# TEXTS AND READINGS IN MATHEMATICS 56

## Iranian Mathematics Competitions
### 1973-2007

## Texts and Readings in Mathematics

# Iranian Mathematics Competitions
# 1973-2007

**Bamdad R. Yahaghi**
**University of Golestan, Iran**

 HINDUSTAN
BOOK AGENCY

Published in India by
Hindustan Book Agency (India)
P 19 Green Park Extension
New Delhi 110 016
India

email: info@hindbook.com
Http://www.hindbook.com

ISBN 978-81-85931-99-9

*beh vajh-e marhamat ey saakenaane sadr-e jalaal,*
*ze rooy-e Hafez o een aastaaneh yaad aareed.*

**Dedicated to all those who teach people to think.**

# Contents

Foreword     xi

Preface     xiii

Acknowledgement     xv

**Part 1.   Problems**     1
- 1.1. First Competition, University of Tehran, March 1973     2
- 1.2. Second Competition, Shiraz (former Pahlavi) University, March 1974     4
- 1.3. Third Competition, (former Jondi Shapour) University of Ahwaz, March 1975     7
- 1.4. Fourth Competition, University of Tabriz, March 1976     9
- 1.5. Fifth Competition, Sharif (former Aryamehr) University of Technology, March 1977     10
- 1.6. Sixth Competition, The University of Isfahan, March 1978     12
- 1.7. Seventh Competition, Ferdowsi University of Mashhad, March 1980     13
- 1.8. Eighth Competition, Shiraz University, March 1984     14
- 1.9. Ninth Competition, Tehran Teacher Training (Tarbiat Moallem) University, September 1985     15
- 1.10. Tenth Competition, University of Sistan and Baluchestan, March 1986     17
- 1.11. Eleventh Competition, The University of Birjand, March 1987     19
- 1.12. Twelfth Competition, Guilan University, March 1988     20
- 1.13. Thirteenth Competition, University of Tehran, March 1989     21
- 1.14. Fourteenth Competition, The University of Isfahan, March 1990     23
- 1.15. Fifteenth Competition, Ferdowsi University of Mashhad, March 1991     24
- 1.16. Sixteenth Competition, Razi (Rhazes or Rasis) University of Kermanshah, March 1992     26
- 1.17. Seventeenth Competition, Shahid Beheshti (former National) University, March 1993     27
- 1.18. Eighteenth Competition, Sharif University of Technology, March 1994     28
- 1.19. Nineteenth Competition, University of Kerman, March 1995     29
- 1.20. Twentieth Competition, Sharif University of Technology, February 1996     30

1.21. Twenty-First Competition, University of Tehran, March 1997          30
1.22. Twenty-Second Competition, University of Ahwaz, March 1998          31
1.23. Twenty-Third Competition, Sharif University of Technology,
      March 1999                                                         39
1.24. Twenty-Fourth Competition, Khajeh Nasir Toosi University of
      Technology, May 2000                                               40
1.25. Twenty-Fifth Competition, IKIU of Qazvin, May 2001                 41
1.26. Twenty-Sixth Competition, IASBS of Zanjan, May 2002                43
1.27. Twenty-Seventh Competition, Bu-Ali Sina (Avecina) University
      of Hamedan, May 2003                                               44
1.28. Twenty-Eighth Competition, Sharif University of Technology,
      May 2004                                                           46
1.29. Twenty-Ninth Competition, University of Mazandaran in
      Babolsar, May 2005                                                 48
1.30. Thirtieth Competition, Tafresh University, May 2006                50
1.31. Thirty-First Competition, Ferdowsi University of Mashhad, May
      2007                                                               51


Part 2.  Solutions                                                       55
  2.1.  First Competition                                                56
  2.2.  Second Competition                                               61
  2.3.  Third Competition                                                70
  2.4.  Fourth Competition                                               76
  2.5.  Fifth Competition                                                80
  2.6.  Sixth Competition                                                90
  2.7.  Seventh Competition                                              94
  2.8.  Eighth Competition                                               99
  2.9.  Ninth Competition                                               105
  2.10. Tenth Competition                                               118
  2.11. Eleventh Competition                                            124
  2.12. Twelfth Competition                                             128
  2.13. Thirteenth Competition                                          135
  2.14. Fourteenth Competition                                          144
  2.15. Fifteenth Competition                                           153
  2.16. Sixteenth Competition                                           159
  2.17. Seventeenth Competition                                         163
  2.18. Eighteenth Competition                                          166
  2.19. Nineteenth Competition                                          171
  2.20. Twentieth Competition                                           177
  2.21. Twenty-First Competition                                        184
  2.22. Twenty-Second Competition                                       188
  2.23. Twenty-Third Competition                                        192
  2.24. Twenty-Fourth Competition                                       197
  2.25. Twenty-Fifth Competition                                        209
  2.26. Twenty-Sixth Competition                                        215
  2.27. Twenty-Seventh Competition                                      218
  2.28. Twenty-Eighth Competition                                       224

2.29. Twenty-Ninth Competition                                     230
2.30. Thirtieth Competition                                        235
2.31. Thirty-First Competition                                     241

**Part 3.   Problem Index**                                        249
First Competition                                                  250
Second Competition                                                 250
Third Competition                                                  251
Fourth Competition                                                 251
Fifth Competition                                                  251
Sixth Competition                                                  252
Seventh Competition                                                252
Eighth Competition                                                 252
Ninth Competition                                                  253
Tenth Competition                                                  253
Eleventh Competition                                               254
Twelfth Competition                                                254
Thirteenth Competition                                             254
Fourteenth Competition                                             255
Fifteenth Competition                                              255
Sixteenth Competition                                              256
Seventeenth Competition                                            256
Eighteenth Competition                                             256
Nineteenth Competition                                             257
Twentieth Competition                                              257
Twenty-First Competition                                           257
Twenty-Second Competition                                          257
Twenty-Third Competition                                           258
Twenty-Fourth Competition                                          258
Twenty-Fifth Competition                                           258
Twenty-Sixth Competition                                           259
Twenty-Seventh Competition                                         259
Twenty-Eighth Competition                                          259
Twenty-Ninth Competition                                           260
Thirtieth Competition                                              260
Thirty-First Competition                                           260

Appendix A.   Historical Introduction And The Winners List         263
A.1. Iranian University Students Mathematics Competitions – a
     historical introduction                                       263
A.2. Winners of the Competitions                                   267

Index                                                              279

# Foreword

This volume offers a collection of problems with solutions in different branches of mathematics given in the annual student competitions of the Iranian Mathematical Society (IMS) during the past 31 years.

After the establishment of numerous new Iranian universities and mathematics departments in the 1960's, there was a growing need for collaboration among these departments. Thus the first national conference was held in Shiraz in 1970. The Iranian Mathematical Society was born just after this meeting, and was registered in 1971.

One of the important activities of the IMS is holding university student competitions which were inaugurated in 1972 by our Society's General Assembly. The first such competition was held in 1973 and the 31st held in 2007 at the Ferdowsi University of Mashhad.

This collection includes all the problems of these contests with nice and interesting solutions presented by a former brilliant student of Sharif University of Technology and one of the winners of these competitions, Dr. Bamdad Yahaghi, who now holds a Ph.D. in Mathematics.

I am sure that this excellent problem book, which is published simultaneously with its Persian version, would be a very helpful reference for all students of mathematics around the world.

A.R. Medghalchi
President of the IMS

# Preface

*taa khod raa be tcheezi nadaadi be kol-leeyat,*
*aan tcheez sa'b-o doshvaar meenomaayad.*
*tchoon khod raa be kol-leeyat be tcheezi daadi,*
*deegar doshvaari namaanad.*
[Until you devote yourself to a task totally,
that task looks hard and unreachable.
As you devote yourself to a task totally,
there remains no difficulty.]
*–Shamsoddin Tabrizi*

International mathematical competitions have gained great popularity in recent years. They provide the first test for a young person's mathematical prowess, and success in these competitions often translates into gaining admittance to the best research institutions for graduate work. For this reason some colleges and even high schools have organized special classes for training mathematically talented students for these competitions. The purpose of this text is to provide a selection of mathematical problems that are not only suitable for special college level (and occasionally even high school level) courses designed for these competitions, but which can significantly improve the level of mathematical sophistication of students who are intrigued by and envision a career in mathematics in general.

I was asked by the Iranian Mathematical Society (IMS) to prepare a problem book on the basis of college level competitions in Iran. Since Iranian students have done quite well in international competitions and their success to some extent reflects the training that they received in special courses, I decided to make the book available to a wider audience. The problems (and their solutions) that are presented in this book are from national mathematical competitions in Iran at college level from 1973 to 2007. I provided my own solutions to most of the problems and also utilized solutions from the files of the IMS, which contained approximately 40% of all the solutions. It might be worth mentioning that there is only one problem in this book that I have not been able to resolve yet. That is Problem 3(c) of 1.3.1.

I should point out that I have not edited the problems of the competitions and they are direct translations from Persian into English. Unfortunately, there were some typos and mistakes in the original version. The errors have been corrected and except for trivial typographical ones, the corrections are so indicated as footnotes throughout Chapter One. Some comments on the problems also appear in footnotes.

xiii

I am indebted, and hence express my deepest gratitude, to the colleagues who contributed in different ways to these competitions and to this book — either as members of the Scientific Committee for the competitions, or by proposing problems or assisting in finding elegant solutions. Unfortunately, some of those colleagues have left us; may they rest in peace. I am thankful to my dear friend Dr. Hossein Hajiabolhassan who constantly encouraged me and helped me with the preparation of the book. I would like to thank Professor Jonathan Borwein for reading the manuscript and Professors Heydar Radjavi, and Mehrdad Shahshahani for reading parts of the manuscript, and for making helpful comments. The idea of making a problem index for the book was suggested by Professor Borwein. My warm thanks go to Professor Rajendra Bhatia, the founding editor of TRIM, for his kind efforts and to the anonymous referee for reading the manuscript carefully and for making constructive comments. I also would like to thank Professor Alireza Jamali, Professor Ebad Mahmoodian, the past President of the IMS, Dr. Rashid Zaare-Nahandi, and Messrs. M. Shokoohi and M. Abdi-Zadeh and the administrative assistants at the IMS's office for their assistance in gaining access to the existing solution files of the IMS. Last but not least, I am grateful to Ms. Anahita Samie for drawing the figures for this book except Figure 3, which was drawn by my dear friend Shaahin Amiri-Sharifi.

Bamdad R. Yahaghi
bbaammddaadd55@gmail.com

# Acknowledgement

A number of friends and colleagues took interest in this work and their contributions are implicit in some elegant solutions to the problems. While inadvertently I may have left out the names of some, the contributions of
Saeed Akbari (the second solution of Problem 7 of 1.9.3),
Kasra Alishahi (the third solution of Problem 1(b) of 1.24.2),
Rajendra Bhatia (the first proof of the lemma presented in Solution 3 of 2.3.1),
Hossein Hajiabolhassan (Problems: 2 of 1.21.2, 2 of 1.23.2, and 6 of 1.28.1),
Hossein Hajiabolhassan and the late Mojtaba Mehrabadi (Problem 2 of 1.14.2),
Ramin Mohammadalikhani (the second solution of Problem 2 of 1.13.1),
Ali Mohammadian (Problems: 5 of 1.25.1 and 6 of 1.27.2),
Omid Naghshineh Arjmand (the second proof of the lemma presented in Solution 3 of 2.3.1, Problem 3(b) of 1.3.1, the lemma presented in Solution 3 of 2.18.1, the second solution of Problem 6 of 1.20),
are duly acknowledged. Needless to say, I am responsible for any shortcomings or errors.

# Part 1

# Problems

## 1.1. First Competition, University of Tehran, March 1973

**1.1.1. Analysis. 1.** A function $\phi$ that is a pointwise limit of continuous real functions is called a *Baire function*.

(a) If $f$ is a real function of the real variable $x$ whose derivative exists everywhere, prove that the function $f'$ is a Baire function.

(b) By giving an example show that every Baire function is not necessarily the derivative of a function.

**2.** Let's call the set of all $n \times n$ real matrices $\mathcal{M}$. The set $\mathcal{M}$ is a metric space if we view the elements of it as $n^2$-dimensional vectors equipped with the Euclidean norm . That is, for every matrix $A = (a_{ij})$, we define the norm of $A$ as follows

$$\|A\| = \left( \sum_{1 \leq i,j \leq n} a_{ij}^2 \right)^{\frac{1}{2}}.$$

Prove that the set of invertible matrices is
(a) open.
(b) disconnected.

**3.** Suppose that the function $f$ is defined on the half-line $(0, +\infty)$ by

$$f(x) = \begin{cases} 0 & x \notin \mathbb{Q}, \\ \frac{1}{p+q} & x = \frac{p}{q}, p, q \in \mathbb{N}, \gcd(p,q) = 1. \end{cases}$$

(a) Show that the limit of this function exists at any point of $(0, +\infty)$.

(b) At what points of $(0, +\infty)$ is the function $f$ continuous? Prove your claim.

**1.1.2. Algebra. 1.** Let $R$ be a commutative ring with identity element.
(a) Prove that an ideal $P$ is prime if and only if

$$\forall a \in R, \forall b \in R : ab \in P \Longrightarrow a \in P \text{ or } b \in P.$$

(b) Prove that every maximal ideal is a prime ideal.

• **Hint.** In a ring $R$, the product of two ideals $T_1$ and $T_2$ is defined as

$$\left\{ \sum_{\text{finite}} xy : x \in T_1, y \in T_2 \right\}.$$

An ideal $M \neq R$ is called maximal if for any ideal $T$, we have

$$M \subseteq T \Longrightarrow T = R.$$

An ideal $P$ is prime if and only if for any two ideals $T_1$ and $T_2$, from $T_1 T_2 \subseteq P$, it follows that $T_1 \subseteq P$ or $T_2 \subseteq P$.

**2.** Suppose that every element of a group $G$ satisfies the equality $x^2 = e$. Prove that the group $G$ is abelian (i.e., commutative).

**3.** Let $E$ and $F$ be two isomorphic sets (that is, there is a one-to-one correspondence between them), and that $f$ is a function from $\mathcal{P}(E)$ into $\mathcal{P}(F)$ satisfying the following three conditions, where $\mathcal{P}$ stands for the power set operation.
(a) $f(\emptyset) = \emptyset$.

(b) $\forall X \in \mathcal{P}(E), \forall Y \in \mathcal{P}(E) : f(X \cup Y) = f(X) \cup f(Y)$.
(c) $\forall X \in \mathcal{P}(E) : \operatorname{card} X \leq \operatorname{card} f(X)$.
($\operatorname{card} A \leq \operatorname{card} B$ means that there exists a one-to-one mapping from $A$ into $B$.)

3.1. Prove that $E$ and $f(E)$ are isomorphic. Moreover, if $X$ and $Y$ are two finite subsets of $E$ such that $X$ is isomorphic to $f(X)$ and $Y$ is isomorphic to $f(Y)$, show that $X \cup Y$ and $f(X \cup Y)$, and $X \cap Y$ and $f(X \cap Y)$ are isomorphic sets, respectively, and that for any two such sets we have $f(X \cap Y) = f(X) \cap f(Y)$.

3.2. Prove that if $E$ is finite and normal, then there exists a normal subset $X_0 \neq \emptyset$ of $E$ such that for any other normal subset $X$ of $E$, we have

$$X \cap X_0 = \emptyset \text{ or } X_0 \subseteq X.$$

(A subset $A$ of $E$ which is isomorphic to $f(A)$ is called a normal subset of $E$.)

**1.1.3. General. 1.** A triangle with sides $a, b, c$ such that $a < b < c$ is given. Set

$$S = \max\left\{\frac{a}{b}, \frac{b}{c}, \frac{c}{a}\right\} \min\left\{\frac{a}{b}, \frac{b}{c}, \frac{c}{a}\right\}.$$

Show that $S \geq 1$.

**2.** Show that the decimal fraction $0.123456789101112131415\ldots$, which is formed by putting consecutive positive integers next to one another, is not periodic.

**3.** Find two distinct real or complex numbers in such a way that each of which is the cube of the other.

**4.** Let $x, y$ be two positive real numbers with $x + y = 1$. Prove that

$$x^x + y^y \geq \sqrt{2}.$$

**1.1.4. Differential Equations.** Find the general solution of the following differential equation

$$\frac{d^2 x}{dt^2} + a^2 x = f(t),$$

where $a$ is a constant and $f$ is a continuous real function.

**1.1.5. Probability and Statistics.** Let $X_1, \ldots, X_{10}$ be ten independent random variables with the probability density functions $f_i(X_i) = \dfrac{e^{-\mu_i} \mu_i^{X_i}}{X_i}$, where $\mu_i = i$ $(i = 1, \ldots, 10)$ and that the codomain of the variable $X_i$ is $\{0, 1, 2, \ldots\}$. Set

$$\overline{X} = \frac{1}{10} \sum_{i=1}^{10} X_i.$$

What is the probability that $\overline{X}$ equals one?

**1.1.6. Topology.** Let $\mathbb{Q}$ be the set of all rational numbers. For each $q \in \mathbb{Q}$, the set $]q, +\infty[$ is denoted by $A_q$. If $T$ is the set consisting of $\mathbb{R}$ and $\emptyset$ and all $A_q$'s, show that $T$ is not a topology on $\mathbb{R}$.
- **Hint.** Consider the sets of the form $A_q$ with $q > \sqrt{2}$ and $q \in \mathbb{Q}$.

## 1.2. Second Competition, Shiraz (former Pahlavi) University, March 1974

**1.2.1. Analysis. 1.** Let $f$ be a real continuous function with nonnegative values on the closed interval $[0, 1]$. Set

$$u_n = \left( \int_0^1 (f(x))^n \, dx \right)^{\frac{1}{n}}, \quad M = \sup_{0 \le x \le 1} f(x).$$

(a) Assuming that $0 < \varepsilon < 1$, prove that there exist numbers $\alpha, \beta$ subject to $0 \le \alpha < \beta \le 1$ such that $M(1 - \varepsilon) \le f(x) \le M$ for all $x$ belonging to the open interval $]\alpha, \beta[$.

(b) Prove that $\lim_{n \to +\infty} u_n = M$.

**2.** Evaluate

$$D = \inf_{f \in \mathcal{F}} \left( \sup_{0 \le t \le 1} |1 - f(t)| + \int_0^1 |1 - f(t)| dt \right),$$

where $\mathcal{F}$ is the vector space of all continuous functions from $[0, 1]$ into $\mathbb{R}$ whose elements take the value zero at zero. Describe the geometrical interpretation of the number $D$.

**3.** Assuming that $C$ is the Cantor set, show that

$$C + C = [0, 2].$$

- **Hint. 1.** If $A$ and $B$ are two subsets of real numbers, by definition,

$$A + B := \{a + b : a \in A, b \in B\}.$$

2. The Cantor set consists of all real numbers between zero and one whose ternary expansions do not have any "one".

**1.2.2. Algebra. 1.** Let $F$ be a field with $n$ elements. Prove that for every integer $m$ with $m \geq 1$, we have

$$\sum_{x \in F} x^m = \begin{cases} n-1 & n-1 \mid m, \\ 0 & n-1 \nmid m. \end{cases}$$

- **Hint.** $F^* = F \setminus \{0\}$ is a cyclic group.

**2.** Let $A$ be a ring.
- **Definition 1.** An element $z \in A$ is called right quasi-regular if there is a $z' \in A$ such that $z + z' - zz' = 0$.
- **Definition 2.** An element $z \in A$ is called left quasi-regular if there is a $z' \in A$ such that $z + z' - z'z = 0$.
- **Definition 3.** An element $z \in A$ is called quasi-regular if there is a $z' \in A$ such that $z + z' - zz' = z + z' - z'z = 0$.

(a) An element $z \in A$ is quasi regular if an only if it is right and left quasi-regular.

(b) If $A$ has an identity element, then the identity element is not right quasi-regular. And if $x$ is a right quasi-regular element, then $1 - x$ is invertible[1] in $A$.

**3.** Let $E$ be a finite-dimensional vector space over a field $K$ and $\phi$ a homomorphism from the ring $K[x]$ into the ring $\mathcal{L}(E)$ with the hypothesis that $\phi(1)$ is the identity element of $\mathcal{L}(E)$.

(a) If $P$ and $D$ belong to $K[x]$ and $D$ is a divisor of $P$, prove that

$$\ker\big(\phi(P)\big) \supset \ker\big(\phi(D)\big).$$

(b) If $R$ is another element of $K[x]$ and $D$ is the greatest common divisor of $P$ and $R$, prove that

$$\ker\big(\phi(D)\big) = \ker\big(\phi(P)\big) \cap \ker\big(\phi(R)\big).$$

- **Hint.** By $K[x]$, we mean the ring of all polynomials in one indeterminate $x$ with coefficients coming from $K$ and $\mathcal{L}(E)$ is the ring of all endomorphisms of $E$. For any $u$ belonging to $\mathcal{L}(E)$, $\ker u$ means the kernel of the linear transformation $u$.

**1.2.3. General. 1.** Find the minimum and the maximum number of "Wednesdays occurring on the 13th of the month" which might happen in a solar year, according to the Persian calendar that is.

**Remark.** Recall that there are 12 months in the Persian calendar, each of the first six months has 31 days, each of the sixth-eleventh months has 30 days, and the last month has 29 days except that in a leap year it has 30 days.

**2.** Prove that

$$\left(\frac{x+y+z+t}{4}\right)^n \leq \frac{x^n + y^n + z^n + t^n}{4},$$

where $x, y, z, t$ are positive real numbers and $n$ is a natural number.

---

[1] "invertible" must read "right invertible"!

**3.** A particle, moving along a straight line, goes one unit of distance in one unit of time in such a way that the speed of it at the initial point as well as the final point is zero. Prove that [the absolute value of] the particle acceleration at some point on its path is greater than or equal to four.

**4.** We know that in Iran the plate number of every car is a five-digit number none of whose digits is zero. In big cities, like Tehran, in addition to a number a [Persian] letter is also used to characterize the plate, e.g., "alef", "be", "pe", etc. By explaining your reasoning, find the total number of the car plates that contain the letter "dal".

**Remark.** The Persian alphabet has 32 letters.

**1.2.4. Probability and Statistics.** Let $X$ and $Y$ be two random variables subject to the following conditions.

$$\text{var}(X) = \text{var}(Y) = \sigma^2, \ \text{cov}(X, Y) = \lambda,$$

where $\sigma$ and $\lambda$ are given real numbers. Define the random variables $U$ and $V$ by $U = X + Y$ and $V = X - Y$.

(a) Find the covariance of $U$ and $V$.

(b) Are $U$ and $V$ independent? If so, prove it; if not, by an example justify your answer.

**1.2.5. Topology.** Consider a nonempty topological space $(E, T)$ and two nonempty subsets $A$ and $K$ of $E$ with $K \supset A$. Suppose that $B$ and $C$ are two closed subsets of the topological space $(K, T_K)$ such that

$$A \subset B \cup C, \ A \cap C \neq \emptyset, \ A \cap B \neq \emptyset.$$

Prove that if $A$ is connected, then $A \cap B \cap C \neq \emptyset$.

• **Hint.** By $T_K$ we mean the induced topology which is induced by $T$ on $K$. The problem can be proved by contradiction.

**1.2.6. Differential Equations.** Find the general solution of the following differential equation subject to the two cases $|x| > 1$ and $|x| < 1$.

$$3(x^2 - 1)y^2 y' + xy^3 = x^3 + x^2 - x - 1.$$

## 1.3. Third Competition, (former Jondi Shapour) University of Ahwaz, March 1975

**1.3.1. Analysis. 1.** Let $f$ be a real function whose domain is $[a, b]$. If $f''$ exists on $[a, b]$ and is positive, then for all $\xi \in [a, b]$,[2] there exists a point $x_0 \in [a, b]$ such that

$$f'(\xi) = \frac{f(a) - f(x_0)}{a - x_0} \text{ or } f'(\xi) = \frac{f(b) - f(x_0)}{b - x_0}.$$

**2.** A sequence $(a_n)_{n=1}^{+\infty}$ of numbers in the closed interval $[0, 1]$ is given such that the elements of the sequence are all distinct. The goal is to find a continuous function $f$ from $[0, 1]$ into $[0, 1]$ such that $f(a_n) = a_{n+1}$. Prove that
   (a) in general, the problem has no solution.
   (b) if the sequence is increasing or decreasing, the problem has a solution.

**3.** In each part of this problem, you can use the preceding parts.
   (a) If the complex number $z$ is a root of the following equation with complex coefficients

$$x^p + c_1 x^{p-1} + \cdots + c_{p-1}x + c_p = 0,$$

then

$$\exists\, k \in \{1, \ldots, p\} : |z| \leq 2\sqrt[k]{|c_k|}.$$

   (b) The following two equations with complex coefficients are given.

$$x^p + c_1 x^{p-1} + \cdots + c_{p-1}x + c_p = 0,$$
$$x^p + c_1' x^{p-1} + \cdots + c_{p-1}'x + c_p' = 0.$$

If the positive numbers $K$ and $\delta$ satisfy the following

$$\forall i = 1, \ldots, p : \left( |c_i| < K^i, |c_i - c_i'| < K^i \delta \right),$$

then for any root $z_j$ of the first equation there exists a root $z_k'$ of the second equation such that

$$\left| z_j - z_k' \right| < 2K \sqrt[p]{\delta}.$$

   (c) Suppose that the real numbers $K$ and $\alpha$ are such that $K > 0$ and $0 < \alpha \leq 1$. Also suppose that the complex valued functions $b_1, \ldots, b_p$ of a real variable $t$ belonging to the compact interval $[t_1, t_2]$ are such that

$$|b_i(t)| < K^i, \ |b_i(t) - b_i(t')| < K^i |t - t'|^\alpha,$$

for all $t, t' \in [t_1, t_2]$ and each $i = 1, \ldots, p$. Prove that if the complex valued function $f$ is continuous on the closed interval $[t_1, t_2]$ and that it satisfies the equation

$$f^p + b_1 f^{p-1} + \cdots + b_{p-1}f + b_p = 0,$$

then

$$\left| f(t_2) - f(t_1) \right| < 4pK \sqrt[p]{(t_2 - t_1)^\alpha}.$$

   • **Hint.** If $F : I \to U$ is a continuous function from an interval into an open subset of the complex plane, the image of $F$ lies in one of the connected components of $U$.

---

[2] "$\xi \in [a, b]$" must read "$\xi \in (a, b)$"!

**1.3.2. Algebra. 1.** Let $M$ be a module on a ring $K$ and $M_1$ and $M_2$ two submodules of $M$. Set

$$M_3 = M_1 + M_2, \quad M_4 = M_1 \cap M_2.$$

(a) Show that the quotient modules $\dfrac{M_3}{M_1}$ and $\dfrac{M_2}{M_4}$ are isomorphic.

(b) If $K$ is a commutative field and $M_1$ and $M_2$ are finite-dimensional, show that

$$\dim M_1 + \dim M_2 = \dim M_3 + \dim M_4.$$

**2.** Prove that a necessary and sufficient condition for a group to be commutative is that the following function is a homomorphism of groups.

$$f : G \to G, \ f(x) = x^2.$$

**3.** Suppose that for any element $y$ of a semigroup $S$ with a right identity element (that is, $xe = x$ for all $x \in S$), there exists a $\bar{y}$ with the property that $\bar{y}y = e$.

(a) By giving an example, show that $S$ might not be a group.

(b) If the right identity element is unique, prove that $S$ is a group.

**1.3.3. General. 1.** The real function $F(x, y)$ where $x$ and $y$ are two real variables is considered. Suppose that this function has the following two properties.[3]

$$\forall x, y \in \mathbb{R} : \ F(x, y) = 0 \iff x = y,$$
$$\forall x, y, z \in \mathbb{R} : \ F(y, x) \le F(x, z) + F(z, y).$$

(a) Prove that $F(x, y) \ge 0$ for all $x, y \in \mathbb{R}$.

(b) Prove that $F(x, y) = F(y, x)$ for all $x, y \in \mathbb{R}$.

**2.** Without using mathematical induction, prove that

$$\sum_{m=0}^{n} \left( \frac{n!}{m!(n-m)!} \right)^2 = \frac{(2n)!}{(n!)^2},$$

for all $n \in \mathbb{N}$.

**3.** Two flat mirrors $\Delta$ and $\Delta'$ which are perpendicular to a plane $P$ and two points $M, N \in P$ are given. From the point $M$, shine a beam of light on the mirror $\Delta$, in the plane $P$, in such a way that the reflected light beam after hitting the mirror $\Delta'$ passes through the point $N$.

---

[3] "$\forall x, y \in \mathbb{R} : \ F(x, y) = 0 \implies x = y$" is redundant!

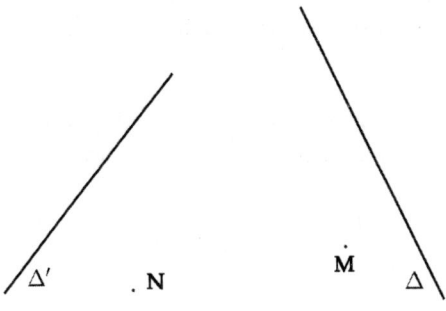

Figure 1

## 1.4. Fourth Competition, University of Tabriz, March 1976

**1.4.1. Analysis. 1.** In this problem $\mathbb{R}$ denotes the real line and $\mathbb{R}^2$ the Euclidean plane. Prove that if $f$ is a continuous and one-to-one function from $\mathbb{R}$ into $\mathbb{R}^2$, its inverse is not necessarily continuous.

**2.** Let $(\alpha_n)_{n=1}^{+\infty}$ be a sequence of nonnegative real numbers such that

$$\lim_{n \to +\infty} \alpha_n = 0.$$

Prove that there are infinitely many number of indices $n$ such that $\alpha_m \leq \alpha_n$ for all $m$ greater than or equal to $n$.

**1.4.2. Algebra. 1.** Let $A$ be a commutative ring and $B$ an ideal of $A$. Set

$$R(B) = \{x \in A : \exists r \in \mathbb{N} \ni x^r \in B\}.$$

We call $R(B)$ the radical of $B$. We call the ideal $B$ of a commutative ring $A$ a prime ideal if

$$\forall a, b \in A, ab \in B \implies a \in B \text{ or } b \in B.$$

(a) Prove that $R(B)$ is an ideal of $A$.
(b) Prove that if the ideal $B$ is prime, then so is the ideal $R(B)$.
(c) If we set $A = \mathbb{Z}$, and if $a$ is an element of $\mathbb{Z}$ and[4]

$$B = \langle a \rangle = a\mathbb{Z} = aA,$$

prove that there are prime numbers $p_1, \ldots, p_r$ such that

$$R(B) = \langle p_1 p_2 \cdots p_r \rangle.$$

**2.** Prove that in a vector space $E$, whose dimension is at least 2, for any two vectors $x$ and $y$ which are not linearly dependent, one can find an automorphism $u : E \to E$ such that $u(x) = x$ and $u(y) = x + y$. From this, conclude that if an endomorphism $f : E \to E$ commutes with any automorphism $v$ (i.e., $f \circ v = v \circ f$), then for every vector $x \in E$, the two vectors $x$ and $f(x)$ are linearly

---

[4] Obviously, we must have $a \neq 0, \pm 1$.

dependent whence $f$ is just the scalar product of the identity automorphism by a fixed scalar (i.e., $f = \lambda id_E$).

**3.** Let $R$ be an integral domain with identity. Suppose that every descending chain of the ideals of $R$ terminates. (that is, for every chain of ideals of $R$ as follows

$$I_1 \supseteq I_2 \supseteq \cdots \supseteq I_k \supseteq \cdots ,$$

there exists a positive integer $n$ such that $I_m = I_n$ for all $m \geq n$.) Prove that $R$ is a field.

    • **Hint.** Assume that $x$ is a nonzero element in $R$, investigate the ideals $x^i R$ $(i = 1, 2, \ldots)$.

**1.4.3. General. 1.** Evaluate the following integral.

$$I = \int_0^{\frac{\pi}{2}} \frac{\sin^n x}{\cos^n x + \sin^n x} dx, \ n \in \mathbb{N}.$$

**2.** We have three boxes $A$ , $B$, and $C$, of which two are empty and one has a prize in it. We choose one of these boxes at random (say, for example $B$); it is plain that the probability that the chosen box contains the prize is $\frac{1}{3}$ because of the three boxes only one contains the prize. Now, suppose that of the two not-chosen boxes $A$ and $C$, we open the one that does not contain the prize (say, for example $A$). Therefore, the prize is in one of the boxes $B$ and $C$.

    (a) What is the probability that the chosen box contains the prize?
    (b) What is the probability that the other box, i.e., $C$, contains the prize?[5]

## 1.5. Fifth Competition, Sharif (former Aryamehr) University of Technology, March 1977

**1.5.1. Analysis. 1.** If two functions are integrable on the interval $(0, 1)$ in the Riemann sense, is it true that the composition of the two functions is integrable on the interval $(0, 1)$ in the Riemann sense? Explain your reasoning.

**2.** Let $\mathbb{R}$ be the set of all real numbers and $f$ a continuous function from $\mathbb{R}$ into $\mathbb{R}$ that does not assume any value more than twice. Prove that $f$ assumes at least a value exactly once.

**3.** Let $A$ be a subset of the real numbers $\mathbb{R}$. A point $p$ of $A$ is called a congestion point of $A$ if every neighborhood of $p$ of the form $(p - \varepsilon, p + \varepsilon)$ contains uncountably many points of $A$. Prove that all but countably many points of $A$ are congestion points of $A$ ($A$ is uncountable).

---

[5] This problem is also known as the Monty Hall problem.

**1.5.2. Algebra. 1.** Let $V$ be a finite-dimensional vector space over $\mathbb{C}$ (the complex numbers). If $A : V \to V$ is a linear transformation such that $A \neq I$, $A^2 \neq I$, $A^3 = I$, find the eigenvalues of $A$. Extend this to the case where $A^k = I$. Do the eigenvalues of $A$ form a group under multiplication? Find a necessary and sufficient condition for this set to form a group.

**2.** Let $G$ be a group and $a$ and $b$ two elements of it satisfying the following relations.

$$a \neq 1, b \neq 1, aba^{-1} = b^2, a^7 = 1,$$

where 1 is the identity element of the group. Find the orders of $a$ and $b$.

**3.** Let $R$ be a ring with identity element such that every $x \in R$ satisfies the following relation: $x^3 + 2x^2 + x = 0$.
 (a) Prove that $2x = 0$ for all $x \in R$.
 (b) Prove that $R$ is a commutative ring.

**1.5.3. General. 1.** Let $m, n \in \mathbb{N}$. Prove that the number $\dfrac{(mn)!}{m!(n!)^m}$ is an integer.

**2.** If $S = \{z_1, \ldots, z_k\}$ is a subset of the complex numbers, define the set $C(S)$ by

$$C(S) := \{z = \alpha_1 z_1 + \cdots + \alpha_k z_k | \alpha_i \geq 0, \alpha_1 + \cdots + \alpha_k = 1\}.$$

If $f$ is a polynomial of degree greater than or equal to two and $A$ and $A'$ denote the sets of the roots of the equations $f(z) = 0$ and $f'(z) = 0$, respectively, where $f'$ denotes the derivative of $f$, prove that

$$C(A) \supset C(A').$$

**3.** Consider the trajectory of a billiard ball $B$ which moves on an ellipse-shaped billiard table $\xi$. Suppose that the angle of incidence is equal to the angle of reflection. Assuming that the initial ball trajectory does not intersect the line segment joining the the two foci, show that there exists another ellipse $\eta$ which is confocal with the ellipse $\xi$ such that the ball trajectory is always tangent to $\eta$.
 • **Hint.** Recall that if $PL$ and $PL'$ are two tangents from a point $P$ to an ellipse, then the angles $FPL$ and $F'PL'$ are equal. ($F$ and $F'$ are the foci of the ellipse.)

## 1.6. Sixth Competition, The University of Isfahan, March 1978

**1.6.1. Analysis. 1.** In the $xy$-plane a point is called rational if both of its coordinates are rational. Prove that if the center of a given circle in the plane is not rational, then there are at most two rational points on the circle.

**2.** Suppose that in a metric space $M$ a sequence $(f_n)_{n=1}^{+\infty}$ of continuous functions is uniformly convergent to a function $f$. Prove that for every sequence $(x_n)_{n=1}^{+\infty}$ converging to a point $x \in M$, we have

$$\lim_{n \to +\infty} f_n(x_n) = f(x).$$

**3.** If $f$ is a continuous function on $\mathbb{R}$ and moreover $\int_0^{+\infty} |f(x)| \, dx < +\infty$, prove that

$$\lim_{n \to +\infty} \int_0^{+\infty} \left| f\left(x + \frac{1}{n}\right) - f(x) \right| dx = 0.$$

**1.6.2. Algebra. 1.** Let $G$ be a group with $|G| = p^n a$, where $p$ is a prime and $p$ and $a$ are relatively prime. If $G$ has subgroups $A$ and $B$ satisfying the following conditions

$$|A| = p^n, |B| = p^m, 0 < m \le n, B \nsubseteq A,$$

prove that $AB$ cannot be a subgroup of $G$.

**2.** Let $A_1, \ldots, A_n$ be mutually commuting $m \times m$ matrices such that $A_i^2 = 0$ for all $1 \le i \le n$. If $m < 2^n$, prove that $A_1 A_2 \cdots A_n = 0$.

**3.** If $R$ is a ring with identity $(1 \ne 0)$ and $e$ and $f$ are two commuting elements of $R$ such that $e^2 = e$ and $f^2 = f$, prove that

$$(e - f)^n = 0 \implies e = f,$$

for all positive integers $n$.

**1.6.3. General. 1.** Let $n_1, n_2, \ldots, n_k$ be $k$ integers of which $m_1, m_2, \ldots, m_k$ is a permutation. Prove that

$$|n_1 - m_1| + |n_2 - m_2| + \cdots + |n_k - m_k|$$

is an even integer.

**2.** An $n \times n$ matrix whose elements are nonnegative integers is given. If the sum of the elements on any row and any column corresponding to any nonzero element of the matrix is at least $n$, prove that the sum of all elements of this matrix is no less than $\frac{n^2}{2}$.

**3.** Suppose 1,000,000 points inside a circle are given. Can one find a straight line not passing through any of the points that divides the circle into two sections each of which containing 500,000 of the points?

## 1.7. Seventh Competition, Ferdowsi University of Mashhad, March 1980

**1.7.1. Analysis. 1.** Is there a closed set $S \subsetneq \mathbb{R}^2$ such that for each $x \in \mathbb{R}^2 \setminus S$ there are exactly two points in $S$ as the closest point of $S$ to $x$?

**2.** For each positive rational number $r$ $(r \in \mathbb{Q}^+)$, suppose $I_r$ is an open interval such that $r < s \implies \bar{I}_r \subset I_s$. The function $f$ on the set $A = \bigcup_{r \in \mathbb{Q}^+} I_r$ is defined by

$$f(x) = \inf \{r : x \in I_r\}.$$

Show that $f$ is continuous.

**3.** Suppose that the sequence $(x_n)_{n=1}^{+\infty}$ is defined by

$$x_n = \frac{n+1}{2^{n+1}} \sum_{k=1}^{n} \frac{2^k}{k}, \quad n = 1, 2, \ldots.$$

Prove that $\lim_{n \to +\infty} x_n$ exists and find its value.

**1.7.2. Algebra. 1.** (a) If in a ring $R$ the element $e$ is a left identity element (i.e., $ex = x, \forall x \in R$) that is unique, show that $e$ is the identity element of this ring (i.e., $ex = xe = x, \forall x \in R$).

(b) If zero is the only nilpotent element of the ring $R$, show that for every idempotent element $a$ of $R$ we have

$$ax = xa, \quad \forall x \in R.$$

• **Note.** The element $b$ of a ring $R$ is called nilpotent if there exists a positive integer $n$ such that $b^n = 0$; $b$ is called idempotent if $b^2 = b$.

**2.** Let $G$ be a group and $G_1$ a normal subgroup of $G$ such that the groups $\frac{G}{G_1}$ and $G_1$ are commutative. Prove that for any arbitrary subgroup $H$ of $G$, there is a commutative subgroup $H_1$ of $H$ such that $H_1$ is a normal subgroup of $H$ and $\frac{H}{H_1}$ is commutative.

**3.** An $n$-dimensional vector space $V$ and a linear transformation $\theta : V \to V$ are given. Consider the powers of $\theta$, i.e., $\theta^0 = 1, \theta, \theta^2, \ldots$. Prove that there is a nonzero integer $s$ such that

$$V = \mathrm{im}(\theta^s) \oplus \ker(\theta^s).$$

## 1.8. Eighth Competition, Shiraz University, March 1984

**1.8.1. Analysis. 1.** Let $F$ and $G$ be two closed subsets of $\mathbb{R}^n$, where $\mathbb{R}$ denotes the real numbers endowed with its ordinary topology. Suppose that $f : F \to \mathbb{R}$ and $g : G \to \mathbb{R}$ are two continuous functions which coincide on $F \cap G$.

(a) Prove that there is a continuous function $h : F \cup G \to \mathbb{R}$ such that $h$ is an extension of both of the functions $f$ and $g$.

(b) By giving an example, show that if $F$ and $G$ are not closed, the assertion in (a) is not correct.

**2.** Suppose that $(a_n)_{n=1}^{+\infty}$ is a decreasing sequence of real numbers and that $\sum_{n=1}^{+\infty} a_n$ is convergent. Prove that $\sum_{n=1}^{+\infty} n(a_n - a_{n+1})$ is convergent.[6]

**3.** Let the function $g$ be continuously differentiable on $[0, 1]$. Prove that

$$\lim_{n \to +\infty} \int_0^1 x^n dg(x) = 0.$$

**1.8.2. Algebra. 1.** In the multiplicative group of nonsingular $2 \times 2$ matrices with real entries, let

$$a = \begin{pmatrix} 2 & 0 \\ 0 & 1 \end{pmatrix}, \quad b = \begin{pmatrix} 1 & 1 \\ 0 & 1 \end{pmatrix}.$$

If $H$ is the subgroup generated by $b$, show that $aHa^{-1}$ is a proper subgroup of $H$.

**2.** Let $R$ be an infinite integral domain (i.e., a commutative ring with identity without divisors of zero). If $R$ has finitely many units, prove that $R$ has infinitely many maximal ideals. (A unit is an element which has a multiplicative inverse.)

**3.** Let $A = (a_{ij})$ be an $n \times n$ matrix over the field of real numbers such that for all $i$ we have $\sum_{j=1}^{n} a_{ij} = a$. If $A^2 = I$ ($I$ is the identity matrix), find $a$.

---

[6] For a counterpart of this problem for sequences of real functions, see Problem 2 of 1.12.1.

**1.8.3. General. 1.** Without using derivative, find the minimum of the three variable real function

$$f(x, y, z) = x^2 + 4y^2 + z^2 - 6x + 4y.$$

**2.** Five people have 719 rials altogether. If each person has an integer amount of money, that the money of no two people is equal, and that the ratio of a person's money with respect to that of any other person with less amount of money is an integer, determine how much money each person has.

**3.** A bus which must pass through the city $A$ passes through the crossroad $B$ of the city with the probability of $\frac{1}{3}$. If the traffic-light of the crossroad is, consecutively, 30 seconds red and 30 seconds green, find the average of the stop time of the bus at the crossroad.

## 1.9. Ninth Competition, Tehran Teacher Training (Tarbiat Moallem) University, September 1985

**1.9.1. Analysis. 1.** On the set of real numbers $\mathbb{R}$, define the equivalence relation $\sim$ as follows

$$a \sim b \iff a - b \in \mathbb{Q},$$

where $\mathbb{Q}$ is the set of rational numbers. Prove that every equivalence class of $\sim$ is dense in $\mathbb{R}$.

**2.** Prove that if $f$ is a continuous and one-to-one function from $\mathbb{R}$ into $\mathbb{R}$, then the inverse of $f$ (from $f(\mathbb{R})$ onto $\mathbb{R}$) is continuous as well.

**3.** Let $f$ be a continuous and increasing function from $[a, b]$ into $[a, b]$ and $f(a) = a$. Prove that if $E = \{x | a \leq x \leq b, f(x) \geq x\}$, then $f(E) = E$.

**4.** Let the function $f$ be defined on $[0, 1]$ by

$$f(x) = \begin{cases} 0 & x \notin \mathbb{Q}, \\ \frac{1}{q} & x \in \mathbb{Q}, x = \frac{p}{q}, p, q \in \mathbb{N}, \gcd(p, q) = 1. \end{cases}$$

Prove that $\int_0^1 f(x)dx$ exists.

**5.** Let $f : [0, 1] \to [0, 1] \times [0, 1]$ be a function. Prove that the function $f$ can have any two properties of the following three properties but cannot have more than two properties.

Continuity, Injectivity, Surjectivity.

**1.9.2.  Algebra. 1.** Let $G$ be a finite group and $p$ the smallest prime dividing the order of $G$. Prove that every subgroup of index $p$ in $G$ is a normal subgroup of $G$.

**2.** Let $\mathbb{Z}_3$ denote the field of integers mod 3. Define, explicitly, an isomorphism from the field $\dfrac{\mathbb{Z}_3[x]}{(x^2+1)}$ onto the field $\dfrac{\mathbb{Z}_3[x]}{(x^2+x+2)}$.

**3.** Find the greatest common divisor of the two polynomials $4x^4 - 2x^2 + 1$ and $-3x^3 + 4x^2 + x + 1$ in $\mathbb{Z}_7[x]$.

**4.** A quadratic extension $K \supseteq F$ such that $\mathrm{ch}(F) \neq 2$ is given (here, $\mathrm{ch}(F)$ denotes the characteristic of the field $F$). Prove that there exists an element $y \in K$ such that $y^2 \in F$ and that $\{1, y\}$ is a basis for $K$ over $F$.

**5.** Let $F$ be a field with characteristic 2, $V$ an $n$-dimensional vector space over $F$, and $T : V \to V$ a linear transformation such that $T^2 = I$. Set $W = \{v \in V : Tv = v\}$. Prove that $\dim(W) \geq n/2$. (Here, $I$ denotes the identity transformation on $V$.)

**1.9.3.  General. 1.** Evaluate the determinant of an $n \times n$ matrix whose diagonal entries are all equal to $r$ and whose off-diagonal entries are all equal to $\lambda$.

**2.** A husband and a wife, working in one of the poultry farms of the country, have posed a problem on the number of their chickens as follows. If they sell 75 of their chickens, their chicken food will run out twenty days after the normal time. But if they buy another one hundred chickens, their chicken food will run out fifteen days before the normal time. Find the number of chickens in the chicken farm.

**3.** Let $\alpha$ be a real number. Prove that there is no positive continuous function $f : [0, 1] \to \mathbb{R}$ such that

$$\int_0^1 f(x)dx = 1, \quad \int_0^1 xf(x)dx = \alpha, \quad \int_0^1 x^2 f(x)dx = \alpha^2.$$

**4.** Find the number of the solutions of the following equation in the set of positive integers.

$$x_1 + x_2 + \cdots + x_m = n,$$

where $m$ and $n$ are natural numbers with $m < n$.

**5.** On a square grid paper the following figure is drawn. Can you cut the paper along the lines and divide it into two pieces in such a way that putting the pieces next to one another forms a chessboard?

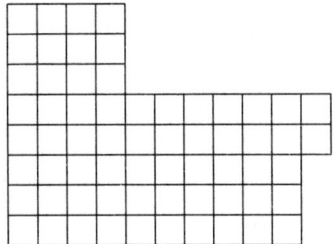

**6.** The start hour in a factory is 8 o'clock in the morning. A worker has estimated that, by car, s/he goes the distance from home to work within 10 to 20 minutes (with uniform distribution).

(a) If this worker leaves his/her home at $7\frac{3}{4}$, what is the probability that s/he does not get to work on time?

(b) If 15 minutes is needed to have breakfast at the work place, determine the latest time that this worker can leave home to get to work on time and with the probability of 75% to have time to eat breakfast.

**7.** Let $c_1, \ldots, c_n$ be $n$ real numbers. Consider the matrix $(c_i c_j)_{n \times n}$. Evaluate $\det\left(I + (c_i c_j)\right)$, where $I$ is the identity matrix of size $n$.

## 1.10. Tenth Competition, University of Sistan and Baluchestan, March 1986

**1.10.1 Analysis. 1.** The real function $f$ satisfying the following conditions is given.

(a) $f(1) = 1$.

(b) $f'(x) = \dfrac{1}{x^2 + \left(f(x)\right)^2}$ for all $x \geq 1$.

Prove that $\lim_{x \to +\infty} f(x)$ exists and is less than $1 + \frac{\pi}{4}$.

**2.** The continuous function $f : [a, b] \to \mathbb{R}$ satisfying the following conditions is given.

(a) $f(a) = f(b) = 0$.

(b) The second derivative of $f$ on the interval $(a, b)$ exists [and is bounded]. Show that

$$\int_a^b |f(x)|\, dx \leq M \frac{(b-a)^3}{12},$$

where $M = \sup\{|f''(x)| : x \in (a, b)\}$.

• **Hint.** Use the auxiliary function $g(t) = (t-a)(t-b)f(x) - (x-a)(x-b)f(t)$.

**3.** Prove that

$$\lim_{n \to +\infty} \int_0^3 \frac{x^2(1-x)x^n}{1+x^{2n}}\, dx = 0.$$

**1.10.2. Algebra. 1.** Prove that the additive group of $\mathbb{R}$ has no maximal subgroup.

**2.** Let $A, B, C, D$ be ideals in a unital ring $R$ such that

$$A + B = A + C = A + D = R. \qquad (*)$$

If $M = B \cap C \cap D$, prove that $A + M = R$.

Give an example of a unital ring $R$ having ideals $A, B, C, D$ which satisfy $(*)$.

**3.** Let $E$ be an extension field of $F$ and $\alpha \in E$ algebraic on $F$ of degree $n$. If $m < n$ and $\gcd(n, m!) = 1$, then show that $F(\alpha) = F(\alpha^m)$.

**4.** Prove that no set of nilpotent matrices can span $M_n(F)$, where $M_n(F)$ is the vector space of all $n \times n$ matrices over the field $F$.

**1.10.3. General. 1.** Consider a test in which the probability of success (S) is equal to $p$ and the probability of failure (F) is equal to $q = 1-p$. Perform this test two times. If the results turn out to be $FS$ or $SF$, set the random variable $X$ to be 0 or 1, respectively. If not, do the test another two more times. Again if the results of the two tests are $FS$ or $SF$, we take the random variable $X$ to be 0 or 1, respectively; and if not, we perform the test another two more times... Prove that $P\{X = 0\} = P\{X = 1\} = \frac{1}{2}$.

**2.** Show that every closed set $A$ in the plane with the property that $A^\circ = \emptyset$ is the boundary of an open set in the plane. (Here, $A^\circ$ denotes the interior of the set $A$.)

**3.** Prove that every solution of the differential equation $(*)$ is a solution of the integral equation $(**)$ and vice versa.

$$x'' = f(t, x), \quad x(0) = x_0, \quad x'(0) = y_0, \qquad (*)$$

where $f(t, x)$ is a continuous function in a region $D$ which contains the point $(0, x_0)$.

$$x(t) = x_0 + y_0 t + \int_0^t (t - s) f(s, x(s)) ds. \qquad (**)$$

**4.** Prove that the product of $r$ consecutive natural numbers is divisible by $r!$ and from there conclude that $\dfrac{n!}{r!(n - r)!}$ is always an integer.

## 1.11. Eleventh Competition, The University of Birjand, March 1987

**1.11.1. Analysis. 1.** The function $f : [0,1] \to \mathbb{R}$ is differentiable and $f$ and $f'$ have no common zero. Prove that the set of zeros of $f$ in $[0,1]$ is finite. (Recall that a solution of the equation $f(x) = 0$ is called a zero of the function $f$.)

**2.** (a) Show that the sequence of functions $f_n(x) = x^{x^{\cdot^{\cdot^{x}}}}$ is pointwise convergent to a function $f$ on $[1, e^{\frac{1}{e}}]$ which satisfies the functional equation $f(x) = x^{f(x)}$.

• **Hint.** Prove that for all $a \in [1, e^{\frac{1}{e}}]$ the sequence $\left(f_n(a)\right)_{n=1}^{+\infty}$ is increasing and bounded. Then, in view of the fact that the function $g(x) = a^x$ is continuous on $\mathbb{R}$, show that $f(a) = a^{f(a)}$.

(b) Show that the function $f$ is one-to-one. Then using the inverse of $f$, conclude that $f$ is continuous. And then, prove that the sequence $(f_n)_{n=1}^{+\infty}$ converges uniformly on $[1, e^{\frac{1}{e}}]$.

**3.** Suppose that for every $n \in \mathbb{N}$, $\varphi_n : [-1,1] \to \mathbb{R}$ has second order derivative and that $\sup \left\{ \left| \varphi_n''(x) \right| : n \in \mathbb{N}, x \in [-1,1] \right\} < +\infty$, prove that $\sum_{n=1}^{+\infty} a_n$ is convergent, where $a_n = \frac{1}{n} \left| \int_{-1}^{1} \varphi_n(t) \cos n\pi t \, dt \right|$.

**1.11.2. Algebra. 1.** Suppose that the ring $R$ has exactly two two-sided ideals. Prove that if there is an element $u$ in $R$ for which $ux = x$ for all $x \in R$, then $R$ is a ring with identity and $u = 1_R$.

**2.** Prove that $\mathbb{A}$, the field of algebraic numbers, is not a finite extension of $\mathbb{Q}$.

**3.** Let $G$ be a finite group with the following property: for every two elements $x$ and $y$ of $G$ with $x \neq e$ and $y \neq e$, where $e$ is the identity element of $G$, there exists an automorphism $\theta \in \text{Aut}(G)$ such that $y = \theta(x)$. Prove that there is a prime number $p$ such that

$$G \cong \mathbb{Z}_p \oplus \mathbb{Z}_p \oplus \cdots \oplus \mathbb{Z}_p.$$

**1.11.3. General. 1.** The matrix $A$ is chosen at random from the set of all $2 \times 2$ matrices with entries from $\mathbb{Z}$. What is the probability that the determinant of $A$ is an even number?

**2.** A real number $c$ is given. Show that if one of the roots of the equation

$$x^3 - \frac{3}{4}x + c = 0$$

is in the closed interval $[-1,1]$, then all of the roots of this equation are in the interval $[-1,1]$.

**3.** Let $X$ be a set with $n$ elements and $C$ a family of the subsets of $X$ which satisfies the following conditions.

(i) If $A, B \in C$, then $A \cup B \in C$.

(ii) If $A \in C$, then $A^c := X \setminus A \in C$.

(iii) $\emptyset \in C$.

Show that the number of the elements of $C$ is equal to $2^k$ where $k \leq n$.

## 1.12. Twelfth Competition, Guilan University, March 1988

**1.12.1. Analysis. 1.** Let $f : \mathbb{R} \to \mathbb{R}$ be a continuous function. Moreover, suppose that there is an $M > 0$ such that for all $x, y \in \mathbb{R}$

$$|f(x) - f(y)| \geq M|x - y|.$$

Show that the function $f$ is one-to-one and surjective.

**2.** Let $(f_n)_{n=1}^{+\infty}$ be a decreasing sequence of nonnegative functions on a nonvoid set $S$. That is,

$$\forall n \in \mathbb{N}: \ f_n : S \longrightarrow \mathbb{R}, f_n \geq f_{n+1}, f_n \geq 0.$$

Prove that the series $\sum_{n \geq 1} f_n$ converges uniformly on $S$ if and only if the series $\sum_{n \geq 1} n(f_n - f_{n+1})$ does.[7]

**3.** Let $g : [0, 1] \to \mathbb{R}$ be a Riemann integrable function. Prove that

$$\int_0^1 \left( \int_x^1 g(t)dt \right) dx = \int_0^1 tg(t)dt.$$

**1.12.2. Algebra. 1.** Let $G$ be a group of order $2p$, where $p$ is an odd prime. Prove that $G$ has one and only one subgroup of order $p$. Also prove that $G$ has $p$ subgroups of order 2, or one subgroup of order 2. In the latter case, show that $G$ is a cyclic group.

**2.** Let $G$ be an abelian group of odd order and $\phi$ a homomorphism of $G$ of order 2. Show that every element $g \in G$ can, uniquely, be written as $g = xy$ where $\phi(x) = x$ and $\phi(y) = y^{-1}$.

**3.** Let $R$ be a unital ring with characteristic 2 such that for every $x \neq 1$ and $y \neq 1$, we have $xy^2 = xy$. Show that $R$ is commutative.

**4.** Suppose that $R$ is a unital ring, that every ideal of $R$ is principal, and that $f : R \to R$ is a surjective homomorphism. Prove that $f$ is an isomorphism.

**5.** Let $A$ be a $3 \times 3$ invertible matrix over a field $F$ such that $\det A = 1$ and $\operatorname{tr}(A) = \operatorname{tr}(A^{-1}) = 0$. Prove that $A^3 = I$.

**6.** If $\alpha$ and $\beta$ are two distinct roots of the equation

$$1 + x + \frac{x^2}{2!} + \cdots + \frac{x^p}{p!} = 0,$$

---

[7] This problem, as stated, is not correct! More precisely, for the "if part" of the problem, we need to assume further that the sequence $(f_n)_{n=1}^{+\infty}$ converges uniformly to zero on $S$.

where $p$ is a prime with $p > 2$, show that $\alpha - \beta$ and $\alpha + \beta$ and $\alpha\beta$ are irrational numbers.

**1.12.3. General. 1.** A circle $C$ centered at $O$ is given and the center of a square is on the circle. If the area of the square is not greater than half of the area of the circle, prove that one of the vertices of the square is inside the circle.

**2.** 1700 people have participated in a true-false test. Knowing that 15 questions have been given in the test, that none of the participants has answered two consecutive questions correctly, and that all of them have answered all questions, do there exist two equal answer sheets?

**3.** Evaluate

$$1 - C_n^2 + C_n^4 - C_n^6 + \cdots,$$

where $n \in \mathbb{N}$, $C_n^k := \dfrac{n!}{k!(n-k)!}$ if $k = 0, \ldots, n$, and $C_n^k := 0$ otherwise.

## 1.13. Thirteenth Competition, University of Tehran, March 1989

**1.13.1. Analysis. 1.** Let $f : \mathbb{R} \to \mathbb{R}$ be (Riemann) integrable and $f(x+y) = f(x) + f(y)$ for all $x, y \in \mathbb{R}$. Show that there exists a number $c$ such that $f(x) = cx$ for all $x \in \mathbb{R}$.

**2.** Show that if $a_n, b_n$ are in $\mathbb{R}$ for all $n \in \mathbb{N}$, $(a_n + b_n)b_n \neq 0$, and that both series $\displaystyle\sum_{n=1}^{+\infty} \frac{a_n}{b_n}$ and $\displaystyle\sum_{n=1}^{+\infty} \left(\frac{a_n}{b_n}\right)^2$ are convergent, then so is the series $\displaystyle\sum_{n=1}^{+\infty} \frac{a_n}{a_n + b_n}$.

**3.** Let $(f_n)_{n=1}^{+\infty}$, with $f_n : [0,1] \longrightarrow \mathbb{R}$, be a sequence of differentiable functions such that $\|f_n'\|_\infty \leq 1$ for all $n \in \mathbb{N}$. Show that if for any continuous function $g : [0,1] \longrightarrow \mathbb{R}$, we have

$$\lim_{n\to+\infty} \int_0^1 f_n g = 0,$$

then the sequence $(f_n)_{n=1}^{+\infty}$ converges uniformly to zero on $[0,1]$.

**1.13.2. Algebra. 1.** Let $R$ be a unital, commutative, and uncountable ring such that for every ideal $0 \neq I \lhd R$, the quotient ring $R/I$ is countable. Prove that $R$ is an integral domain.

**2.** Let $R$ be a subring of $M_n(\mathbb{Q})$. If

$$\begin{pmatrix} \alpha & 0 \\ 0 & \alpha \end{pmatrix} \in R \text{ for all } \alpha \in \mathbb{Q}, \qquad (*)$$

prove that every left (or right) ideal of $R$ is finitely generated. By finding a left ideal in the ring

$$T = \left\{ \begin{pmatrix} z & q_1 \\ 0 & q_2 \end{pmatrix} \mid z \in \mathbb{Z}, q_1, q_2 \in \mathbb{Q} \right\},$$

which is not finitely generated, show that the condition $(*)$ cannot be dropped.

**3.** Give an example of a group $G$ which contains two elements $a$ and $b$ such that $a$ and $b$ are of order two but $ab$ is of order infinity.

**4.** Let $G = \mathrm{GL}_2(\mathbb{Z}_3)$ (the multiplicative group of all $2 \times 2$ invertible matrices on $\mathbb{Z}_3$), $K = Z(G)$ denote the center of $G$, and

$$H = \left\{ \begin{pmatrix} a & b \\ 0 & c \end{pmatrix} : a, b, c \in \mathbb{Z}_3, ac \neq 0 \right\}.$$

(a) Prove $K \leq H \leq G$ and that the order of $H$ is 12.
(b) Prove

$$K = \bigcap_{x \in G} x^{-1} H x.$$

(c) Prove

$$\frac{G}{K} \cong S_4,$$

where $S_4$ denotes the group of all permutations on four letters.

**5.** $A = (a_{ij})$ is an $n \times n$ matrix, where

$$a_{ij} = \begin{cases} \delta_{i,n} & j = 1, \\ \delta_{i,j-1} & j > 1. \end{cases}$$

Suppose that $\xi$ is an $n$th root of unity in $\mathbb{C}$. Set

$$\nu(\xi) = \begin{pmatrix} 1 \\ \xi \\ \xi^2 \\ \vdots \\ \xi^{n-1} \end{pmatrix}.$$

Prove that $\nu(\xi)$ is an eigenvector of $A$ and find its corresponding eigenvalue.

**1.13.3. General. 1.** A regular $n$-gon is inscribed in a circle with radius one. Choose an arbitrary point on the circle and find the squares of the distances from the point $M$ to the vertices of the regular $n$-gon. Prove that the sum of these values is $2n$.

**2.** Find all polynomials $f(x)$ with rational coefficients with the property that $f(x)$ is irrational whenever $x$ is irrational.

**3.** Let $n$ parallel lines in the space that do not lie in a plane be given. Show that there are at least $n$ distinct planes each of which passes through at least two lines of these given lines ($n > 2$).

## 1.14. Fourteenth Competition, The University of Isfahan, March 1990

**1.14.1. Analysis. 1.** Let $f : \mathbb{R} \longrightarrow \mathbb{R}$ be twice differentiable on $\mathbb{R} \setminus \{x_0\}$ for some $x_0 \in \mathbb{R}$. If $f'(x) < 0 < f''(x)$ on $x < x_0$, and $f'(x) > 0 > f''(x)$ on $x > x_0$, then $f$ is not differentiable at $x_0$.

**2.** Let $g : \mathbb{R} \longrightarrow \mathbb{R}$ be a continuous function with the property that $g(x) > 0$ for $x \neq 0$ and $g(0) = 0$. Let $f : \mathbb{R} \longrightarrow \mathbb{R}$ be a uniformly continuous and bounded function such that $g \circ f$ is integrable on $\mathbb{R}$. Prove that $\lim_{x \to \infty} f(x) = 0$.

**3.** Let $f : \mathbb{R} \longrightarrow \mathbb{R}$ be a continuous function and $[a, b]$ and $[c, d]$ two closed intervals such that $[a, b] \subset f([c, d])$. Prove that there exists a subinterval $[r, s]$ of $[c, d]$ such that $f([r, s]) = [a, b]$.

**1.14.2. Algebra. 1.** Prove that if $G$ is a finite $p$-group (p a prime number), then $G' \neq G$, where $G'$ is the commutator (a.k.a. derived) subgroup of $G$.

**2.** Let $G$ be a finite group and $K$ a normal subgroup of $G$ of order $p$, where $p$ is the smallest prime dividing $|G|$. Prove that $K$ is a subgroup of $Z(G)$, where $Z(G)$ denotes the center of the group.

**3.** Give an example of a ring $R$ having two elements which have the greatest common divisor in $R$ but do not have the least common multiple in $R$.

**4.** Prove that if $F$ is a field and $n$ an integer greater than one, then $x^n y + x^{n-1} + 1$ is irreducible in $F[x, y]$. What can be said in case $n = 1$?

**5.** Let $V$ be a finite-dimensional vector space over a field $F$ and $V_1$ and $V_2$ two subspaces of $V$ with $\dim V_1 = \dim V_2$. Prove that there exists a subspace $U$ of $V$ such that

$$U \oplus V_1 = U \oplus V_2.$$

**1.14.3. General. 1.** Let $S$ be a set with $n$ elements, and

$$A = \{A_1, \ldots, A_n\}$$

a family containing $n$ distinct subsets of $S$. Show that there exists an element $x \in S$ such that the sets

$$A_1 \cup \{x\}, \ldots, A_n \cup \{x\}$$

are distinct.

**2.** A publisher is to exhibit 1369 titles of its published books. The books are to be exhibited as follows. Each day 100 titles are to be placed on the exhibition table and that no two titles are placed on the table more than once. Determine the maximum number of the days of the publisher exhibition.

**3.** Find all nonzero real numbers $a_1, \ldots, a_n$ with the following property:

$$\sum_{i=1}^{n} a_i^m = \sum_{i=1}^{n} a_i, \quad m = 1, \ldots, n+1.$$

## 1.15. Fifteenth Competition, Ferdowsi University of Mashhad, March 1991

**1.15.1. Analysis. 1.** The real function $f$ is defined on $[0, +\infty)$ and is increasing. The function $\varphi$ is defined on $[0, +\infty)$ by

$$\varphi(x) = \int_0^x f(t)dt.$$

(a) Prove that for all $x, y \geq 0$

$$\varphi\left(\frac{x+y}{2}\right) \leq \frac{1}{2}\Big(\varphi(x) + \varphi(y)\Big).$$

(b) Conclude that $\varphi$ is convex.

**2.** The real function $g$ is continuous on $[0, 1]$ and $g(0) = 0$. Define the sequence $(f_n)_{n=1}^{+\infty}$ of functions on $[0, 1]$ by

$$f_n(x) = \frac{g(x)(\sin x)^n}{1 + nx}.$$

Prove that the sequence $(f_n)_{n=1}^{+\infty}$ is uniformly convergent on $[0, 1]$.

**3.** A function $f$ from $(0, +\infty)$ into $(0, +\infty)$ is defined; it is a one-to-one correspondence and for all $x, y \in (0, +\infty)$, we have

$$2xy \leq xf(x) + yf^{-1}(y).$$

($f^{-1}$ is the inverse of $f$.)
   (a) Show that for all $x, y \in (0, +\infty)$,

$$\frac{y-x}{y}f(x) \leq f(y) - f(x) \leq \frac{y-x}{x}f(y).$$

(b) Conclude that there exists a $c \in \mathbb{R}$ such that $f(x) = cx$ for all $x \in (0, +\infty)$.

**1.15.2. Algebra. 1.** Let $G$ be a group of order $p^{\alpha}m$, where $p$ is a prime, $a \in \mathbb{N}$, $p \nmid m$, and that $G$ has exactly $(1 + p)$ Sylow $p$-subgroups. Show that

$$\left| \bigcap_{i=1}^{p+1} P_i \right| = p^{\alpha - 1},$$

where $P_i$'s are the Sylow $p$-subgroups of $G$.
- **Hint.** $[G : A \cap B] \leq [G : A][G : B]$.

**2.** Let $R$ be a unital ring such that every left ideal of it is also a right ideal of $R$. Prove that the intersection of all prime ideals of $R$ is equal to the set of the nilpotent elements of $R$.

**3.** Show that for every $n \times n$ matrix $A$, there is an $n \times n$ matrix $B$ such that $AB$ is an idempotent. [8]

**1.15.3. General. 1.** Find a necessary and sufficient condition for the following property: the product of two integers is divisible by the sum of them.

**2.** Let $f$ be nonnegative on $[0, 1]$ and $\int_0^1 f(x)dx = 1$. Prove that

$$\int_0^1 \left( x - \int_0^1 uf(u)du \right)^2 f(x)dx \leq \frac{1}{4}.$$

**3.** A train has $n$ wagons. Any passenger who is boarding the train (independent of others), chooses a wagon at random to get on.
  (a) What is the probability that at least one passenger gets on every wagon?
  (b) Using (a), evaluate the following sum.

$$\binom{n}{1}1^p - \binom{n}{2}2^p + \binom{n}{3}3^p - \cdots + (-1)^{n-1}\binom{n}{n}n^p,$$

where $1 \leq p \leq n$.

---

[8] This problem, as stated, is trivial! To make the problem nontrivial, prove that for every $n \times n$ matrix $A$ with entries from a field $F$, there is a matrix $B \in M_n(F)$ such that $AB$ is an idempotent whose rank is equal to that of $A$ (see Solution 3 of 2.15.2).

## 1.16. Sixteenth Competition, Razi (Rhazes or Rasis) University of Kermanshah, March 1992

**1.16.1.  Analysis. 1.** Let $g : [0, 1] \longrightarrow \mathbb{R}$ be a continuous function with $g(1) = 0$. If $f_n(x) = x^n g(x)$, show that the sequence $(f_n)_{n=1}^{+\infty}$ converges uniformly on $[0, 1]$.

**2.** The function $f : \mathbb{R} \longrightarrow \{0, 1, \ldots, 9\}$ is defined by

$$f(x) = \begin{cases} a_2 & x \neq [x], \\ 9 & x = [x], \end{cases}$$

where $[x]$ denotes the integral part of $x$ and $a_2$ is the second digit of the decimal expansion of $x - [x]$.
   (a) Prove that $f$ is periodic and determine its period.
   (b) If $c$ is the period of $f$, evaluate the following integral

$$\int_0^c x \, df(x).$$

(If the decimal expansion of a number ends in zero, reduce the last nonzero digit by one and change all other digits on the right of it to 9.)

**3.** Let $f : \mathbb{R} \longrightarrow \mathbb{R}$ be a uniformly continuous function. Prove that there are positive numbers $a$ and $b$ such that $|f(x)| \leq a|x| + b$ for all $x \in \mathbb{R}$.

**1.16.2. Algebra. 1.** Let $G$ be a finite nonabelian group, and $A, B$ two distinct abelian subgroups of $G$ such that

$$[G : A] = [G : B] = p,$$

where $[G : A]$ and $[G : B]$ denote the indexes of $A$ and $B$ in $G$, respectively, and $p$ is the smallest prime dividing $|G|$. Prove that $\mathrm{Inn}(G) \cong \mathbb{Z}_p \times \mathbb{Z}_p$, where $\mathrm{Inn}(G)$ stands for the set of all inner isomorphisms of $G$.

**2.** Let $R$ be a ring and $r \in R$ be such that $r - r^2$ is nilpotent. Prove that $R$ has a nonzero idempotent element whenever $r$ is not nilpotent.

**3.** Let $A = (a_{ij}) \in M_n(\mathbb{Q})$, where $a_{ij} = \gcd(i, j)$. Is $A$ invertible? Why?

## 1.17. Seventeenth Competition, Shahid Beheshti (former National) University, March 1993

**1.17.1. Analysis. 1.** Suppose that the function $f$ is continuous on the interval $[a, b]$ and that it is differentiable on $(a, b)$. Also suppose that the graph of $f$ is not a line segment. Prove that there exists a $c \in (a, b)$ such that

$$|f'(c)| > \left|\frac{f(b) - f(a)}{b - a}\right|.$$

**2.** If $p_1(x), p_2(x), p_3(x), p_4(x)$ are real polynomials in the indeterminate $x$, prove that

$$\int_{-1}^{x} p_1(t)p_3(t)dt \int_{-1}^{x} p_2(t)p_4(t)dt - \int_{-1}^{x} p_1(t)p_4(t)dt \int_{-1}^{x} p_2(t)p_3(t)dt$$

is divisible by $(x + 1)^4$.

**3.** Let $X$ be a metric space and $f : X \longrightarrow \mathbb{R}$ a continuous function. Show that the set $\{x \in X : f(x) = 0\}$ is an open subset of $X$ if and only if there exists a continuous function $g : X \longrightarrow \mathbb{R}$ such that $f = gf^2$.

**4.** Let $f : [0, 1] \longrightarrow \mathbb{R}$ be a continuous function such that $f(0) = f(1)$. Prove that there are two points $a, b$ satisfying the following: $0 < a \le b \le 1$, $b - a = \frac{1}{2}$, and $f(b) = f(a)$.

**1.17.2. Algebra. 1.** Let $G$ be a finite group and $H \le G$ be such that

$$\forall\, x \in G \;\; \big(\text{if } x \notin H \Longrightarrow H \cap (x^{-1}Hx) = \{e_G\}\big).$$

Prove that $[G : H]$ and $|H|$ are relatively prime.

**2.** Let $R$ be the ring of all $n \times n$ matrices over a field $F$, and $R[x, y]$ the ring of all polynomials in two indeterminates $x, y$ with coefficients from $R$. (note: $\forall a \in R : ax = xa, ay = ya, xy = yx$.) Suppose $f, g \in R[x, y]$ are such that $fg = 1$. Determine $gf$.

**3.** Let $V$ be a finite-dimensional vector space over a field $F$ and $T : V \longrightarrow V$ a linear transformation. Determine the dimension of $\ker T \cap T(V)$ in terms of the ranks of the powers of $T$.

## 1.18. Eighteenth Competition, Sharif University of Technology, March 1994

**1.18.1. Analysis. 1.** Let $(a_n)_{n=0}^{+\infty}$ be a sequence of positive reals and $A_n = a_0 + \cdots + a_n$. If $\lim_{n \to +\infty} A_n = +\infty$ and $\lim_{n \to +\infty} \dfrac{a_n}{A_n} = 0$, show that the radius of convergence of the power series $\sum_{n=0}^{+\infty} a_n x^n$ is one.

**2.** Let $f, g$ be continuous and periodic functions whose periods are one. Show that

$$\lim_n \int_0^1 f(x)g(nx)dx = \left( \int_0^1 f(x)dx \right) \left( \int_0^1 g(x)dx \right).$$

**3.** For any real number $x$, write the binary expansion of $x - [x]$ as follows

$$x - [x] = \sum_{k=1}^{\infty} \frac{a_k}{2^k}.$$

Show that the function $g : \mathbb{R} \longrightarrow \mathbb{R}$ defined by $g(x) = \limsup_{n \to \infty} \dfrac{\sum_{j=1}^{n} a_j}{n}$ maps every nonempty interval onto $[0, 1]$ (i.e., for every nonempty interval $I$, we have $g(I) = [0, 1]$). Using this, construct an open map from $\mathbb{R}$ into $\mathbb{R}$ that is not continuous.

**1.18.2. Algebra. 1.** Let $G$ be a finite group and $K$ a normal subgroup of $G$ such that $K$ as a group is simple and the square of its order does not divide the order of $G$. Prove that if $H$ is a subgroup of $G$ such that $H \cong K$, then $H = K$.

**2.** The following ring is given.

$$R = \begin{pmatrix} \mathbb{Z} & \mathbb{Q} \\ 0 & \mathbb{Q} \end{pmatrix} = \left\{ \begin{pmatrix} a & b \\ 0 & c \end{pmatrix} : a \in \mathbb{Z}, b, c \in \mathbb{Q} \right\}$$

Prove that every ascending chain of right ideals of $R$ is necessarily stable; that is, if

$$I_1 \subseteq I_2 \subseteq I_3 \subseteq \cdots, \quad \forall j \in \mathbb{N} : I_j \lhd_r R,$$

then there is an $n \in \mathbb{N}$ such that $I_n = I_{n+1} = I_{n+2} = \cdots$.

**3.** Let $V$ be a finite-dimensional vector space over a field $F$. If $\{V_i\}_{i \in I}$ is a set of proper subspaces of $V$ such that $\dim V_i = \dim V_j$ for all $i, j \in I$ and if $|I| < |F|$, prove that there is a proper subspace $U$ of $V$ such that

$$V = V_i \oplus U,$$

for all $i \in I$.

## 1.19. Nineteenth Competition, University of Kerman, March 1995

**1.19.1. Analysis. 1.** If $f : [0, 1] \longrightarrow [0, 1]$ is a continuous function, show that the equation

$$2x - \int_0^x f(t)dt = 1$$

has only one root in $[0, 1]$.

**2.** Let $f : (0, +\infty) \longrightarrow \mathbb{R}$ be a continuously differentiable function such that $\lim_{x \to +\infty} (f(x) + hf'(x)) = 0$ for some $h \in \mathbb{R}^+$. Show that $\lim_{x \to +\infty} f(x) = \lim_{x \to +\infty} f'(x) = 0$.

**3.** Let $A$ be the set of all continuous real functions on $[0, 1]$ and[9], for each $n \in \mathbb{N}$, $E_n$ denote the set consisting of all $f \in A$ for which there is an $a \in [0, 1]$ such that

$$|f(x) - f(a)| \leq n|x - a|,$$

for all $x \in [0, 1]$.
   (a) Show that $E_n$'s are closed sets with empty interior.
   (b) Conclude that there is a continuous function on $[0, 1]$ which is nowhere differentiable.

**1.19.2. Algebra. 1.** Let a group $G$ be given. Suppose that the set

$$\mathcal{A} = \left\{ N \triangleleft G : \frac{G}{N} \cong G \right\}$$

is ordered by inclusion and that $\mathcal{A}$ has a maximal element. Prove that the trivial group $\{e\}$ is the only maximal element of $\mathcal{A}$.

**2.** Let $R$ be a ring with identity. Set $S = \sum J$, where $J$ can be any minimal left ideal of $R$ (in case such ideals are non-existent, set S=0).
   (a) Prove that $S$ is a two-sided ideal in $R$ (i.e., $S \triangleleft R$).
   (b) If $S \neq 0$ and the product of any two nonzero two-sided ideals of $R$ is a nonzero two-sided ideal of $R$, then $S = \bigcap I$, where $I$ can be any nonzero two-sided ideal of $R$.

**3.** A square matrix $A = (a_{ij})_{n \times n}$ over a field $F$ is given. Suppose that there are scalars $x_i, y_i \in F$ such that $a_{ij} = x_i + y_j$ $(1 \leq i, j \leq n)$. Prove that rank$(A) \leq 2$.

**4.** Let $F$ be a field, $n \in \mathbb{N}$, and $A \in M_n(F)$ with rank$(A) = 1$. Prove that

$$\det(I + A) = 1 + \text{tr}(A).$$

---

[9] The set, in fact the algebra, $A$ is assumed to be endowed with the uniform metric of $A$, which is induced by the uniform norm of $A$, usually denoted by $||.||_\infty$

## 1.20. Twentieth Competition, Sharif University of Technology, February 1996

**1.** Let $\mathbb{C}$ be the field of complex numbers and
$$\{(x,y),(z,t),(x',y'),(z',t')\} \subseteq \mathbb{C}^2.$$
Prove that there are scalars $\alpha, \beta \in \mathbb{C}$ which are not simultaneously zero such that the vectors $\theta = \alpha(x,y) + \beta(z,t)$ and $\theta' = \alpha(x',y') + \beta(z',t')$ are linearly dependent.

**2.** The sequence $(a_n)_{n=1}^{+\infty}$ of nonnegative real numbers satisfies the following property:
$$1 + a_{m+n} \le (1 + a_m)(1 + a_n),$$
for all $m, n \in \mathbb{N}$. Prove that the sequence $(x_n)_{n=1}^{+\infty}$ defined by $x_n = \sqrt[n]{1 + a_n}$ is convergent.

**3.** Let $G$ be a group such that for all $\sigma \in \mathrm{Aut}(G)$ and for all $x \in G$ we have $\sigma(x) = x$ or $\sigma(x) = x^{-1}$. Prove that $G$ is solvable.

**4.** The function $f : (\frac{1}{4}, 1) \longrightarrow \mathbb{R}$ for all $x \in (\frac{1}{4}, 1)$ satisfies the following
$$x^{f(x)} = f(x).$$
Prove that $f$ is uniformly continuous on $(\frac{1}{4}, 1)$.

**5.** Let $R$ be a commutative ring with identity and with the following properties:
  (a) The intersection of all of its nonzero ideals is nontrivial.
  (b) If $x$ and $y$ are zero divisors in $R$, then $xy = 0$.
  Prove that $R$ has exactly one nontrivial ideal.

**6.** Let $f, g : [0, +\infty) \longrightarrow \mathbb{R}$ be two functions. Suppose that $f$ is decreasing and $\lim_{x \to +\infty} f(x) = 0$ and that $g$ is a periodic function such that $\int_0^p g = 0$, where $p$ is the period of $g$. Show that $\int_0^{+\infty} fg$ is convergent.

## 1.21. Twenty-First Competition, University of Tehran, March 1997

### 1.21.1. Analysis. 1. A set $S \subset \mathbb{R}$ and a function $f : S \longrightarrow \mathbb{R}$ are given.
  (a) Assume that $\alpha$ is a limit point of the set $S$ and that for every $\varepsilon > 0$ the set $\{x \in S : |f(x)| \ge \varepsilon\}$ is finite. Show that $\lim_{x \to \alpha} f(x) = 0$.
  (b) Assume that $S$ is compact and that for every limit point $\alpha$ of $S$, we have $\lim_{x \to \alpha} f(x) = 0$. Show that the set $\{x \in S : |f(x)| \ge \varepsilon\}$ is finite for all $\varepsilon > 0$.

**2.** Let $g : \mathbb{R} \longrightarrow \mathbb{R}$ be a continuous function satisfying the following
$$\lim_{x \to +\infty} (g(x + t) - g(x)) = 0,$$
for all $t \in \mathbb{R}$.

(a) Suppose that $K$ is a compact subset of $\mathbb{R}$. Show that
$$\forall \varepsilon > 0 \ \exists M > 0 \ni x > M \ \Rightarrow \ |g(x+t) - g(x)| < \varepsilon \ (\forall \ t \in K).$$

(b) Use (a) to show that
$$\lim_{x \to +\infty} \left( \int_x^{x+1} g(u)du - g(x) \right) = 0, \quad \lim_{x \to +\infty} \frac{g(x)}{x} = 0.$$

**3.** Let $f : [0,1] \longrightarrow \mathbb{R}$ be a bounded function whose limit from the left exists at any point in $[0,1]$. Prove that $f$ is Riemann integrable on $[0,1]$.

**1.21.2. Algebra. 1.** Let $R$ be a ring and $A$ and $B$ ideals of $R$ with $A \subseteq B$. By definition, $A$ is said to be *small* in $B$ and we write $A \overset{s}{\subseteq} B$ if for every ideal $C$ of $R$ for which $A + C = B$, then $C = B$. Show that if $A \overset{s}{\subseteq} B$ and $C \overset{s}{\subseteq} D$ and $A \cap C = B \cap D = \{0\}$, then $A + C \overset{s}{\subseteq} B + D.$[10]

**2.** Let $G$ be a group and $H$ a subgroup of $G$ such that $H \leq Z(G)$, where $Z(G)$ denotes the center of $G$. Prove that if $[G : H] = p^2$, where $p$ is prime, then the derived subgroup of $G$ is cyclic.

**3.** Let $A$ be a nonzero real $n \times n$ matrix such that
$$A = (a_{ij})_{1 \leq i,j \leq n}, \quad a_{ik}a_{jk} = a_{kk}a_{ij}, \ \forall i, j, k.$$

Prove that
(a) $\mathrm{tr}(A) \neq 0$.
(b) The matrix $A$ is symmetric.
(c) The characteristic polynomial of $A$ is equal to $x^{n-1}(x - \mathrm{tr}(A))$.

## 1.22. Twenty-Second Competition, University of Ahwaz, March 1998

**1.22.1. Analysis. 1.** Let $f : [0,1] \longrightarrow \mathbb{R}$ be a differentiable function with $f(0) = f(1) - 1 = 0$. Show that for each $n \in \mathbb{N}$, there are $x_1, \ldots, x_n \in [0,1]$ such that $\displaystyle\sum_{i=1}^{n} \frac{1}{f'(x_i)} = n$.

**2.** Let $X$ be a metric space, $f : X \longrightarrow \mathbb{R}$ a continuous function, and $(f_n)_{n=1}^{+\infty}$ a sequence of continuous nonnegative functions on $X$.

(a) If $\sum_{n=1}^{+\infty} f_n = f^2$, then there exists a sequence $(g_n)_{n=1}^{+\infty}$ of continuous real functions on $X$ such that $f_n = g_n f$ for all $n \in \mathbb{N}$.

(b) Prove that $\sum_{n=1}^{+\infty} g_n = f$ provided that $\sum_{n=1}^{+\infty} g_n$ is uniformly convergent on $X$ and that the interior of $f^{-1}(\{0\})$ is the empty set.

---

[10] The hypothesis "$A \cap C = B \cap D = \{0\}$" is redundant!

**3.** Let $(f_n)_{n=1}^{+\infty}$ be a sequence of analytic functions from the region $D \subset \mathbb{C}$ into $\mathbb{C}$ such that $f_n$ uniformly converges to $f$ on $D$. Prove that if $\gamma$ is a simple closed curve inside $D$, then $f$ is analytic inside and on the curve $\gamma$ and that for all $z_0$ inside $\gamma$, we have

$$\lim_n \int_\gamma \frac{f_n(z)}{z - z_0} dz = \int_\gamma \frac{f(z)}{z - z_0} dz.$$

**1.22.2. Algebra. 1.** Let $G$ be a group, $H \leq K \leq G$, $[G : K]$ an odd number, and $[K : H] = 2$. Also suppose that there is a $k \in K$ of order 2 which is not conjugate to any element of $H$ in $G$. Prove that $G$ has a subgroup of index 2.

**2.** In a commutative ring $R$ (which is not necessarily unital), the ideal $M$ is called maximal if $R^2 \not\subseteq M$ and $M$ is not contained in any ideal other than itself and $R$. Now, we define a subring $J$ of $R$ as follows. If $R$ has no maximal ideal, set $J = R$. If not, set $J$ to be the intersection of all maximal ideals of $R$. Prove that $J$ has no maximal ideal.

**3.** Let $V$ be an $n$-dimensional vector space over a field $F$ ($n \in \mathbb{N}$) and $S, T : V \longrightarrow V$ two linear transformations such that the characteristic polynomial of one of them is irreducible over $F$. If $L = TS - ST \neq 0$, then the rank of $L$ is greater than one.

**1.22.3. General. 1.** If $\lim\limits_{x \to +\infty} \left(\dfrac{b+x}{b-x}\right)^x = 9$, then $b$ is equal to

(a) 3.      (b) $3^e$.      (c) $\ln 3$.      (d) $\ln 9$.

**2.** If $\lim\limits_{x \to 0} \dfrac{1}{2bx - \sin x} \displaystyle\int_0^x \frac{t^2}{\sqrt{t-a}} dt$ exists and is nonzero, then which of the following is admissible?

(a) $a < 0$, $b = \frac{1}{2}$.
(b) $a = 0$, $b = \frac{1}{2}$.
(c) $a > 0$, $b = \frac{1}{2}$.
(d) $a = -1$, $b = \frac{-1}{2}$.

**3.** Which of the following is the general solution of the differential equation $e^x(1+x)dx - (xe^x - ye^y)dy = 0$?

(a) $x^2 e^{x-y} + y^3 = c$.
(b) $y^2 + 2xe^{x-y} = c$.
(c) $y^2 e^{x-y} + x^2 = c$.
(d) $xe^{x-y} + \frac{1}{2}y^2 = 0$.

**4.** If $A$ and $B$ are two ends of the diagonal of a cube with side length 3, and an ant wants to travel from $A$ to $B$, what is the smallest distance in meters that the ant needs to travel?

(a) $3\sqrt{3}$.　　(b) $3 + 3\sqrt{3}$.　　(c) $3\sqrt{6}$.　　(d) $3\sqrt{5}$.

**5.** If $f(x) = \sum_{n=1}^{+\infty} \dfrac{(x+2)^n}{2^n(n+1)}$, then which one is the domain of $f$?

(a) $[-4, 0)$.
(b) $[-4, 0]$.
(c) $(-4, 0]$.
(d) None of (a), (b), and (c).

**6.** If $x^3 + y^3 = xy + 1$ and in a neighborhood of $(1, 1)$, $y$ is a function of $x$, which one is $y'(1)$?

(a) $-1$.　　(b) $\frac{-2}{3}$.　　(c) $1$.　　(d) $\frac{3}{2}$.

**7.** A particle, initially located in the origin, is discretely moving along the $x$-axis one unit to the right or one unit to the left with probability $p$ and $q = 1-p$, respectively. The probability that after $2k$ moves ($k \geq 5$), the particle is 10 units far away from the origin is

(a) $C_{2k}^{k+5} p^k q^k (p^{10} + q^{10})$.
(b) $C_{2k}^{k+5} p^{k+5} q^{k-5}$.
(c) $C_{2k}^{k+5} p^{k-5} q^{k+5}$.
(d) $C_{2k}^{k+5} p^{k-5} q^{k-5} (p^{10} + q^{10})$.

**8.** An arrow, as shown in the figure, is moving according to the following law. If the arrow is put at a point $B$, then it jumps as much as the length of the arc $AB$ in the trigonometric direction. At how many point(s) can one put the arrow so that after 36 seconds it hits the point $A$?

(a) Exactly at one point.
(b) Exactly at 18 points.
(c) Exactly at 35 points.
(d) At no point.

**9.** Consider the function $f(x) = \begin{cases} x^2 & x \text{ irrational} \\ 0 & x \text{ rational} \end{cases}$. The point(s) at which $f$ is differentiable is/are

(a) the two points $x = 0, 1$.
(b) exactly one point.
(c) rational points.
(d) no point.

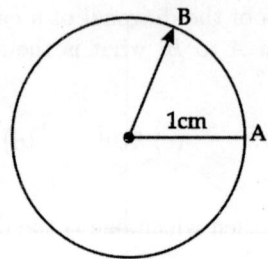

Figure 2

**10.** Consider the series $\displaystyle\sum_{n=0}^{+\infty} \frac{\cos(x^n)}{1+n^2}$.

(a) The series converges only at zero.
(b) The series converges everywhere.
(c) The series converges at nonzero points.
(d) The series converges absolutely at one point.

**11.** If we consider the sequence $(x_n)_{n=1}^{+\infty}$ $(x_1 \neq 0)$ of real numbers satisfying the following recursive relation $x_{n+1} = \dfrac{x_n}{1+x_n}$, then

(a) $x_n \to 1$ as $n \to +\infty$.
(b) $(x_n)_{n=1}^{+\infty}$ has a subsequence converging to one.
(c) $(x_n)_{n=1}^{+\infty}$ has a subsequence converging to zero.
(d) $(x_n)_{n=1}^{+\infty}$ is divergent.

**12.** Which of the following propositions is correct?

(a) $\lim_n \int_0^n \frac{dx}{1+x^2}$ exists and is positive.
(b) $\lim_n \int_0^n \frac{dx}{1+x^2}$ exists and is negative.
(c) $\lim_n \int_0^n \frac{dx}{1+x^2}$ is divergent.
(d) $f(x) = \frac{1}{1+x^2}$ is not integrable on the interval $[0, n]$.

**13.** If the function $f : \mathbb{R} \longrightarrow \mathbb{R}$ is differentiable and its derivative is bounded, then

(a) the function $f$ is bounded.
(b) the function $f$ is increasing.
(c) the function $f$ is uniformly continuous.
(d) the function $f$ is decreasing.

**14.** If the function $f : [0, 1] \longrightarrow [0, \frac{1}{2}]$ is continuous, then

(a) the graph of $f$ intersects the line $y = 2x$.
(b) there exists only one point $x_0$ for which $f(x_0) = x_0$.
(c) $f(x) \neq 0$ for all $x \in [0, 1]$.
(d) the equation $f(x) = x$ has at most two solutions.

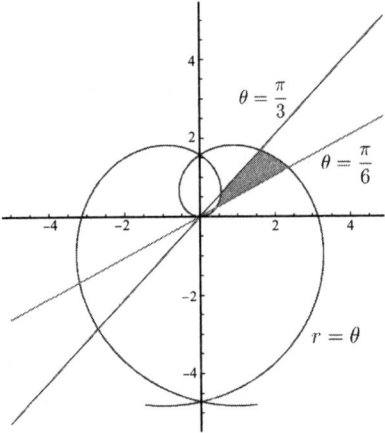

Figure 3

**15.** If $f(x) = \dfrac{|x|}{1 + |x|}$ for all $x \in \mathbb{R}$, then

(a) the function $f$ is increasing.
(b) $f(x + y) = f(x) + f(y)$.
(c) $f(f(x + y)) \le f(f(x)) + f(f(y))$.
(d) the function $f$ is continuous only at zero and $f(x + y) \le f(x) + f(y)$.

**16.** Below is drawn the graph of the polar curve $r = \theta$. Which one is the area of the shaded region?

(a) $\int_{\frac{\pi}{6}}^{\frac{\pi}{3}} \theta^2 d\theta$.

(b) $\int_{-\frac{\pi}{3}}^{-\frac{\pi}{6}} \frac{1}{2}\theta^2 d\theta$.

(c) $\int_{\frac{2\pi}{3}}^{\frac{5\pi}{6}} \frac{1}{2}\theta^2 d\theta - \int_{\frac{\pi}{6}}^{\frac{\pi}{3}} \frac{1}{2}\theta^2 d\theta$.

(d) $\int_{\frac{5\pi}{3}}^{\frac{11\pi}{6}} \frac{1}{2}\theta^2 d\theta - \int_{\frac{\pi}{6}}^{\frac{\pi}{3}} \frac{1}{2}\theta^2 d\theta$.

**17.** The value of $\int_0^1 x^5 e^x dx$ is

(a) $-44e$.　　(b) $120 + 44e$.　　(c) $44e$.　　(d) $e$.

**18.** If $y = \displaystyle\int_{\sin x}^{x^2} (x + t)dt$, then $\dfrac{dy}{dx}$ at $x = 0$ is equal to

(a) 5.　　(b) 0.　　(c) 2.　　(d) $-1$.

**19.** For what values of $p$, does the integral $\displaystyle\int_1^2 \dfrac{dx}{x(\ln x)^p}$ converge?

(a) $p = 1$.　　(b) $p > 1$.　　(c) $p < 1$.　　(d) $p = 2$.

**20.** $\int_0^{+\infty} \dfrac{e^{-ax} - e^{-bx}}{x} dx$ $(0 < a < b)$ is equal to

(a) $\ln(\frac{a}{b})$.     (b) $\ln(\frac{b}{a})$.     (c) $\ln(ab)$.     (d) $e^{ab}$.

**21.** Which of the following statements is not correct?

(a) A power series might converge at both of the end points of its interval of convergence.
(b) A power series can converge conditionally or diverge at any of the end points of its interval of convergence.
(c) A power series can converge absolutely at one of the end points of its interval of convergence and converge conditionally at the other end point.
(d) A power series can converge conditionally at one of the end points of its interval of convergence and diverge at the other end point.

**22.** Using the spherical coordinates to compute the volume of the ellipsoid, the Jacobian of the transformation is equal to

(a) $\rho^2 \sin(\varphi)$.     (b) $\rho \sin(\varphi)$.     (c) $abc\rho^2 \sin(\varphi)$.     (d) $abc$.

**23.** Letting $F(x, y, z) = (x - y)\overrightarrow{i} + (y - z)\overrightarrow{j} + (x - y)\overrightarrow{k}$ $(-1 \le x, y, z, \le 1)$ be a vector field, how much is the flux of the field $F$ through the surface of a cube of side length two centimeters?

(a) 8 cm$^3$.     (b) 16 cm$^3$.     (c) 1 cm$^3$.     (d) 14 cm$^3$.

**24.** For the vector field $F(x, y, z) = f(x, y, z)\overrightarrow{i} + g(x, y, z)\overrightarrow{j} + h(x, y, z)\overrightarrow{k}$, we have Curl$(F) = 0$ but $\oint_C F.dR \ne 0$, where $C : x^2 + y^2 = 1$. Then

(a) div$(F) < 0$.
(b) the domain of $F$ is connected.
(c) the domain of $F$ is simply connected.
(d) the domain of $F$ is not connected.

**25.** The family of orthogonal curves to the curves $r = c(1 + \cos\theta)$ is

(a) $r = c(1 - \sin\theta)$.
(b) $r = c\sin\theta$.
(c) $r = c(1 + \sin\theta)$.
(d) $r = c(1 - \cos\theta)$.

**26.** The general solution of the differential equation $y'' - 2y' - 3y = 64xe^{-x}$ is

(a) $y = c_1e^{3x} + c_2e^{-x} - e^{-x}(8x^2 + 4x + 1)$.
(b) $y = c_1e^{3x} + c_2e^{-x}$.
(c) $y = c_1e^{3x} + c_2e^{-x} - 8x^2e^{-x}$.
(d) $y = c_1e^{3x} + c_2e^{-x} - e^{-x}(8x^2 + 1)$.

**27.** Using the Gamma function for the definition of the factorial of numbers, $(\frac{1}{2})!$ is

a) $\sqrt{\pi}$.  (b) $\frac{\sqrt{\pi}}{2}$.  (c) $\frac{1}{\sqrt{\pi}}$.  (d) $\sqrt{\frac{\pi}{2}}$.

**28.** The equation $x^4 + x^2 - 1 = 0$ has roots as follows.

(a) Two positive roots, two negative roots.
(b) One positive root, one negative root, two nonreal roots.
(c) No real root.
(d) Four real roots.

**29.** The equation $f(x) = e^{-x} - \sin x = 0$ has a real root in the interval $[0.4, 0.8]$. The appropriate fixed point iteration function to find this root is

(a) $g(x) = x + e^{-x} - \sin x$.
(b) $g(x) = -\ln(\sin(x))$.
(c) $g(x) = \sin^{-1}(e^{-x})$.
(a) $g(x) = \frac{1}{2}(2x + e^{-x} - \sin x)$.

**30.** In the bisection method, the number of times needed in order to obtain accuracy within $\varepsilon$ is equal to

(a) $1 + \dfrac{\log \frac{b-a}{\varepsilon}}{\log 2}$.
(b) $\log(b-a) - \log(\varepsilon)$.
(c) cannot be determined.
(d) $1 + \dfrac{\log(b-a)}{\log 2}$.

**31.** In the iteration method, $P_{n+1} = g(P_n)$ $(n \geq 0)$ where $g(x) = x - \dfrac{f(x)}{f'(x)}$, as the multiplicity of the root $x = P$ increases,

(a) $g'(P)$ approaches zero.
(b) the speed of the convergence does not change.
(c) $g'(P)$ gets very close to one.
(d) the method is not convergent.

**32.** Which of the following propositions is correct?

(a) If $\sum_{n=1}^{+\infty} \frac{a_n}{a_{n+1}}$ converges, then so does the series $\sum_{n=1}^{+\infty} a_n$.
(b) If $\sum_{n=1}^{+\infty} a_n$ converges, then $\lim_n na_n = 0$.
(c) If $\lim_n na_n = 0$, then the series $\sum_{n=1}^{+\infty} a_n$ is convergent.
(d) If $\sum_{n=1}^{+\infty} \frac{a_{n+1}}{a_n}$ converges, then so does the series $\sum_{n=1}^{+\infty} a_n$.

**33.** Which one is the equation of the plane which includes the line $2y - 2 = 2x = z$ and is parallel to the line $\begin{cases} x - y = 0 \\ z = 0 \end{cases}$?

(a) $y - x = 1$.
(b) $y = 2x$.
(c) $z - 2y - 2x = 2$.

(d) $y - x = -1$.

**34.** Which of the following statements is a more complete definition of "algorithm"?

(a) "Algorithm" is a method for solving a problem.
(b) "Algorithm" is a logical method together with a terminating condition for solving a problem.
(c) "Algorithm" is a logical procedure without ambiguity, which includes a finite set of stages that are related to one another, together with a terminating condition.
(d) "Algorithm" just means method.

**35.** Every digit in octal expansion is equivalent to

(a) one digit in decimal expansion.
(b) five digits in binary expansion.
(c) two digits in hexadecimal expansion.
(d) three digits in binary expansion.

**36.** $A$ is infinite if and only if

(a) $A$ is countable.
(b) $\text{card}(A \setminus B) = \text{card}(A)$ for all $B \subseteq A$.
(c) $A$ is equivalent to any of its infinite subsets.
(d) for all natural numbers $n$, $A$ includes a subset which is equivalent to an $n$-set.

**37.** Let $A$ be an ordered set and $B \subseteq C \subseteq A$. Which of the following claims is true?

(a) If $\sup(A)$ and $\sup(C)$ exist, then $\sup(B) \leq \sup(C)$.
(b) If $\sup(B)$ and $\sup(C)$ exist, then $\sup(C) \leq \sup(B)$.
(c) If $\sup(B)$ and $\sup(C)$ exist, then $\inf(B) = \inf(C)$.
(d) There is no relation between $\sup(B)$ and $\sup(C)$.

**38.** Which of the following is equivalent to Zorn's Lemma?

(a) Every subset of an ordered set having an upper bound has sup.
(b) There is a choice function for every nonempty family of nonempty subsets of an arbitrary set $X$.
(c) If $\text{card}(A) \leq \text{card}(B) \leq \text{card}(P(A))$, then $\text{card}(A) = \text{card}(B)$ or $\text{card}(B) = \text{card}(P(A))$.
(d) Every bounded subset of an ordered set has a maximal element.

**39.** The integral $\displaystyle\int_{-a}^{a} \frac{\sin x dx}{e^x + e^{-x}}$ is equal to

(a) 2.
(b) 0.
(c) $2 \int_{0}^{a} \frac{\sin x dx}{e^x + e^{-x}}$.

(d) It does not exist.

**40.** If $f(a + b - x) = f(x)$ for all $x \in [a, b]$, then $\int_a^b x f(x) dx$ is equal to

(a) $\frac{a-b}{2} \int_a^b f(x) dx$.

(b) $\int_a^b f(x) dx$.

(c) $\frac{a+b}{2} \int_a^b f(x) dx$.

(c) $(a + b) \int_a^b f(x) dx$.

## 1.23. Twenty-Third Competition, Sharif University of Technology, March 1999

**1.23.1. Analysis. 1.** Let $\mathcal{X} = \mathcal{C}_b(\mathbb{R})$ be the space of all continuous bounded functions $g : \mathbb{R} \longrightarrow \mathbb{R}$ which is endowed with the norm $\|g\| = \sup_{t \in \mathbb{R}} |g(t)|$. For a given function $f \in \mathcal{X}$, define $f_\alpha : \mathbb{R} \longrightarrow \mathbb{R}$ by $f_\alpha(t) = f(t + \alpha)$, where $\alpha \in \mathbb{R}$. Prove or disprove the following propositions.

(a) If $\{f_\alpha : \alpha \in \mathbb{R}\}$ is compact, then $f$ is uniformly continuous on $\mathbb{R}$.

(b) If $f$ is uniformly continuous on $\mathbb{R}$, then the set $\{f_\alpha : \alpha \in \mathbb{R}\}$ is compact in $\mathcal{X}$.

**2.** Let $\alpha \in \mathbb{R}$ and $\alpha \neq 2k\pi$ for all $k \in \mathbb{Z}$. Find the sum of the series $\sum_{n=1}^{+\infty} \frac{e^{in\alpha}}{n}$ and justify your answer. Ditto for $\sum_{n=1}^{+\infty} \frac{\sin(n\alpha)}{n}$ and $\sum_{n=1}^{+\infty} \frac{\cos(n\alpha)}{n}$.

**3.** Let $f, f_1, f_2, \ldots$ be real continuous functions on the interval $[a, b]$ whose derivatives are also continuous on $[a, b]$. If the sequence $(f_n)_{n=1}^{+\infty}$ converges to $f$ pointwise and the sequence $(f'_n)_{n=1}^{+\infty}$ converges pointwise to a continuous function $g$, show that $f'(x) = g(x)$ for all $x \in [a, b]$. (Note: at the end points $a$ and $b$, only the right and left derivatives are to be considered.)

**1.23.2. Algebra. 1.** Let $G$ be a nonabelian group. Prove that the group of the inner automorphisms of $G$ cannot be nonabelian of order 8.[11]

**2.** Let $R$ be a ring and $H$ the intersection of all nonzero right ideals of $R$. Show that if $H \neq 0$, then $H$ is a two-sided ideal in $R$ and that we have $H^2 = 0$, or else $R$ is a division ring.

**3.** Let $A$ be an $n \times n$ matrix with entries from a field $F$. Prove that if for every matrix $B$ with trace zero, we have $\text{tr}(AB) = 0$, then $A = \lambda I$ for some scalar $\lambda$ in $F$.

---

[11] This problem is wrong!

## 1.24. Twenty-Fourth Competition, Khajeh Nasir Toosi University of Technology, May 2000

**1.24.1. First Day. 1.** Let $f, g : [a, b] \longrightarrow \mathbb{R}$ be continuous and that $g'$ exists on $[a, b]$. Show that if $(f(a) - g'(a))(g'(b) - f(b)) > 0$, then there exists a $c \in (a, b)$ such that $f(c) = g'(c)$.

**2.** Let $D$ be a domain in the complex plane and $u$ a real valued harmonic function on $D$. Show that if the set

$$A = \{(x_0, y_0) \in D : \exists r > 0 \ni \forall (x, y) \in B_r(x_0, y_0) \cap D \Rightarrow u(x, y) \le u(x_0, y_0)\}$$

is nonempty, then $u$ is constant on $D$. Recall that a domain is a connected open set and that

$$B_r(x_0, y_0) = \{(x, y) \in \mathbb{R}^2 : \sqrt{(x - x_0)^2 + (y - y_0)^2} < r\}.$$

**3.** Let $t(n)$ be the smallest prime factor of $n$ for all $1 < n \in \mathbb{N}$. Prove that $t(n) < t(3^n - 2^n)$.

**4.** Let $G$ be an infinite group with the property that for every two infinite subsets $X$ and $Y$ of $G$, there exist $x \in X$ and $y \in Y$ such that $xy = yx$. Prove that if the center of $G$ has finite index in $G$, then $G$ is abelian.

**5.** Let $M(n, d)$ denote the maximal number of $n$-tuples from $\{0, 1\}$ such that every two of which differ at least in $d$ components. (For instance, $M(4, 3) = 2$.) Prove that $M(2d - 1, d) \le 2d$.

**6.** A ruthless governor has made three mathematicians, who have been sentenced to death, play the following game.

Two lookalike boxes, one of which contains two black marbles and a white marble, and the other which contains two white marbles and a black marble, are provided. Each of the convicts picks a marble, which is not to be replaced, from one of the boxes at random. If the picked marble is black, the convict will be executed; if not, s/he will be set free. Assuming that every convict witnesses the choice(s) made by the convict(s) prior to oneself and that the convicts will make the most logical choice, what is the probability that the second person survives? The probability of survival is higher for the third person or for the second person?

**1.24.2. Second Day. 1.** Let $B$ be a nonempty bounded set in $\mathbb{R}^n$ with the property that for each pair of points $x, y$ in $B$, there exists an open ball $U$ such that $U \subseteq B$ and $x, y \in U$.
(a) Prove that $B$ is an open ball.
(b) Show that if we replace $\mathbb{R}^n$ by a complete metric space, then the conclusion of (a) does not necessarily hold.

**2.** Let $f$ be a continuous function on the interval $[a, b]$ such that $\int_a^b f(t)dt \neq 0$. Show that for each $k \in (0, 1)$, there exists a $c \in (a, b)$ such that

$$\int_a^c f(t)dt = k \int_a^b f(t)dt.$$

**3.** Let matrices $A, B, C$ be such that $AB$, $BC$, and $ABC$ make sense. Denoting the rank of any matrix $P$ by $r(P)$, prove that
(a) $r(BC) + r(AB) \leq r(ABC) + r(B)$.
(b) $r(A) + r(B) \leq r(AB) + n$,
where $n$ is the number of columns of the matrix $A$.

**4.** Let $R$ be a ring. Prove that if for any left ideal $I$ of $R$, we have $I^2 = I$, then for any two-sided ideal $K$ of $R$ and any left ideal $I$ of $R$, we have $I \cap K = KI$. (Remark: Note that $R$ is not necessarily unital.)

**5.** In an athletic tournament, $n$ teams have participated and any two teams have played once against one another. Assuming that the game result is to be either win or lose, prove that at the end of the tournament there are two teams whose wins are equal if and only if there are three teams $A, B, C$ such that $A$ has won $B$, $B$ has won $C$, and $C$ has won $A$.

**6.** Two persons are playing the following game.
First, the first person chooses a triple from $\{0, 1\}$ (for instance, $(0, 1, 0)$). The second person, who knows the triple chosen by the first person, chooses a different triple. Then a machine which at random generates 0 with probability $\frac{1}{2}$ and 1 with probability $\frac{1}{2}$ is turned on. We write the numbers generated by the machine from left to right and the player whose chosen triple comes first is to win the game. Show that no matter what choice is made by the first person, there is a choice for the second person so that the probability of winning the game by the second person is greater than $\frac{1}{2}$.

## 1.25. Twenty-Fifth Competition, IKIU of Qazvin, May 2001

**1.25.1. First Day. 1.** Let $G$ be a finite group of order $n$ such that $[G : Z(G)] = 4$. Prove that $8 \mid n$. For any given natural number $n$ such that $8 \mid n$, construct a group with the aforementioned property. (Here, $Z(G)$ denotes the center of $G$.)

**2.** Let $S, T$ be two linear transformations on a vector space $V$. Prove that if $S^2 = S$, $T^2 = T$, $\ker T \subseteq \operatorname{im} S$, and $\operatorname{im} T \subseteq \ker S$, then $T + S$ is the identity transformation.

**3.** Let $f$ be a twice differentiable function such that $f''(t) < 0$ for all $t \in \mathbb{R}$. Prove that if for two real numbers $x$ and $y$ we have $f'(y) + x < f(y + 1)$, then $f(y) > x$.

**4.** Let $f : (a, b) \longrightarrow \mathbb{R}$ be continuously differentiable and
$$\lim_{x \to a^+} f^2(x) = 0, \quad \lim_{x \to b^-} f^2(x) = e - 1.$$
Prove that if $2f(x)f'(x) - f^2(x) \geq 1$ for all $x \in (a, b)$, then $0 < b - a \leq 1$. Give an example in which $b - a = 1$. (Note: $f^2(x) = (f(x))^2$.)

**5.** Seven boxes, on each of which a number from 1 to 7 is written, are at our disposal. Seven marbles numbered 1, seven marbles numbered 2, ..., and seven marbles numbered 7 are provided. We place the marbles in the boxes at random in such a way that each box has exactly seven marbles. Play the following game.

In stage one, a marble is drawn at random from box number one (without replacement). In stage $i$ ($i \geq 2$), a marble is drawn at random from the box whose number is identical to that of the marble drawn in stage $(i - 1)$ (without replacement). The game ends when one is required to draw a marble from an empty box. Find the probability that all the marbles are drawn from all boxes.

**6.** Prove that the number of triangles, with integer sides, each of which having perimeter $n$ is equal to the number of partitions of the number $n$ into the summands 2, 3, and 4 in which the summand 3 appears at least once.

**1.25.2.  Second Day. 1.** Let $R$ be an integral domain and $U(R)$ the multiplicative group of the units of $R$. Prove that every finite subgroup of $U(R)$ is cyclic.

**2.** Let $p$ be an odd prime. Prove that every prime factor of $2^p + 1$ which is different from 3 is of the form $2kp + 1$.

**3.** Let $(X, d)$ be a compact metric space and $f : X \longrightarrow X$ a surjective function. Show that if
$$d(f(x), f(y)) \geq d(x, y),$$
for all $x, y \in X$, then $f$ is continuous.

**4.** Let $f : \mathbb{C} \longrightarrow \mathbb{C}$ be a function with $f(0) = 0$ and such that for all $z \in \mathbb{C}$ and $w \in \{0, 1, i\}$, we have
$$|f(z) - f(w)| = |z - w|.$$
Find the function $f$ explicitly. ($i^2 = -1$.)

**5.** Let $n$ be a an odd number and $A$ an $n \times n$ matrix whose entries are from the set $\{-1, 1\}$. If the product of the entries of the $i$th row is shown by $a_i$ and the product of the entries of the $j$th column is shown by $b_j$, prove that

$$\sum_{i=1}^{n} a_i + \sum_{j=1}^{n} b_j \neq 0.$$

Does the above assertion hold for an even $n$?

**6.** A society is called ideal if for every two distinct members $a$ and $b$ of the society, there exists a member $c$ of it such that $c$ is acquainted with one and only one of $a$ or $b$ (We assume that acquaintance is a symmetric and nonreflexive relation.) Prove that every ideal society of $n$ people ($n \geq 2$) contains an ideal society of $n - 1$ people.

## 1.26. Twenty-Sixth Competition, IASBS of Zanjan, May 2002

**1.26.1. First Day. 1.** Let $f : [a, b] \rightarrow [a, b]$ be a continuous function which is differentiable on $(a, b)$ and $f(a) = a$ , $f(b) = b$. Prove that there exist two distinct points $x_1$ and $x_2$ in $(a, b)$ such that $f'(x_1)f'(x_2) = 1$.

**2.** Suppose that $U = \{z \in \mathbb{C} : |z| < 1\}$ and $D$ is an open set in $\mathbb{C}$ such that $\overline{U} \subseteq D$, and that the function $f$ is analytic on $D$. Also let $f(0) = 1$ and that $|f(z)| > 2$ for all $z$ for which $|z| = 1$. Prove that there exists $z_0 \in U$ such that $f(z_0) = 0$.

**3.** Letting $p$ be an odd prime, prove that
$$p^{p-1} + (2p - 2)^{p-1} \equiv 1 \pmod{p(2p - 2)}.$$

**4.** Let $R$ be a unital commutative ring such that every ideal of it is principal. Prove that if $R$ has only one maximal ideal, then $Rx \subseteq Ry$ or $Ry \subseteq Rx$ for all $x, y \in R$.

**5.** A *k-element cover* for a set $S$ is a collection of $k$ nonempty subsets of $S$ such that the union of it is $S$. An $n$-element cover of the set $S$ is called minimal if no $n - 1$ elements of it cover $S$. If the number of the minimal $k$-element covers of a set with $n$ elements is denoted by $M(n, k)$, prove that
$$M(n, n-1) = \frac{n}{2}(2^n - n - 1).$$

**6.** Let $X$ be a set with $n$ elements and positive integers $m, k$ be such that $m \leq k$ and $m + k \leq n$. Denote the set of all subsets of $X$ with $m$ elements by $X^{\{m\}}$. Suppose that $f : X^{\{m\}} \rightarrow \mathbb{R}$ is a function where $\mathbb{R}$ is the set of real numbers. Prove that if for every member $S$ of $X^{\{k\}}$ we have $\displaystyle\sum_{T \in X^{\{m\}}, T \subseteq S} f(T) = 0$, then $f \equiv 0$.

**1.26.2. Second Day. 1.** Let $\{r_1, r_2, \dots\}$ be the set of rational numbers in $[0, 1]$. For each $x \in [0, 1]$, we let $A_x = \{n \in \mathbb{N} : r_n \leq x\}$ and define the function $f : [0, 1] \to \mathbb{R}$ by

$$f(x) = \sum_{n \in A_x} \frac{1}{2^n}.$$

Prove that $f$ is continuous at any irrational point of the interval $[0, 1]$.

**2.** Let $f : \mathbb{R} \to \mathbb{R}$ be a continuous function and that $(f \circ f \circ f)(x) = x$ for all $x \in \mathbb{R}$. Prove that $f(x) = x$ for all $x \in \mathbb{R}$.

**3.** Let $G$ be a group which has two distinct maximal subgroups, say, $H$ and $K$. Prove that if $H$ and $K$ are abelian and $Z(G) = \{e\}$, then $H \cap K = \{e\}$.

**4.** Let $A$ be an $n \times n$ matrix with entries in a field $F$. Let $\lambda \in F$ be an eigenvalue of $A^n$ corresponding to an eigenvector $x$ such that for all $a_0, a_1, \dots, a_{n-1} \in F$, if $\left(\sum_{i=0}^{n-1} a_i A^i\right) x = 0$, then $a_0 = a_1 = \cdots = a_{n-1} = 0$. Prove that $\left(A^n - \lambda I\right)^n = 0$. [12]

**5.** An extended die is a homogeneous cube on each of whose sides an arbitrary positive integer is written. (These numbers are not necessarily distinct.) Can one design two extended dice (which are not necessarily identical) in such a way that the probability that the sum of the numbers on the rolled dice is the same as those of two ordinary dice? (That is, the probability of getting a sum of 2 is $\frac{1}{36}$, the probability of getting a sum of 3 is $\frac{2}{36}$, etc.) In case, you answer in the affirmative, find all possible answers; and if you answer in the negative, prove your claim.

**6.** A matrix $M_{m \times n} = (m_{ij})$ with real entries is said to be *balanced*, if whenever for all $i, i', j, j'$ we have $1 \leq i < i' \leq m$ and $1 \leq j < j' \leq n$, then $m_{ij} + m_{i'j'} \leq m_{ij'} + m_{i'j}$. Suppose that there are two rows $i_1$ $i_2$ such that by interchanging the rows, the matrix remains balanced. Prove that for all $i_1'$ and $i_2'$, if $i_1 < i_1' < i_2' < i_2$, then by interchanging the rows $i_1'$ and $i_2'$ the matrix remains balanced.

## 1.27. Twenty-Seventh Competition, Bu-Ali Sina (Avecina) University of Hamedan, May 2003

**1.27.1. First Day. 1.** Let $n$ be a natural number greater than 1. Denote the set of all $n \times n$ matrices with real entries by $M_n(\mathbb{R})$. Define the following metric on $M_n(\mathbb{R})$. For $A = (A_{ij})$ and $B = (B_{ij})$,

$$d(A, B) = \max\left\{\left|A_{ij} - B_{ij}\right| : i, j = 1, \dots, n\right\}.$$

Prove that $GL_n(\mathbb{R})$, the set of nonsingular matrices, is an open and disconnected subset of $M_n(\mathbb{R})$.

---

[12] Prove that $A^n - \lambda I = 0$!

**2.** Let $f$ be a function that is analytic on $\mathbb{C}$. Let $L$ and $M$ be two orthogonal lines intersecting at a point $A$ such that $f(L) = L$ and $f(M) = M$. Prove that if $z_1$ and $z_2$ are two complex numbers that are symmetric with respect to $A$, then $f(z_1)$ is the image of $f(z_2)$ with respect to $A$.

**3.** Let $(a_n)_{n=1}^{+\infty}$ be a sequence of real numbers defined as follows

$$a_0 = 0, a_1 = b, a_{n+1} = a_n \sqrt{1 + a_{n-1}^2} + a_{n-1}\sqrt{1 + a_n^2}, n \geq 1.$$

Determine $a_n$ in terms of $b$.

**4.** Let $K$ be a nonempty set and $\emptyset \neq I \subseteq K$. Let $\{A_i : i \in I\}$ be a family of the subsets of $K$. Prove that if

$$\{i : i \notin A_i\} \in \{A_i : i \in I\} \cup \{\emptyset\},$$

then $i \in A_i$ for all $i \in I$.

**5.** Let $A$ be a $3 \times 2$ matrix and $B$ a $2 \times 3$ matrix with complex entries for which

$$AB = \begin{pmatrix} 8 & 2 & -2 \\ 2 & 5 & 4 \\ -2 & 4 & 5 \end{pmatrix}.$$

Determine the matrix $BA$. (Hint: Compute $(AB)^2$.)

**6.** Let $G$ be a group and $H$ a subgroup of it such that for all $x \in G \setminus H$ and all $y \in G$, there is a $u \in H$ such that $y^{-1}xy = u^{-1}xu$. Prove that $H$ is normal in $G$ and $G/H$ is abelian.

**1.27.2. Second Day. 1.** Let $f : \mathbb{R} \to \mathbb{R}$ be a differentiable function, $a, b \in \mathbb{R}$ and $a < b$. If $f(a) = f(b) = 0$, $f'(a) > 0$, and $f'(b) > 0$, prove that $f'$ has at least two roots in the interval $(a, b)$.

**2.** Prove that the interval $[0, 1]$ cannot be written as a union of pairwise disjoint closed intervals each of which having a positive length less than one.

**3.** Let $D$ be a countable subset of the Euclidean plane, i.e., $\mathbb{R} \times \mathbb{R}$. Prove that there is a partition for $D$ into two subsets $X$ and $Y$ such that any line parallel to the $x$-axis intersects $X$ in finitely many points and any line parallel to the $y$-axis intersects $Y$ in finitely many points.

**4.** Let $\{A_1, \ldots, A_n\}$ be a family of finite sets and $S = \bigcup_{i=1}^{n} A_i$. Suppose that a fixed number $k$, $1 \leq k \leq n$, satisfies the following conditions:
 (a) The union of any $k$ elements of the family $\{A_1, \ldots, A_n\}$ is equal to $S$.
 (b) The union of any $k-1$ elements of the family $\{A_1, \ldots, A_n\}$ is not equal to $S$.
 (c) $|S| = \binom{n}{k-1}$
 Find the number of the elements of all of $A_i$'s.

**5.** Let $a, b$ be two natural numbers such that $\gcd(a, b) = 1$. Prove that

$$\text{ord}_{ab}(a + b) = \text{lcm}[\text{ord}_b(a), \text{ord}_a(b)]$$

(Here, by $\text{ord}_n(m)$, we mean the order of $m$ modulo $n$.)

**6.** Let $R$ and $R'$ be two rings whose elements are all idempotents and consider a function $f : R \to R'$ that is one-to-one and onto such that $f(xy) = f(x)f(y)$ for all $x, y \in R$. Prove that $R \cong R'$.

## 1.28. Twenty-Eighth Competition, Sharif University of Technology, May 2004

### 1.28.1. First Day. 
**1.** Prove that any one-to-one and entire function is of the form $f(z) = az + b$, where $a$ and $b$ are fixed complex numbers and $a \neq 0$.

**2.** Let $d_1$ and $d_2$ be the Euclidean metrics on $\mathbb{R}$ and $\mathbb{R} \times \mathbb{R}$, respectively, $A = \{(x, y) \in \mathbb{R} \times \mathbb{R} : x = 0 \text{ or } y = 0\}$, and $d = d_2|_A$ the metric induced by the Euclidean metric of $\mathbb{R} \times \mathbb{R}$ on $A$. Prove that if $f : (\mathbb{R}, d_1) \to (A, d)$ is continuous and surjective, then $f^{-1}\{(0,0)\}$ has at least three elements.[13]

**3.** Let $f$ be an arithmetic function with the property that for every natural number $n$, $\sum_{d|n} f(d) = n^2$. ($\sum_{d|n}$ means sum over all positive divisors of $n$.) If $\phi$ is Euler's totient function, prove that

$$\frac{f(n)}{\phi(n)} = n \prod_{p|n, p \text{ prime}} \left(1 + \frac{1}{p}\right),$$

for all natural numbers $n > 1$.

**4.** Let $G$ be a subgroup of $S_n$ with the property that for all $1 \leq i \leq n$ and $1 \leq j \leq n$, there exists $\sigma \in G$ such that $\sigma(i) = j$. Prove that for all $1 \leq k \leq n$, $G_k \cap Z(G) = \{e\}$ where $G_k = \{\tau \in G : \tau(k) = k\}$, $Z(G)$ denotes the center of $G$, and $e$ is the identity element of $G$.

**5.** Let $A$ be the set of all vectors of $n$-tuples whose entries are zero or one in such a way that in each vector the numbers of one entries is odd. By explaining your reasoning, determine, the maximal number of vectors from the set $A$ such that every two of which share an even number of one entries.

**6.** Let $(P, \leq)$ be a poset (reflexive, antisymmetric, and transitive) with $n$ elements. For every element $i \in P$, set $U_i = \{j : j \in P, j > i\}$ and $L_i = \{j : j \in P, j < i\}$. Suppose that relative to every element $i$ of $P$, there corresponds a real number $X_i$ subject to the following

$$X_i = \begin{cases} \frac{1 - \sum_{j \in L_i} X_j}{n - |U_i|} & \text{if } L_i \neq \emptyset, \\ \frac{1}{n - |U_i|} & \text{if } L_i = \emptyset. \end{cases}$$

Prove that $0 \leq X_i \leq 1$ for all $i \in P$.

---

[13] Prove that $f^{-1}\{(0,0)\}$ has infinitely many elements!

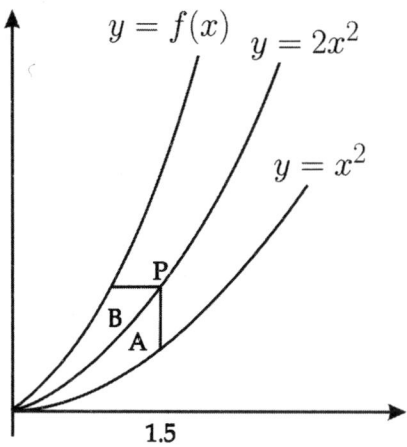

Figure 4

**1.28.2. Second Day. 1.** Let, in Figure 4, the function $f$ be one-to-one and continuous and that for every point $P$ on the curve $y = 2x^2$, the areas of the regions $A$ and $B$ are equal to one another. Determine the function $f$ explicitly.

**2.** Let $f : [a, b] \to (a, b)$ be a continuous function. Prove that for every natural number $n$, there exist a positive number $\alpha$ and $c \in (a, b)$ such that

$$f(c) + f(c + \alpha) + \cdots + f(c + n\alpha) = (n + 1)(c + \frac{n}{2}\alpha).$$

**3.** A matrix $O$ is called orthogonal if $OO^t = O^tO = I$, where $O^t$ denotes the transpose of the matrix $O$ . Let $A$ and $B$ be two orthogonal square matrices with entries in $\mathbb{C}$ and $\det(A) + \det(B) = 0$. Can one conclude that $\det(A + B) = 0$? Why?

**4.** Suppose that $R$ is a unital ring and that there is a natural number $n$ with the property that for all $k \in \{n, n+1, n+2\}$ and for all $x, y \in R$, $(xy)^k = x^k y^k$. Prove that $R$ is commutative.

**5.** Of three people who are suspected to have committed a murder, one is the murderer. The three people are to be interrogated by taking a test which has five questions. If the person being tested is innocent, the probability that the person responds positively to any question is 0.4; and if guilty, the probability of responding positively is 0.8. Of these three people, one who is chosen at random is being interrogated. This person responds positively to four questions and negatively to one question. What is the probability that the person is the murderer?

**6.** Let $X$ be a set and $r$ a natural number. Let $X_r$ be the set of all subsets of $X$ which have $r$ elements. Suppose that $F$ is a subset of $X_r$ with the property that the intersection of every $k$ elements of $F$ is nonempty ($k \geq 2$ is a fixed number). If we set

$$I(F) = \min \{|T| : T \subseteq X, T \cap A \neq \emptyset, \forall A \in F\},$$

prove that $I(F) \leq \frac{r-1}{k-1} + 1$.

## 1.29. Twenty-Ninth Competition, University of Mazandaran in Babolsar, May 2005

**1.29.1. First Day. 1.** Let $f : [0, a] \to \mathbb{R}$ be a continuous and positive function. Prove that

$$\left( \int_0^a f(x)dx \right) \left( \int_0^a \frac{dx}{f(x)} \right) \geq a^2.$$

**2.** Let $(n_i)_{i=1}^{+\infty}$ be an increasing sequence (not necessarily strictly) of natural numbers with $n_1 \geq 2$ such that the series $\displaystyle\sum_{i=1}^{+\infty} \frac{1}{n_1 \cdots n_i}$ converges to a real number $x$. Prove that $x$ is rational if and only if there exists a natural number $\ell$ such that $n_i = n_\ell$ for all $i \geq \ell$.

**3.** Consider the sequence of all natural numbers whose digits consist of 1.

$$1, 11, 111, \ldots$$

Prove that if the natural number $m$ is relatively prime with respect to 30, then infinitely many terms of the above sequence are divisible by $m$.

**4.** Let $R$ be an arbitrary ring (not necessarily unital) that has no nonzero nilpotent two-sided ideal. Prove that every nonzero right ideal in $R$ has an element whose square is not zero.

**5.** Let $\mathbb{Z}$, $\mathbb{E}$, and $\mathbb{O}$ be the set of integers, even integers, and odd integers, respectively. Set

$$X := \{A \in \mathcal{P}(\mathbb{Z}) : \text{both } A \cap \mathbb{E}, A \cap \mathbb{O} \text{ are infinite}\},$$
$$Y := \{A \in \mathcal{P}(\mathbb{Z}) : A \text{ is infinite}\}.$$

We know that there exists a one-to-one correspondence from $X$ into $Y$. Exhibit a law for a surjective function $f : X \to Y$.

**6.** Let $S$ be the $k$-dimensional vector space of all binary (zero and one) sequences of length $n$ over the field $\mathbb{Z}_2$. The distance between two elements $X, Y$ of $S$ are to be defined as the number of entries of $X$ and $Y$ that are different from one another. (More precisely, if $X = (x_1, \ldots, x_n)$ and $Y = (y_1, \ldots, y_n)$, then the distance between $X$ and $Y$ is equal to the number of $i$'s for which

$x_i \neq y_i$.) Suppose that the minimum distance between two distinct elements of $S$ is equal to $d$. Prove that

$$d \leq \frac{n2^{k-1}}{2^k - 1}.$$

**1.29.2. Second Day. 1.** Let $D = \{z \in \mathbb{C} : |z| < 1\}$ and $f : D \to \mathbb{C}$ be an analytic function such that $f(\frac{1}{n}) \in \mathbb{R}$ for all natural numbers $n \geq 2$. Prove that for all natural numbers $n$, $f^{(n)}(0) \in \mathbb{R}$, where $f^{(n)}$ denotes the $n$th derivative of the function $f$.

**2.** (i) Prove that if $X$ is a connected metric space, then for all $\varepsilon > 0$ and every two points $x, y \in X$, there exist an $n \in \mathbb{N}$ and points $x_1, \ldots, x_n \in X$ such that $x_1 = x$, $x_n = y$ and that $d(x_i, x_{i+1}) < \varepsilon$ for all $i < n$.

(ii) Give an example showing that the converse of the assertion of (i) does not hold.

(iii) Prove that the converse of the assertion of (i) holds provided that $X$ is compact.

**3.** Let $G$ be a group and $K$ a subgroup of it.

(i) Prove that $\dfrac{N_G(K)}{C_G(K)}$ is isomorphic to a subgroup of $\mathrm{Aut}(K)$.

(ii) Prove that if $K$ is cyclic and $K \triangleleft G = G'$, then $K \leq \mathrm{Z}(G)$.

**4.** Let $F$ be a field, $M_n(F)$ the set of all $n \times n$ matrices with entries in $F$, $A \in M_n(F)$, and the invertible matrix $P \in M_n(F)$ be such that $P^{-1}AP$ is upper triangular. Prove that every two invariant subspaces of $A$ are comparable with respect to inclusion if and only if there exist a $\lambda \in F$ and a nilpotent matrix $N \in M_n(F)$ with $N^{n-1} \neq 0$ such that $A = \lambda I + N$.

**5.** Two people, named $A$ and $B$, are playing a coin-flipping game as follows. They throw their coins, if the result of flipping the two coins turns out to be the same, $A$ is going to win both coins; if not, $B$ is going to take both coins. Suppose that $A$ has $m$ coins and $B$ has $n$ coins. On average, how many times should the game be played till one of the players runs out of the coins?

**6.** Let $C$ be the Cantor set. Prove that $C - C = [-1, 1]$. (Hint: it is worth mentioning that $C - C = \{x - y : x, y \in C\}$ and that $C$ is equal to the set of all numbers in $[0, 1]$ whose ternary expansions have only 0 or 2.)

## 1.30. Thirtieth Competition, Tafresh University, May 2006

**1.30.1. First Day. 1.** Let $f : [0, +\infty) \longrightarrow \mathbb{R}$ be a continuous function and $\lim_{x \to +\infty} f(x) = 1$. Evaluate the following limit.

$$\lim_{n \to +\infty} \int_{1385}^{2006} f(nx)dx.$$

**2.** Let $m \in \mathbb{N}$, $c \in \mathbb{C}$, $a_j \in \mathbb{C}$, and $|a_j| = 1$ for all $1 \leq j \leq m$. If

$$\lim_{n \to +\infty} \sum_{j=1}^{m} a_j^n = c,$$

then $c = m$ and $a_j = 1$ for all $1 \leq j \leq m$.

**3.** Let $R$ be a commutative ring with identity and $a$ an element of $R$ with $a^3 - a - 1 = 0$. Prove that if $J$ is an ideal of $R$ such that the quotient ring $R/J$ has at most four elements, then $J = R$.

**4.** Let $p, q$ be prime numbers such that $p = 2q + 1$ and $q \equiv 1 \pmod 4$. Prove that 2 is a primitive root modulo $p$.

**5.** Prove that

$$\sum_{|k| < \sqrt{m}} \binom{2m}{m + k} \geq 2^{2m-1},$$

for all $m \in \mathbb{N}$ with $m \geq 1$

**Hint.** By the Chebyshev Inequality if $X$ is a random variable whose mean and variance are $\mu$ and $\sigma^2$, respectively, then $P(|X - \mu| \geq \lambda\sigma) \leq \frac{1}{\lambda^2}$ for all $\lambda > 0$.

**6.** A financial group has $n$ members each of whom has a number of coins. Let $k \in \mathbb{N}$ be a fixed positive integer. A business among four people, say, $p_1$, $p_2$, $p_3$, and $p_4$, of the $n$ people, who are chosen arbitrarily, can be done subject to the following conditions:

(a) (sum of the number of coins owned by $p_3$ and $p_4$) − (sum of the number of coins owned by $p_1$ and $p_2$) + $2k > 0$,

(b) each of $p_1$ and $p_2$ has at least $k$ coins.

If the conditions (a) and (b) are met, then a business is done as follows. Each of $p_3$ and $p_4$ earns exactly $k$ coins which are lost by each of $p_1$ and $p_2$. Prove that after a finite number of doing business, the condition (a) or (b) in the above does not hold for any four people of the $n$ people.

**1.30.2. Second Day. 1.** Let $X$ be a separable metric space, that is, $X$ has a countable dense subset. Let $f : X \longrightarrow \mathbb{R}$ be a function for which $\lim_{x \to a} f(x)$ exists for all $a \in X$. Prove that the set of points at which $f$ is discontinuous is at most countable.

**2.** Suppose that $f(z) = \sum_{n=0}^{+\infty} a_n z^n$ defines a nonconstant analytic function, where the radius of the convergence of the series is equal to $R > 0$. Prove that the closest distance from a zero of $f$ to the origin is at least $\frac{R|a_0|}{M+|a_0|}$, where $M = M(R) = \sup_{|z|=R} |f(z)|$.

**3.** Let $G$ be a group with the property that the order of any element of $G'$, the derived subgroup of $G$, is finite. Prove that the set of all elements of $G$ whose orders are finite is a subgroup of $G$.

**4.** Let $K$ be a field, $F$ a subfield of $K$, $n \in \mathbb{N}$, and $A$ an $F$-algebraic $n \times n$ matrix with entries in $K$ such that $\operatorname{rank}(A) = \operatorname{rank}(A^2)$. Prove, firstly, that $K^n = \operatorname{im} A \oplus \ker A$ and, secondly, that there is a polynomial $f \in F[x]$ such that $E := f(A)$ is an idempotent matrix satisfying $E(x + y) = x$ for all $x \in \operatorname{im} A$ and $y \in \ker A$.

**5.** For an arbitrary subset $C$ of the natural numbers, set $C \oplus C := \{x+y|x,y \in C, x \neq y\}$. Prove that there is a unique partition of the set of natural numbers into two subsets $A$ and $B$ such that none of $A \oplus A$ and $B \oplus B$ contains a prime.

**Hint.** Recall that by Bertrand's conjecture (a.k.a. Bertrand's principle) for every natural number $n$ there exists at least a prime number $p$ such that $n < p \leq 2n$.

**6.** Let $C$ be a circle whose perimeter is equal to one and $0 < \alpha < \frac{1}{2}$. The distance between two points of the circle is defined to be the length of the shortest arc joining the two points. Let $T \subseteq C$ and $T = I_1 \cup \cdots \cup I_m$, where $m \in \mathbb{N}$ and $I_j$'s are disjoints arcs of the circle ($1 \leq j \leq m$). Prove that if the distance between every two points of $T$ is less than or equal to $\alpha$, then

$$\sum_{j=1}^{m} \ell(I_j) \leq \alpha,$$

where $\ell(I_j)$ denotes the length of the arc $I_j$.

## 1.31. Thirty-First Competition, Ferdowsi University of Mashhad, May 2007

**1.31.1. First Day. 1.** We have colored a circle by two colors. Does there necessarily exist a monocolored triangle inscribed in the circle which is
   (a) an equilateral triangle?
   (b) a right triangle?
   (c) an isosceles triangle?

**2.** The Minkowski sum of two sets $A, B \subseteq \mathbb{R}^d$ is defined as follows

$$A + B = \{a + b \in \mathbb{R}^d | a \in A, \; b \in B\}.$$

Prove that if $A$ is bounded and $B$ is closed, then

$$(A + B)' = (A' + B) \cup (A + B'),$$

where $A'$ stands for the set of the limit points of the set $A$.

**3.** Suppose that there is a group $G$ which has exactly $n$ subgroups of index 2 ($n$ is a natural number.). Prove that there exists a finite abelian group which has exactly $n$ subgroups of index 2.

**4.** Can one find two biased dice in such a way that the probability of getting a sum of $j$, for all $2 \leq j \leq 12$, for the two simultaneously thrown dice is a number in the interval $\left(\frac{2}{33}, \frac{4}{33}\right)$?

**5.** Show that $\mathbb{R}^2$ has a dense subset whose no three points are collinear.

**6.** Let $A$ be an $n \times n$ matrix with real entries. Prove that

$$\det(A) = \frac{1}{n!} \det \begin{pmatrix} \mathrm{tr}(A) & 1 & 0 & \cdots & \cdots & 0 \\ \mathrm{tr}(A^2) & \mathrm{tr}(A) & 2 & 0 & \cdots & 0 \\ \mathrm{tr}(A^3) & \mathrm{tr}(A^2) & \mathrm{tr}(A) & 3 & \ddots & \vdots \\ \vdots & \vdots & \vdots & \ddots & \ddots & 0 \\ \mathrm{tr}(A^{n-1}) & \mathrm{tr}(A^{n-2}) & \cdots & \cdots & \mathrm{tr}(A) & n-1 \\ \mathrm{tr}(A^n) & \mathrm{tr}(A^{n-1}) & \cdots & \cdots & \mathrm{tr}(A^2) & \mathrm{tr}(A) \end{pmatrix}.$$

**1.31.2. Second Day. 1.** Let $n$ be an odd natural number. Prove that there are $n$ consecutive natural numbers whose sum is a complete square. In this spirit, do there exist twelve natural numbers whose sum is a complete square?

**2.** For what values of the nonzero reals $\alpha$ and $\beta$ the following limit exists (and is finite)?

$$\lim_{x,y \to 0+} \frac{x^{2\alpha} y^{2\beta}}{x^{3\alpha} + y^{3\beta}}$$

**3.** Let $A$ be a nonvoid set and $A^n$ be the set of all ordered $n$-tuples of the elements of $A$. For $\alpha = (\alpha_1, \ldots, \alpha_n)$ and $\beta = (\beta_1, \ldots, \beta_n)$ in $A^n$, define

$d(\alpha, \beta) =$ the number of corresponding components of $\alpha$ and $\beta$ that differ with one another.

For two arbitrary elements $x$ and $y$ in $A^n$, prove that there is a one-to-one correspondence between the following two sets

$$C = \{z \in A^n : d(x, z) < d(y, z)\},$$
$$D = \{z \in A^n : d(y, z) < d(x, z)\}.$$

**4.** Let $R$ be a commutative ring with identity. Prove that the ring $R[x]$ has infinitely many maximal ideals.

**5.** Let $x_1, \ldots, x_{2n}$ be real numbers such that removing any of them the remaining ones can be partitioned into two sets with equal sums ($n \geq 2$). Prove that $x_i$'s are all zero.

**6.** Let $T$ be the union of all line segments with one end at $M = (0, 1)$ and the other end at a point on the $x$-axis having rational coordinates. In other words,

$$T = \{(tq, 1 - t) \in \mathbb{R}^2 | t \in [0, 1], \ q \in \mathbb{Q}\}.$$

(a) Assume that $A, B \in T$ such that $A$, $B$, and $M$ are not collinear. Show that any continuous path from the point $A$ to the point $B$ in the set $T$ has to pass through the point $M$ (by the aforementioned path, we mean a continuous function $\gamma : [0, 1] \to T$ with $\gamma(0) = A$ and $\gamma(1) = B$.).

(b) Prove that every continuous function from $T$ into $T$ has at least a fixed point.

# Part 2

# Solutions

## 2.1. First Competition

**2.1.1. Analysis. 1.** (a) For the given function $f$, let $f_n : \mathbb{R} \to \mathbb{R}$ be defined by $f_n(x) = n\left(f(x + \frac{1}{n}) - f(x)\right)$, where $n \in \mathbb{N}$. It is plain that the sequence $(f_n)_{n=1}^{+\infty}$ converges pointwise to $f'$ on $\mathbb{R}$, finishing the proof of this part.

(b) First, we recall the following which is known as Darboux's Theorem.

**Darboux's Theorem.** *The derivative of any differentiable function on a closed interval has the intermediate value property.*

**Proof.** Suppose that $f : [a, b] \longrightarrow \mathbb{R}$ is a differentiable function, $f'(a) \neq f'(b)$, and $\lambda \in \mathbb{R}$ is between $f'(a)$ and $f'(b)$. We need to show that there exists a $c \in (a, b)$ such that $f'(c) = \lambda$. To this end, if necessary replacing $f$ by $-f$, without loss of generality, we may assume that $f'(a) < f'(b)$. There are two cases to consider.

(i) $f'(a) < \lambda \leq \frac{f(b)-f(a)}{b-a}$.
Define the function $g : [a, b] \longrightarrow \mathbb{R}$ by

$$g(x) = \begin{cases} f'(a) & x = a, \\ \frac{f(x)-f(a)}{x-a} & a < x \leq b. \end{cases}$$

It is readily seen that $g$ is continuous on $[a, b]$. We have $g(a) < \lambda \leq g(b)$. It thus follows from the Intermediate Value Theorem that there exists an $x_0 \in (a, b]$ such that $g(x_0) = \frac{f(x_0)-f(a)}{x_0-a} = \lambda$. Therefore, by the Mean Value Theorem, $\lambda = f'(c)$ for some $c \in (a, b)$, which is what we want.

(i) $\frac{f(b)-f(a)}{b-a} < \lambda < f'(b)$.
In this case, defining the function $g : [a, b] \longrightarrow \mathbb{R}$ by

$$g(x) = \begin{cases} \frac{f(b)-f(x)}{b-x} & a \leq x < b, \\ f'(b) & x = b, \end{cases}$$

the assertion follows in a similar fashion. $\qquad\square$

Define the function $f : \mathbb{R} \to \mathbb{R}$ by

$$f(x) = \begin{cases} 0 & x \leq 0, \\ 1 & x > 0. \end{cases}$$

The function $f$ is a Baire function because the sequence $(f_n)_{n=1}^{+\infty}$ of continuous functions, where $f_n : \mathbb{R} \to \mathbb{R}$ is defined by

$$f_n(x) = \begin{cases} 0 & x \leq 0, \\ nx & 0 < x < \frac{1}{n}, \\ 1 & \frac{1}{n} \leq x, \end{cases}$$

converges to the function $f$ pointwise. Yet, in view of Darboux's Theorem , the function $f$ is not the derivative of any function on $\mathbb{R}$ because $f$ does not have the intermediate value property (a.k.a. Darboux's property ). $\qquad\blacksquare$

**2.** (a) For $X = (x_{ij}) \in \mathcal{M}$, by definition, we have

$$\det(X) = \sum_{\sigma \in S_n} \text{sgn}(\sigma) x_{1\sigma 1} \cdots x_{n\sigma n},$$

where $S_n$ denotes the symmetric group on $\{1, \ldots, n\}$ and $\text{sgn}(\sigma)$ stands for the sign of the permutation $\sigma \in S_n$. Hence, the function $\det : \mathcal{M} \to \mathbb{R}$ is continuous, for $f$ is a polynomial function in $n^2$ variables $x_{ij}$ where $1 \leq i, j \leq n$. We have

$$U(\mathcal{M}) = \det^{-1}(\mathbb{R} \setminus \{0\}),$$

where $U(\mathcal{M})$ denotes the set of invertible matrices in $\mathcal{M}$. It follows that $U(\mathcal{M})$ is an open subset of $\mathcal{M}$ because $\mathbb{R} \setminus \{0\}$ is an open subset of $\mathbb{R}$.

(b) To see that $\mathcal{M}$ is disconnected, just note that

$$U(\mathcal{M}) = \det^{-1}(\mathbb{R}^-) \cup \det^{-1}(\mathbb{R}^+)$$

where $\mathbb{R}^- = (-\infty, 0)$ and $\mathbb{R}^+ = (0, +\infty)$. Thus, in view of the proof of (a), the open sets $\det^{-1}(\mathbb{R}^-)$ and $\det^{-1}(\mathbb{R}^+)$ form a disconnectedness for $U(\mathcal{M})$, finishing the proof. ∎

**3.** (a) We need the following lemma.

**Lemma.** *Let $N \in \mathbb{R}^+$ and $I$ a bounded interval of $\mathbb{R}$. Then, in the interval $I$ there are finitely many $r \in \mathbb{Q}$ whose denominators are less than or equal to $N$.*

**Proof.** Without loss of generality assume that $N$ is a natural number. Just note that, by the Archimedean property of real numbers and the well-ordering principle of natural numbers, the set $\bigcup_{b=1}^{N} \{\frac{a}{b} \in I : a \in \mathbb{Z}\}$ is a finite set, proving the assertion. □

We claim that $\lim_{x \to a} f(x) = 0$ for all $a \in \mathbb{R}^+$. To see this, for a given $\varepsilon > 0$, first pick $N \in \mathbb{N}$ such that $\frac{1}{N} < \varepsilon$. It follows from the lemma above that there exists $\delta > 0$ such that $x = \frac{p}{q} \in \mathbb{Q}$ with $\gcd(p, q) = 1$ and $0 < |x - a| < \delta$ imply that $q > N$. So for this given $\varepsilon > 0$, find $\delta > 0$ as in the above. If $0 < |x - a| < \delta$, there are two cases to consider:
(i) $x = \frac{p}{q} \in \mathbb{Q}$ with $\gcd(p, q) = 1$. Noting that $q > N$, we can write

$$|f(x) - 0| = \frac{1}{p + q} < \frac{1}{q} < \frac{1}{N} < \varepsilon.$$

(ii) $x \notin \mathbb{Q}$. In this case, we have

$$|f(x) - 0| = 0 < \varepsilon.$$

From these two cases, we see that $\lim_{x \to a} f(x) = 0$, which is what we want.

(b) In view of the claim we made in (a), it follows that $f$ is continuous at the irrational points of $(0, +\infty)$ and is discontinuous at the rational points of $(0, +\infty)$. ∎

**2.1.2. Algebra. 1.** (a) "if": Suppose that $I_1, I_2 \lhd R$ with $I_1 I_2 \subseteq R$ and $I_1 \not\subseteq P$. We show that $I_2 \subseteq P$. Choosing an $x_0 \in I_1 \setminus P$ and assuming that $y \in I_2$ is arbitrary, we have $x_0 y \in I_1 I_2 \subseteq P$. Hence, $y \in P$, for $x_0 \in I_1 \setminus P$. As $y$ was arbitrary, we see that $I_2 \subseteq P$, which is what we want.

"only if": Let $\langle a \rangle$ and $\langle b \rangle$ denote the ideals generated by $a$ and $b$ in $R$, respectively. The ring $R$ is commutative with identity element. So we have $\langle a \rangle = aR$, $\langle b \rangle = bR$, and $\langle a \rangle \langle b \rangle = \langle ab \rangle$. Now from $ab \in P$, it follows that $\langle a \rangle \langle b \rangle \subseteq P$. But $P$ is prime. Therefore, $\langle a \rangle \subseteq P$ or $\langle b \rangle \subseteq P$, implying that $a \in P$ or $b \in P$, which is what we want.

(b) Let $M$ be a maximal ideal in $R$. In view of (a), it suffices to show that $M$ is prime. To this end, suppose that $ab \in M$, where $a, b \in R$. We need to show that $a \in M$ or $b \in M$. If $a \notin M$, then $M + \langle a \rangle = R$ because $M$ is maximal and $a \notin M$. In particular, we see that there are $m \in M$ and $x \in R$ such that $1 = m + ax$. This implies that $b = bm + abx \in M$, for $M$ is an ideal and $ab \in M$. ∎

**2.** It follows from the hypothesis that $g^{-1} = g$ for all $g \in G$. We can write
$$ab = (ab)^{-1} = b^{-1} a^{-1} = ba,$$
for all $a, b \in G$. That is, $G$ is abelian, which is what we want. ∎

**3.** (3.1) In view of (c), we can write
$$\mathrm{card} E \leq \mathrm{card} f(E) \leq \mathrm{card} F = \mathrm{card} E.$$
This, together with the Schroder-Bernstein Theorem, implies that $\mathrm{card} E = \mathrm{card} f(E)$. Now, suppose that $X$ and $Y$ are two finite subsets of $E$ such that $\mathrm{card} X = \mathrm{card} f(X)$ and $\mathrm{card} Y = \mathrm{card} f(Y)$. We can write
$$
\begin{aligned}
\mathrm{card} f(X \cup Y) &= \mathrm{card}\big(f(X) \cup f(Y)\big) \\
&= \mathrm{card} f(X) + \mathrm{card} f(Y) - \mathrm{card}\big(f(X) \cap f(Y)\big). \quad (*)
\end{aligned}
$$
On the other hand,
$$\mathrm{card}(X \cup Y) = \mathrm{card} X + \mathrm{card} Y - \mathrm{card}(X \cap Y). \quad (**)$$
But $\mathrm{card}(X \cup Y) \leq \mathrm{card} f(X \cup Y) = \mathrm{card}\big(f(X) \cup f(Y)\big)$. So it follows from $(*)$ and $(**)$ that $\mathrm{card}\big(f(X) \cap f(Y)\big) \leq \mathrm{card}(X \cap Y)$. As $f(X \cap Y) \subseteq f(X) \cap f(Y)$, we can write
$$\mathrm{card}(X \cap Y) \leq \mathrm{card} f(X \cap Y) \leq \mathrm{card}\big(f(X) \cap f(Y)\big) \leq \mathrm{card}(X \cap Y).$$
Therefore,
$$\mathrm{card} f(X \cap Y) = \mathrm{card}\big(f(X) \cap f(Y)\big) = \mathrm{card}(X \cap Y).$$
That is, $X \cap Y$ and $f(X) \cap f(Y)$ are isomorphic sets. Now, since $f(X \cap Y) \subseteq f(X) \cap f(Y)$ and these two sets are finite, it follows that $f(X \cap Y) = f(X) \cap f(Y)$. This equality together with $(*)$ and $(**)$ implies that $\mathrm{card}(X \cup Y) = \mathrm{card} f(X \cup Y)$, finishing the proof of assertion in this part.

(3.2) Set
$$T = \big\{ \mathrm{card}(A) : \emptyset \neq A \subseteq E \text{ and } \mathrm{card}(A) = \mathrm{card} f(A) \big\}.$$

It is obvious that the set $T$ is a nonempty subset of $\mathbb{N}$. It thus follows from the well-ordering principle of $\mathbb{N}$ that the set $T$ has an initial element, say, card$X_0$, where $X_0$ is a nonempty normal subset of $E$. We claim that for any other normal subset $X$ of $E$, we have

$$X \cap X_0 = \emptyset \text{ or } X_0 \subseteq X.$$

Note that $X_0 \cap X \subseteq X_0$. If $X_0 \cap X = X_0$, we obtain $X_0 \subseteq X$; if not, we show that $X_0 \cap X = \emptyset$, completing the proof. Suppose that $X_0 \cap X \subsetneq X_0$. Since $X_0$ is finite, it follows that

$$\text{card}(X_0 \cap X) < \text{card}(X_0).$$

On the other hand, by (3.1), $\text{card}(X_0 \cap X) = \text{card} f(X_0 \cap X)$. We claim that $X_0 \cap X = \emptyset$. Otherwise, we see that $\text{card}(X_0 \cap X) \in T$, yielding $\text{card}(X_0) \leq \text{card}(X_0 \cap X)$, for $\text{card}(X_0)$ is the initial element of $T$. This is a contradiction. Hence, $X_0 \cap X = \emptyset$ and we are done.    ∎

**2.1.3. General. 1.** From $a < b < c$, we see that $\dfrac{a}{b} < 1$, $\dfrac{b}{c} < 1$ and $1 < \dfrac{c}{a}$. Thus,

$$\max\left(\frac{a}{b}, \frac{b}{c}, \frac{c}{a}\right) = \frac{c}{a}, \quad \min\left(\frac{a}{b}, \frac{b}{c}, \frac{c}{a}\right) \neq \frac{c}{a}.$$

There are two cases to consider.
(i) $\min\left(\dfrac{a}{b}, \dfrac{b}{c}, \dfrac{c}{a}\right) = \dfrac{a}{b}$ and (ii) $\min\left(\dfrac{a}{b}, \dfrac{b}{c}, \dfrac{c}{a}\right) = \dfrac{b}{c}$.
In both cases, from the above inequalities, we easily see that $S > 1$.    ∎

**2.** We show that the decimal fraction $a = 0.123456789101112131415\ldots$ cannot be periodic after any digit. To prove this by contradiction, assume that the decimal fraction $a$ is periodic after its first $k$ digits and that its period is $l$. Obviously, the number $10^{k+l}$ occurs somewhere after the $k$th digit of the decimal fraction $a$. As $10^{k+l}$ has $k + l$ zeros in its decimal expansion and that the period of $a$ after the $k$th digit is assumed to be $l$, it follows that the digits of $a$ after the $k$th digit are all zero, which is obviously impossible, settling the proof.    ∎

**3.** We need to solve the following system of equations in $\mathbb{C}$

$$\begin{cases} x = y^3 \\ y = x^3 \end{cases}, \quad x \neq y.$$

It is plain that $x, y \neq -1, 0, 1$, for otherwise $x = y \in \{-1, 0, 1\}$. By substitution, we obtain

$$x^9 = x, x \neq 0 \implies x^8 = 1 \implies x = \omega^j, 0 < j < 7, j \neq 4,$$

where $\omega = \cos\frac{\pi}{4} + i\sin\frac{\pi}{4}$. It is now easily seen that

$$\begin{cases} x = \omega^j \\ y = \omega^{3j} \end{cases}, 0 < j < 7, j \neq 4,$$

is the general solution of the above system of equations.    ∎

**4.** Using the AM-GM Inequality, we can write

$$x^x + y^y \geq 2\sqrt{x^x y^y} = 2e^{f(x)},$$

where $f : (0,1) \to \mathbb{R}$ is defined by $f(x) = \frac{1}{2}(x \ln x + (1-x) \ln(1-x))$. We have $f'(x) = -\frac{1}{2} \ln(\frac{1}{x} - 1)$. Thus, $f'(x) < 0$ on $(0, \frac{1}{2})$ and $f'(x) > 0$ on $(\frac{1}{2}, 1)$, implying that $f$ attains its only local minimum and hence its absolute minimum at $\frac{1}{2}$. Consequently, we have

$$x^x + y^y \geq 2e^{f(x)} \geq 2e^{f(\frac{1}{2})} = \sqrt{2},$$

for all $x, y \in \mathbb{R}^+$ with $x + y = 1$ and that the equality happens if and only if $x = y = \frac{1}{2}$, which is what we want. ∎

### 2.1.4. Differential Equations.
We know that the general solution of a nonhomogeneous linear equation is $y = y_c + y_p$, where $y_c$ is the general solution of the corresponding homogeneous equation and $y_p$ is a particular solution of the nonhomogeneous linear equation $x'' + a^2 x = f(t)$. In view of this, first of all, it is easily seen that $y_c(t) = c_1 \cos at + c_2 \sin at$ $(t \in \mathbb{R})$, where $c_1, c_2 \in \mathbb{R}$, is the general solution of the homogeneous equation. Using the method of variation of parameters, a straightforward calculation shows that the function $y_p$ defined by

$$y_p(t) = \frac{1}{a} \int_{t_0}^{t} f(x) \sin a(t - x) dx \quad (t \in \mathbb{R}),$$

where $t_0 \in \mathbb{R}$ is a fixed arbitrary real number, is a particular solution of the nonhomogeneous linear equation $x'' + a^2 x = f(t)$. Therefore,

$$y(t) = c_1 \cos at + c_2 \sin at + \frac{1}{a} \int_{t_0}^{t} f(x) \sin a(t - x) dx \quad (t \in \mathbb{R}),$$

is the general solution of the equation $x'' + a^2 x = f(t)$. ∎

### 2.1.5. Probability and Statistics.
We need to calculate $P(\sum_{i=1}^{10} X_i = 10)$. To do this, let, first of all, $X$ be a Poisson random variable with parameter $\lambda$ defined by its probability density function as follows

$$f_X(i) = P(X = i) = \begin{cases} \frac{e^{-\lambda} \lambda^i}{i!} & i \in \mathbb{N} \cup \{0\}, \\ 0 & \text{otherwise.} \end{cases}$$

Next, we find the moment generating function of $X$.

$$m_X(t) = E(e^{tX}) = \sum_{X_i} e^{tX_i} f_X(X_i) = \sum_{i=0}^{+\infty} e^{ti} f_X(i)$$

$$= e^{-\lambda} \sum_{i=0}^{+\infty} \frac{(\lambda e^t)^i}{i!} = e^{\lambda(e^t - 1)}.$$

Now, since each $X_i$ is a Poisson random variable with parameter $\mu_i$ and that $X_i$'s are independent random variables, letting $Y = \sum_{i=1}^{10} X_i$, we can write

$$m_Y(t) = E\left(e^{t \sum_{i=1}^{10} X_i}\right) = E\left(\prod_{i=1}^{10} e^{tX_i}\right) = \prod_{i=1}^{10} E\left(e^{tX_i}\right)$$

$$= \prod_{i=1}^{10} m_{X_i}(t) = e^{(\sum_{i=1}^{10} \mu_i)(e^t - 1)}.$$

Therefore, $Y = \sum_{i=1}^{10} X_i$ is a Poisson random variable with parameter $\sum_{i=1}^{10} \mu_i$, whence

$$P\left(\sum_{i=1}^{10} X_i = 10\right) = f_Y(10) = \frac{e^{-\sum_{i=1}^{10} \mu_i}\left(\sum_{i=1}^{10} \mu_i\right)^{10}}{10!}.$$

But $\mu_i = i$. So, we have

$$P\left(\overline{X} = 1\right) = P\left(\sum_{i=1}^{10} X_i = 10\right) = \frac{e^{-55} 55^{10}}{10!}.$$

■

**2.1.6. Topology.** To prove the assertion by contradiction, suppose that $T$ is a topology on $\mathbb{R}$. We must have

$$\bigcup_{\substack{q > \sqrt{2} \\ q \in \mathbb{Q}}} A_q = (\sqrt{2}, +\infty) \in T,$$

implying that $\sqrt{2} \in \mathbb{Q}$, which is a contradiction, settling the proof. ■

## 2.2. Second Competition

**2.2.1. Analysis. 1.** (a) Since $f$ is continuous and the interval $[0,1]$ is compact, there exists an $x_0 \in [0,1]$ such that

$$M = \sup_{0 \le x \le 1} f(x) = \max_{0 \le x \le 1} f(x) = f(x_0).$$

If $M = 0$, there is nothing to prove. If not, as $f$ is continuous at $x_0$, we see that for given $M\varepsilon > 0$, there exists a neighborhood of $x_0$, say, $(\alpha, \beta)$, where $0 \le \alpha < \beta \le 1$, such that

$$|f(x) - f(x_0)| < M\varepsilon$$

whenever $\alpha < x < \beta$. Noting that $f(x_0) = M \ge f(x)$ for all $x \in [0,1]$, in view of the above inequality, we obtain

$$M(1 - \varepsilon) < f(x) \le M,$$

for all $x \in (\alpha, \beta)$, which is what we want.

(b) For given $0 < \varepsilon < 1$, find $0 \leq \alpha < \beta \leq 1$ from (a). We have $0 < M(1 - \varepsilon) \leq f(x) \leq M$ on $[\alpha, \beta]$, for $f$ is continuous on $[0, 1]$. We can write

$$M^n(1 - \varepsilon)^n(\beta - \alpha) \leq \int_\alpha^\beta f^n \leq \int_0^1 f^n \leq M^n.$$

Taking $n$th root, we obtain

$$M(1 - \varepsilon)(\beta - \alpha)^{\frac{1}{n}} \leq \left( \int_0^1 f^n \right)^{\frac{1}{n}} \leq M. \qquad (*)$$

Now taking $\underline{\lim}_n$ and $\overline{\lim}_n$ of both sides of the left and right inequalities above, we obtain

$$M(1 - \varepsilon) \leq \underline{\lim}_n \left( \int_0^1 f^n \right)^{\frac{1}{n}} \leq \overline{\lim}_n \left( \int_0^1 f^n \right)^{\frac{1}{n}} \leq M.$$

Thus, $0 \leq \overline{\lim}_n \left( \int_0^1 f^n \right)^{\frac{1}{n}} - \underline{\lim}_n \left( \int_0^1 f^n \right)^{\frac{1}{n}} \leq M - M(1 - \varepsilon) = M\varepsilon$. As $0 < \varepsilon < 1$ is arbitrary, we conclude that

$$\overline{\lim}_n \left( \int_0^1 f^n \right)^{\frac{1}{n}} = \underline{\lim}_n \left( \int_0^1 f^n \right)^{\frac{1}{n}}.$$

That is, $\lim_n \left( \int_0^1 f^n \right)^{\frac{1}{n}}$ exists. Letting $n \to +\infty$ in $(*)$, we obtain

$$M(1 - \varepsilon) \leq \lim_n \left( \int_0^1 f^n \right)^{\frac{1}{n}} \leq M.$$

Again, $0 < \varepsilon < 1$ being arbitrary, we see that

$$\lim_n \left( \int_0^1 f^n \right)^{\frac{1}{n}} = M = f(x_0) = \sup_{0 \leq x \leq 1} f(x),$$

finishing the proof. ∎

**2.** Note first that

$$\sup_{0 \leq t \leq 1} |1 - f(t)| \geq |1 - f(0)| = 1, \quad \int_0^1 |1 - f(t)| dt \geq 0,$$

and hence

$$\sup_{0 \leq t \leq 1} |1 - f(t)| + \int_0^1 |1 - f(t)| dt \geq 1,$$

for all $f \in \mathcal{F}$. This implies that

$$D = \inf_{f \in \mathcal{F}} \left( \sup_{0 \leq t \leq 1} |1 - f(t)| + \int_0^1 |1 - f(t)| dt \right) \geq 1.$$

We claim that $D = 1$. To see this, just note that for the sequence $(f_n)_{n=1}^{+\infty}$, where $f_n : [0, 1] \to \mathbb{R}$ $(n \in \mathbb{N})$ is defined by

$$f_n(t) = \begin{cases} nt & 0 \leq t \leq \frac{1}{n}, \\ 1 & \frac{1}{n} < t \leq 1, \end{cases}$$

we have

$$\lim_n \left( \sup_{0 \le t \le 1} \left| 1 - f_n(t) \right| + \int_0^1 \left| 1 - f_n(t) \right| dt \right) = \lim_n \left( 1 + \frac{1}{2n} \right) = 1.$$

As for the geometrical interpretation of $D$, we view $\mathcal{F}$ as a subspace of the normed space of all continuous functions on $[0, 1]$, denoted by $\mathcal{C}[0, 1]$, which is equipped with the following norm

$$\|f\| := \sup_{0 \le t \le 1} \left| f(t) \right| + \int_0^1 \left| f(t) \right| dt.$$

If $1 : [0, 1] \to \mathbb{R}$ is defined by $1(x) = 1$ for all $x \in [0, 1]$, we would then have

$$d(\{1\}, \mathcal{F}) = \inf_{f \in \mathcal{F}} \|1 - f\|$$

$$= \inf_{f \in \mathcal{F}} \left( \sup_{0 \le t \le 1} \left| 1 - f(t) \right| + \int_0^1 \left| 1 - f(t) \right| dt \right) = D = 1.$$

That is, the distance from the vector $1 \in \mathcal{C}[0, 1]$ to the subspace $\mathcal{F}$ of $\mathcal{C}[0, 1]$ is equal to $D = 1$. ∎

**3.** It suffices to show that $[0, 2] \subseteq C + C$. To this end, let $x \in [0, 2]$ be arbitrary. If

$$\frac{x}{2} = 0.x_1 x_2 \ldots$$

denotes the ternary expansion of $\frac{x}{2}$ where $x_i \in \{0, 1, 2\}$, then we can write

$$\frac{x}{2} = 0.a_1 a_2 \ldots + 0.b_1 b_2 \ldots,$$

where $a_i, b_i \in \{0, 1\}$ and $x_i = a_i + b_i$ for all $i \in \mathbb{N}$. So we have

$$x = 2(0.a_1 a_2 \ldots) + 2(0.b_1 b_2 \ldots) = 0.a_1' a_2' \ldots + 0.b_1' b_2' \ldots := a + b,$$

where $a_i' = 2a_i$ and $b_i' = 2b_i$, and hence $a_i', b_i' \in \{0, 2\}$ for all $i \in \mathbb{N}$. Therefore, $a, b \in C$ because there is no "one" in their ternary expansions. This finishes the proof. ∎

**2.2.2. Algebra. 1.** To prove the assertion we need the following general lemma.

**Lemma.** *Let $G$ be a finite group with the property that the equation $x^n = e$ has at most $n$ solutions in $G$ for all $n \in \mathbb{N}$, where $e$ is the identity element of $G$. Then the group $G$ is cyclic.*

**Remark.** In case the group $G$ is abelian, the lemma is a quick consequence of the Fundamental Theorem of finite abelian groups .

**First proof.** Let $|G| = n$ and $C$ be a cyclic group with $n$ elements so that $C = \langle a \rangle$, where $\operatorname{ord}(a) = n$. Suppose that for a divisor $d$ of $n$, there is an element $g_d \in G$ such that $\operatorname{ord}(g_d) = d$. It follows that $g_d, g_d^2, \ldots, g_d^{d-1}, g_d^d = e$ are $d$ solutions of the equation $x^d = e$. As $d | n = |G| = |C|$ and $C$ is cyclic, there is an $a_d \in C$ such that $\operatorname{ord}(a_d) = d$. Consequently, $\langle g_d \rangle \cong \langle a_d \rangle$, and hence $C$ has at least as many elements of order $d$ as $G$ has. But $G$ and $C$ both have $n$ elements. Thus, $G$ must have the same number of elements of order $d$ as $C$, for all $d$ dividing $n$. It follows that $G$ is cyclic because $C$ has an element of order $n$, namely the element $a$. So the proof is complete.

**Second proof.** Define the relation $\sim$ on $G$ as follows

$$a \sim b \iff \text{ord}(a) = \text{ord}(b).$$

It is obvious that $\sim$ is an equivalence relation on $G$. Hence, the relation $\sim$ partitions $G$ into its equivalence classes. In other words, we can write

$$G = \bigcup_{i=1}^{k} [g_i],$$

where $k \in \mathbb{N}$ is less than $|G|$, $\{g_1, \ldots, g_k\}$ is a maximal set of nonequivalent elements of $G$, and $[g_i]$ denotes the equivalence class containing the element $g_i$.

Note that if $G$ happens to be a cyclic group of order $n$ so that $G = \langle a \rangle$ with $\text{ord}(a) = n$, we can write

$$n = |G| = \sum_{i=1}^{k} |[g_i]|,$$

where $g_i$'s $(1 \le i \le k)$ are as in the above and $g_i = a^{j_i}$ for some $1 \le j_i \le n$. Let $d_i = \gcd(j_i, n)$. It is easily checked that

$$\text{ord}(a^k) = \frac{\text{ord}(a)}{\gcd(k, \text{ord}(a))} = \frac{n}{\gcd(k, n)} \tag{$*$}$$

for all $1 \le k \le n$. In view of this, we see that $d_i \ne d_{i'}$ whenever $i \ne i'$ and that $a^k \in [g_i]$ if and only if $\frac{n}{\gcd(k,n)} = \frac{n}{d}$ if and only if $\gcd(k, n) = d$ if and only if $\gcd(\frac{k}{d}, \frac{n}{d}) = 1$. Thus, $|[g_i]| = \phi(\frac{n}{d})$, where $\phi$ denotes Euler's totient function . Consequently,

$$n = \sum_{i=1}^{k} |[g_i]| = \sum_{d|n} \phi\left(\frac{n}{d}\right) = \sum_{d|n} \phi(d).$$

Now back to a general finite group $G$, letting $d_i = \text{ord}(g_i)$, we claim that $|[g_i]| = \phi(d_i)$. That $|[g_i]| \ge \phi(d_i)$ follows from the hypothesis that $d_i = \text{ord}(g_i)$ and that, in view of $(*)$, the cyclic group $\langle g_i \rangle$ has exactly $\phi(d_i)$ generators. On the other hand, $|[g_i]| \le \phi(d_i)$, for otherwise there must exist a $g \in G \setminus \langle g_i \rangle$ with $\text{ord}(g) = d_i$, which, in turn, implies that the equation $x^{d_i} = e$ has at least $d_i + 1$ solutions, namely, $g, g_i, g_i^2, \ldots, g_i^{d_i} = e$, which is a contradiction. Thus, $|[g_i]| = \phi(d_i)$, and hence

$$n = |G| = \sum_{i=1}^{k} \phi(d_i) = \sum_{d|n} \phi(d).$$

This, in particular, implies that $G$ has an element of order $n$. Therefore, $G$ is a cyclic group of order $n$, which is what we want.  $\square$

**Corollary.** *Let $D$ be a division ring and $G$ a finite abelian subgroup of the multiplicative group of $D$, i.e., $D^* = D \setminus \{0\}$. Then, $G$ is a cyclic group. In particular, the multiplicative group of any finite field is cyclic.* □

By the corollary above, there exists an $a \in F^*$ such that $F^* = \langle a \rangle$. Letting $b = a^m$, we can write

$$S := \sum_{x \in F} x^m = \sum_{i=0}^{n-2} (a^i)^m = \sum_{i=0}^{n-2} b^i.$$

If $n - 1 | m$, then, as $a^{n-1} = 1$, we have $b = 1$. This together with the above equality implies that $\sum_{x \in F} x^m = n - 1$. If $n - 1 \nmid m$, then $b^m \neq 1$. So we have $S = bS = 1 + b + \cdots b^{n-1}$, yielding $(b-1)S = 0$ which, in turn, implies that $S = 0$, proving the assertion. ∎

**2.** (a) "$\Longleftarrow$" This is obvious.
  "$\Longrightarrow$" Assume that $z$ is right and left quasi-regular. Thus, there exist $z', z'' \in A$ such that

$$z + z' - z'z = z + z'' - zz'' = 0. \qquad (*)$$

We can write

$$(z + z' - z'z)z'' = 0 \implies zz'' + z'z'' - z'(z + z'') = 0 \implies zz'' = z'z.$$

This together with $(*)$ implies that $z' = z''$. That is, $z$ is quasi-regular, which is what we want.

(b) Proceed by contradiction. The identity element of $A$ being right quasi-regular means there exists $1' \in A$ such that $1 + 1' - 1' = 0$, yielding $1 = 0$, which is impossible. Likewise, the identity element is not left quasi-regular either.
  As pointed out in the footnote of this problem, from the element $x$ being right quasi-regular one can only conclude that $1 - x$ is right invertible. First, let $x$ be right quasi-regular. So there exists $x' \in A$ such that $x + x' - xx' = 0$. From this, we obtain $(1 - x)(1 - x') = 1$. That is, $1 - x$ is right invertible. In fact, as this argument is reversible, in a similar fashion, one can prove that the element $x$ is right (resp. left) quasi-regular if and only if $1 - x$ is right (resp. left) invertible. We now show that in general from $x$ being right quasi-regular one cannot conclude that $1 - x$ is invertible. To this end, in view of the aforementioned comment, it suffices to show that in general from $x$ being right quasi-regular one cannot conclude that $x$ is left quasi-regular as well. To see this, let $F$ be a field of characteristic zero and $F[x]$ the ring of all polynomials with coefficients from the field $F$. View $F[x]$ as a vector space over $F$ and use $A$ to denote $\mathcal{L}(F[x])$, the ring of all linear transformations from $F[x]$ into $F[x]$. Let $I$ denote the identity transformation, $D$ the differentiation linear transformation, i.e. $D(f_0 + f_1 x + \cdots + f_n x^n) = f_1 + 2f_2 x + \cdots + nf_n x^{n-1}$, and

$I_1$ and $I_2$ denote the linear transformations defined by

$$I_1(f_0 + f_1 x + \cdots + f_n x^n) = 1 + f_0 x + \frac{f_1}{2} x^2 + \cdots + \frac{f_n}{n+1} x^{n+1},$$

$$I_2(f_0 + f_1 x + \cdots + f_n x^n) = 2 + f_0 x + \frac{f_1}{2} x^2 + \cdots + \frac{f_n}{n+1} x^{n+1}.$$

Clearly, $DI_1 = DI_2 = I$. That is, $D$ has two right inverses, and hence $D$ is not invertible. This implies that $I - D$ is right quasi-regular but it is not left quasi-regular. Because otherwise $I - (I - D) = D$ will become invertible, which is impossible. ∎

**3.** (a) Since $D$ divides $P$, there exists $Q \in K[x]$ such that $P = QD$. We conclude that $\phi(P) = \phi(Q)\phi(D)$ because $\phi$ is a ring homomorphism. It is now obvious that $\ker \phi(D) \subseteq \ker \phi(P)$, which is what we want.

(b) Suppose that $D = \gcd(P, R)$, where $P, R \in K[x]$. Since $K[x]$ is a Euclidean ring, it follows that there exist $M, N \in K[x]$ such that $D = MP + NR$. As $\phi$ is a ring homomorphism, we conclude that $\phi(D) = \phi(M)\phi(P) + \phi(N)\phi(R)$. This implies $\ker \phi(R) \cap \ker \phi(P) \subseteq \ker \phi(D)$. On the other hand, since $D = \gcd(P, R)$, in view of (a), we see that $\ker \phi(D) \subseteq \ker \phi(P)$ and $\ker \phi(D) \subseteq \ker \phi(R)$, yielding $\ker \phi(D) \subseteq \ker \phi(R) \cap \ker \phi(P)$. Therefore, $\ker \phi(D) = \ker \phi(R) \cap \ker \phi(P)$, finishing the proof. ∎

**2.2.3. General. 1.** If the first day up to the seventh day of the first month of the Persian calendar, called "Farvardin", corresponds to 1 up to 7, respectively, then the thirteenth day of the first month corresponds to 6. From this convention, we will obtain the 12-tuple

$$(6, 2, 5, 1, 4, 7, 3, 5, 7, 2, 4, 6)$$

whose $i$th component is corresponded to the thirteenth day of the month $i$ of the Persian calendar ($1 \leq i \leq 12$). For instance, the number 1 being the fourth component of the above 12-tuple means that the thirteenth day of the fourth month of the Persian calendar occurs on the same day of the week as does the first day of the first month of the calendar. For $1 \leq i \leq 7$, let $n(i)$ denote the number of times that the number $i$ occurs in the above 12-tuple. Obviously,

$$n(i) = \begin{cases} 1 & i \in \{1, 3\}, \\ 2 & i \notin \{1, 3\}. \end{cases}$$

As the first day of "Farvardin" can occur on any day of the week, we see that "Wednesday" can correspond to any of the numbers $1, 2, \ldots, 7$. Therefore, in the Persian calendar, the minimum and the maximum number of "Wednesdays occurring on the 13th of the month" is 1 and 2, respectively. ∎

**2.** Recall that a real function $f$ on an interval $I$ is called *convex* if

$$f(\lambda x + (1 - \lambda)y) \leq \lambda f(x) + (1 - \lambda)f(y),$$

for all $x, y \in I$ and $0 \leq \lambda \leq 1$. It is well-known that a function $f : I \to \mathbb{R}$ which is twice differentiable is convex if and only if $f'' \geq 0$ on $I$. From this, it follows that the function $g_n : (0, +\infty) \to \mathbb{R}$ defined by $g_n(x) = x^n$, where $n \in \mathbb{R}^+$, is convex if and only if $n \geq -1$. In particular, $g_n$ is convex whenever $n \in \mathbb{N}$. This yields $g_n\left(\frac{x+y}{2}\right) \leq \frac{g_n(x)+g_n(y)}{2}$ for all $n \in \mathbb{N}$ and $x, y \in \mathbb{R}^+$. So, we can write

$$\left(\frac{x+y+z+t}{4}\right)^n = g_n\left(\frac{x+y+z+t}{4}\right) \leq \frac{1}{2}g_n\left(\frac{x+y}{2}\right) + \frac{1}{2}g_n\left(\frac{z+t}{2}\right)$$

$$\leq \frac{g_n(x)+g_n(y)+g_n(z)+g_n(t)}{4} = \frac{x^n+y^n+z^n+t^n}{4},$$

for all $n \in \mathbb{N}$ and $x, y, z, t \in \mathbb{R}^+$, which is what we want. ∎

**3.** We prove the following proposition of which the assertion is a quick consequence. Note that if the particle were moving along a straight line, then the tangential acceleration of the motion would be equal to the acceleration of the motion along the straight line.

*A particle moving along a curve goes $s_0$ unit(s) of distance in $t_0$ unit(s) of time in such a way that the instantaneous speed of it at the initial point as well as the final point is zero. Prove that the absolute value of the particle tangential acceleration at some point on its path is greater than or equal to $\frac{4s_0}{t_0^2}$* unit of distance (unit of time)² .

Assume that the initial and final moments of the motion are $0$ and $t_0$, respectively. If the equation of the motion is given by a vector function $x : [0, t_0] \to \mathbb{R}^3$, it follows that $x$ is twice differentiable, $x'(0) = x'(t_0) = 0$, and $s(0) = 0$, $s(t_0) = s_0$, where $x' = \frac{dx}{dt}$ denotes the instantaneous velocity vector and $s(t) = \int_0^t \|x'(\tau)\| d\tau$ the length of the path at time $t$. It is plain that $s'(0) = s'(t_0) = 0$. Also, recall that the tangential acceleration of the motion, denoted by $a_t$, is, by definition, the time derivative of the instantaneous speed, i.e., $a_t = \frac{dv}{dt} = \frac{d^2s}{dt^2} = s''(t)$. As $s$ is continuous, we see from the Intermediate Value Theorem that there is a moment $0 < c < t_0$ at which we have $s(c) = \frac{s(t_0)+s(0)}{2} = \frac{s(t_0)}{2}$. There are two cases to consider.

(i) $0 < c < \frac{t_0}{2}$.

In this case, using the extension of the Mean Value Theorem for second order derivatives, we obtain a moment $t_1$ with $0 < t_1 < c$ such that

$$s(c) = s(0) + s'(0)(c-0) + s''(t_1)\frac{(c-0)^2}{2!} = s''(t_1)\frac{c^2}{2},$$

from which, we get $\frac{s(t_0)}{2} = s''(t_1)\frac{c^2}{2}$, and hence $s''(t_1) = \frac{s(t_0)}{c^2}$. This together with the hypothesis that $0 < c < \frac{t_0}{2}$ implies that $s''(t_1) \geq \frac{4s(t_0)}{t_0^2} > 0$, yielding $|s''(t_1)| \geq \frac{4s(t_0)}{t_0^2} = \frac{4s_0}{t_0^2}$, which is what we want.

(ii) $\frac{t_0}{2} \leq c < t_0$.

Again using the extension of the Mean Value Theorem for second derivatives, we obtain a moment $t_1$ with $\frac{t_0}{2} < t_1 < t_0$ such that

$$s(c) = s(t_0) + s'(t_0)(c-t_0) + s''(t_1)\frac{(c-t_0)^2}{2!},$$

from which, we obtain $\frac{s(t_0)}{2} = s(t_0) + s''(t_1)\frac{(c-t_0)^2}{2}$, and hence $s''(t_1) = \frac{-s(t_0)}{(c-t_0)^2} <$ 0. This together with $\frac{t_0}{2} \le c < t_0$ implies that $|s''(t_1)| = \frac{s(t_0)}{(c-t_0)^2} \ge \frac{4s(t_0)}{t_0^2} = \frac{4s_0}{t_0^2}$, which is what we want. ∎

**4.** Since the digits can be selected independent of the letter, using the product rule of combinatorics, it is obvious that the total number of the car plates that contain the letter "dal" is equal to

$$9 \times 9 \times 9 \times 9 \times 9 \times 1 = 59049,$$

as desired. ∎

### 2.2.4. Probability and Statistics. (a) We can write

$$
\begin{aligned}
\mathrm{cov}(U,V) &= E(UV) - E(U)E(V) \\
&= E((X+Y)(X-Y)) - E(X+Y)E(X-Y) \\
&= E(X^2 - Y^2) - ((E(X))^2 - (E(Y))^2) \\
&= E(X^2) - (E(X))^2 - (E(Y^2) - (E(Y))^2) \\
&= \mathrm{var}(X) - \mathrm{var}(Y) = \sigma^2 - \sigma^2 = 0.
\end{aligned}
$$

Therefore, $\mathrm{cov}(U,V) = 0$.

**Note:** As is obvious now the condition $\mathrm{cov}(X,Y) = \lambda$ was redundant.

(b) Not necessarily. Suppose that $U, V$ are two random variables with $\mathrm{cov}(U,V) = 0$, which are not independent. If we let $X = \frac{1}{2}(U+V)$ and $Y = \frac{1}{2}(U-V)$, we would have

$$
\mathrm{var}(X) = \frac{1}{4}\big(\mathrm{var}(U) + \mathrm{var}(V) + 2\mathrm{cov}(U,V)\big) = \frac{1}{4}\big(\mathrm{var}(U) + \mathrm{var}(V)\big),
$$

$$
\mathrm{var}(Y) = \frac{1}{4}\big(\mathrm{var}(U) + \mathrm{var}(V) - 2\mathrm{cov}(U,V)\big) = \frac{1}{4}\big(\mathrm{var}(U) + \mathrm{var}(V)\big).
$$

Therefore, $\mathrm{var}(X) = \mathrm{var}(Y) = \frac{1}{4}\big(\mathrm{var}(U) + \mathrm{var}(V)\big) = \sigma^2$. Note that the random variables $X + Y = U$ and $X - Y = V$ are not independent by our hypothesis. To give a concrete example, let $U$ and $V$ be two random variables defined by their joint distribution table as follows.

| U \ V | 1 | 2 | 3 | |
|---|---|---|---|---|
| 1 | 0.1 | 0.1 | 0.1 | 0.3 |
| 2 | 0.1 | 0.1 | 0.1 | 0.4 |
| 3 | 0.1 | 0.1 | 0.1 | 0.3 |
| | 0.3 | 0.3 | 0.4 | |

It is easily verified that

| UV | 1 | 2 | 3 | 4 | 6 | 9 |
|---|---|---|---|---|---|---|
| | 0.1 | 0.2 | 0.2 | 0.1 | 0.3 | 0.1 |

From the above table, we obtain $E(UV) = 4.2$. On the other hand, we have $E(U) = 2$ and $E(V) = 2.1$, implying that $E(UV) = E(U)E(V)$, and hence $\text{cov}(U, V) = 0$. But $P(U = 1, V = 1) = 0.1$ and $P(U = 1) = P(V = 1) = 0.3$, yielding $P(U = 1, V = 1) \neq P(U = 1).P(V = 1)$. In other words, $U$ and $V$ are not independent, which is what we wanted. ∎

**2.2.5. Topology.** Proceed by contradiction. As $A \subseteq B \cup C$, we have $A = (A \cap B) \cup (A \cap C)$. Also, by the contradiction hypothesis, $(A \cap B) \cap (A \cap C) = A \cap B \cap C = \emptyset$. But $A \cap B \neq \emptyset$ and $A \cap C \neq \emptyset$. Hence, it suffices to show that $A \cap B$ and $A \cap C$ are closed sets in the induced topology of $A$, for $A$ would then be disconnected, a contradiction. To this end, note that $B$ and $C$ are closed in $(K, T_K)$. Thus, there exist closed sets $B_1$ and $C_1$ in $(E, T)$ such that $B = B_1 \cap K$ and $C = C_1 \cap K$, from which, we obtain

$$A \cap B = A \cap B_1, A \cap C = A \cap C_1.$$

Now as $B_1$ and $C_1$ are closed in $E$, it follows that $A \cap B$ and $A \cap C$ are closed in $A$, which is what we want. ∎

**2.2.6. Differential Equations.** Two cases to consider.
(i) $|x| > 1$.
Rewrite the equation as

$$(y^3)' + \frac{x}{x^2 - 1} y^3 = \frac{(x^2 - 1)(x + 1)}{x^2 - 1} = x + 1.$$

Multiplying both sides of the equation by

$$\mu(x) = \exp\left( \int \frac{x dx}{(x^2 - 1)} \right) = \sqrt{|x^2 - 1|} = \sqrt{x^2 - 1},$$

we obtain

$$\frac{d}{dx} \left( y^3 \sqrt{x^2 - 1} \right) = (x + 1)\sqrt{x^2 - 1},$$

from which, we get

$$y^3 \sqrt{x^2 - 1} = \int x\sqrt{x^2 - 1} dx + \int \sqrt{x^2 - 1} dx.$$

Using integration by parts and dividing both sides of the above equality by $\sqrt{x^2 - 1}$, we see that the general solution of the equation is

$$y^3 = \frac{1}{3}(x^2 - 1) + \frac{x}{2} + \frac{1}{2\sqrt{x^2 - 1}} \ln|x + \sqrt{x^2 - 1}| + \frac{c}{\sqrt{x^2 - 1}},$$

where $c$ is an arbitrary fixed constant.
(ii) $|x| < 1$.
In this case, note that $|x^2 - 1| = 1 - x^2$ yields $\mu(x) = \sqrt{1 - x^2}$. From this, a similar argument as in (i) shows that the general solution of the equation is

$$y^3 = \frac{-1}{3}(x^2 - 1) + \frac{x}{2} + \frac{1}{2\sqrt{1 - x^2}} \arcsin x + \frac{c}{\sqrt{1 - x^2}},$$

where $c$ is an arbitrary fixed constant. ∎

## 2.3. Third Competition

**2.3.1. Analysis. 1. Remark.** It is worth mentioning that the assertion holds under the weaker hypothesis that $f''$ is nonnegative on $(a, b)$ or $f'$ is monotonic on $[a, b]$. Also in the statement of the problem "$\xi \in [a, b]$" must read "$\xi \in (a, b)$" and that "$\xi \in (a, b)$" cannot be weakened, because for the function $f : [0, 1] \to \mathbb{R}$ with $f(x) = x^2$ and $\xi \in \{0, 1\}$, there is no $x_0 \in (0, 1)$ such that

$$f'(\xi) = \frac{f(0) - f(x_0)}{0 - x_0} \text{ or } f'(\xi) = \frac{f(1) - f(x_0)}{1 - x_0}.$$

Since $f'' > 0$ on $[a, b]$, it follows that $f'$ is strictly increasing on $[a, b]$. By the Mean Value Theorem, there exists $a < \xi_0 < b$ such that

$$f'(\xi_0) = \frac{f(b) - f(a)}{b - a}.$$

Now, if $f'(\xi) = f'(\xi_0)$, there is nothing to prove. If not, there are two cases to consider.

(i)

$$f'(\xi) < f'(\xi_0) = \frac{f(b) - f(a)}{b - a}.$$

Define the function $g : [a, b] \to \mathbb{R}$ by

$$g(x) = \begin{cases} f'(a) & x = a, \\ \frac{f(a) - f(x)}{a - x} & a < x \leq b. \end{cases}$$

It is plain that

$$\lim_{x \to a^+} g(x) = \lim_{x \to a^+} \frac{f(a) - f(x)}{a - x} = f'(a) = g(a).$$

Hence, $g$ is continuous at $a$ from the right, and therefore it is continuous on $[a, b]$. As $f'$ is strictly increasing on $[a, b]$, we obtain $f'(a) < f'(\xi)$. So we can write

$$g(a) = f'(a) < f'(\xi) < f'(\xi_0) = g(b).$$

Thus, it follows from the Intermediate Value Theorem that there exists an $a < x_0 < b$ such that

$$f'(\xi) = g(x_0) = \frac{f(a) - f(x_0)}{a - x_0},$$

which is what we want.

(ii)

$$f'(\xi) > f'(\xi_0) = \frac{f(b) - f(a)}{b - a}.$$

The proof, which is omitted, is almost identical to that of (i) except that we need to define the function $g : [a, b] \to \mathbb{R}$ by

$$g(x) = \begin{cases} \frac{f(b) - f(x)}{b - x} & a \leq x < b, \\ f'(b) & x = b. \end{cases}$$

This settles the proof.                                           ∎

**2.** (a) Set $B = (\mathbb{Q} \cap [0,1]) \setminus \{\frac{1}{n} : n \in \mathbb{N}\}$. Plainly, we can write $B = \{b_n : n \in \mathbb{N}\}$. Define the sequence $(a_n)_{n=1}^{+\infty}$ by

$$a_n = \begin{cases} \frac{1}{n} & n \text{ odd}, \\ b_n & n \text{ even}. \end{cases}$$

We claim that for this sequence $(a_n)_{n=1}^{+\infty}$, there is no continuous function $f$ satisfying $f(a_n) = a_{n+1}$ for all $n \in \mathbb{N}$. Suppose to the contrary that $f : [0,1] \to [0,1]$ is a continuous function such that $f(a_n) = a_{n+1}$ for all $n \in \mathbb{N}$. By proving that $f = 0$ on $(0,1]$, and hence on $[0,1]$, we obtain a contradiction, proving the claim. For a given $x \in (0,1]$, pick a subsequence $(a_{n_i})_{i=1}^{+\infty}$ such that $\lim_{i \to +\infty} a_{n_i} = x$. As $x \neq 0$, if necessary by passing to a subsequence, we may assume that $a_{n_i} \notin \{\frac{1}{n} : n \in \mathbb{N}\}$ for all $i \in \mathbb{N}$. This implies that $n_i$'s are all even numbers, from which, we see that $(n_i + 1)$'s are all odd numbers and hence

$$f(x) = \lim_{i \to +\infty} f(a_{n_i}) = \lim_{i \to +\infty} a_{n_i+1} = \lim_{i \to +\infty} \frac{1}{n_i + 1} = 0,$$

as desired.

(b) We sketch the proof for the case in which the sequence $(a_n)_{n=1}^{+\infty}$ is increasing. In case the sequence is decreasing, the proof can be carried out in a similar fashion. Without loss of generality, if necessary by redefining the sequence $(a_n)_{n=1}^{+\infty}$, we may assume that $a_1 = 0$. Let $a_\infty = \lim_{n \to +\infty} a_n$. Define the function $f$ on $[0,1]$ by

$$f(x) = \begin{cases} \frac{a_{n+2} - a_{n+1}}{a_{n+1} - a_n}(x - a_n) + a_{n+1} & a_n \leq x < a_{n+1}, n \in \mathbb{N}, \\ x & a_\infty \leq x \leq 1. \end{cases}$$

It is not difficult to see that $f : [0,1] \to [0,1]$ is continuous and that $f(a_n) = a_{n+1}$ for all $n \in \mathbb{N}$, as desired. ∎

**3.** (a) We need the following proposition.

**Proposition.** *Let the complex number $z$ be a root of the following equation with complex coefficients*

$$x^p + c_1 x^{p-1} + \cdots + c_{p-1} x + c_p = 0,$$

*where $p \in \mathbb{N}$. If $\lambda_k > 0$ for all $1 \leq k \leq p$ and $\sum_{k=1}^{p} \frac{1}{\lambda_k} \leq 1$, then*

$$|z| \leq \max_{1 \leq k \leq p} \sqrt[k]{\lambda_k |c_k|}.$$

Under the hypothesis of the proposition, taking $\lambda_k = 2^k$ for all $1 \leq k \leq p$, we obtain

$$|z| \leq 2 \max_{1 \leq k \leq p} \sqrt[k]{|c_k|},$$

proving the assertion.

In order to prove the proposition, we need the following lemma.

**Lemma.** *Let the complex numbers $z_1, \ldots, z_p$ ($p \in \mathbb{N}$) be the roots of the following equation with complex coefficients*

$$x^p + c_1 x^{p-1} + \cdots + c_{p-1} x + c_p = 0.$$

*Set $r_0 = \max_{1 \le i \le p} |z_i|$. If $r > 0$ and*

$$r^p > |c_1| r^{p-1} + \cdots + |c_p|,$$

*then $r_0 < r$.*

**First proof.** Let $z \in \mathbb{C}$ with $|z| \ge r$ be arbitrary. We can write

$$
\begin{aligned}
\left| \frac{f(z)}{z^p} \right| &= \left| 1 + \frac{c_1}{z} + \cdots + \frac{c_p}{z^p} \right| \\
&\ge 1 - \left| \frac{c_1}{z} \right| - \cdots - \left| \frac{c_p}{z^p} \right| \\
&\ge 1 - \frac{|c_1|}{r} - \cdots - \frac{|c_p|}{r^p} > 0.
\end{aligned}
$$

Thus, $f(z) \ne 0$, and hence $r_0 < r$, as desired.

**Second proof.** Let $k(x) := x^p - |c_1| x^{p-1} - \cdots - |c_p|$. By the hypothesis, $k(r) > 0$, which is equivalent to $\frac{k(r)}{r^p} = 1 - \sum_{i=1}^p \frac{|c_i|}{r^i} > 0$. Thus, for all $x \ge r$, we have

$$1 - \sum_{i=1}^p \frac{|c_i|}{x^i} \ge 1 - \sum_{i=1}^p \frac{|c_i|}{r^i} > 0.$$

Consequently,

$$
\begin{aligned}
k'(x) &= (x^p)' \left( \frac{k(x)}{x^p} \right) + x^p \left( \frac{k(x)}{x^p} \right)' \\
&= p x^{p-1} \left( 1 - \sum_{i=1}^p \frac{|c_i|}{x^i} \right) + x^p \sum_{i=1}^p \frac{i|c_i|}{x^{i+1}} > 0,
\end{aligned}
$$

for all $x \ge r$. Thus, $k$ is strictly increasing on the interval $[r, +\infty)$. Now, let $z \in \mathbb{C}$ be such that $|z| \ge r$. It follows that $k(|z|) \ge k(r) > 0$. So, we can write

$$
\begin{aligned}
|z^p + c_1 z^{p-1} + \cdots + c_p| &\ge |z|^p - |c_1||z|^{p-1} - \cdots - |c_p| \\
&= k(|z|) \ge k(r) > 0.
\end{aligned}
$$

That is, $z^p + c_1 z^{p-1} + \cdots + c_p \ne 0$ whenever $|z| \ge r$. This yields $r_0 = \max_{1 \le i \le p} |z_i| < r$, which is what we want.

**Third proof.** Define $f(z) = z^p + c_1 z^{p-1} + \cdots + c_{p-1} z + c_p$, $g(z) = z^p$, and $h(z) = c_1 z^{p-1} + \cdots + c_p = f(z) - g(z)$ on $\mathbb{C}$. Note that $f(z) \ne 0$ for all $z \in \mathbb{C}$ with $|z| = r$, for otherwise

$$r^p = |z^p| = |-c_1 z^{p-1} - \cdots - c_p| \le |c_1||z|^{p-1} + \cdots + |c_p|,$$

from which, we obtain $r^p \le |c_1| r^{p-1} + \cdots + |c_p|$, which is a contradiction. For all $z$ on the circle $|z| = r$, we have

$$|h(z)| \le |c_1| r^{p-1} + \cdots + |c_p| < r^p = |g(z)|.$$

So it follows from Rouché's Theorem that the entire functions $g(z) = z^p$ and $f(z) = h(z) + g(z)$ have the same number of zeros inside the circle $|z| = r$. Consequently, $f$ has $p$ zeros inside the circle $|z| = r$, implying that $r_0 < r$, which is what we want. $\qquad\square$

**Proof of Proposition.** Set $r_1 = \max_{1 \leq k \leq p} \sqrt[k]{\lambda_k |c_k|}$, where $\lambda_k$'s in $\mathbb{R}^+$ are such that $\sum_{k=1}^{p} \frac{1}{\lambda_k} \leq 1$. For all $r > r_1$ and for each $k = 1, \ldots, p$, we obviously have $\frac{1}{\lambda_k} > \frac{|c_k|}{r^k}$. Thus,

$$1 \geq \sum_{k=1}^{p} \frac{1}{\lambda_k} > \sum_{k=1}^{p} \frac{|c_k|}{r^k},$$

which obtains

$$r^p > |c_1| r^{p-1} + \cdots + |c_p|,$$

and hence $r_0 < r$ in view of the above lemma. As $r > r_1$ was arbitrary, we conclude that $r_0 \leq r_1 = \max_{1 \leq k \leq p} \sqrt[k]{\lambda_k |c_k|}$, which is what we want. $\qquad\square$

(b) Let $f(z) := z^p + c_1 z^{p-1} + \cdots + c_{p-1} z + c_p$, $g(z) := z^p + c'_1 z^{p-1} + \cdots + c'_{p-1} z + c'_p$, and $h(z) := g(z + z_j)$. First, we prove that the absolute value of the product of the roots of $h$, i.e., $|h(0)|$, is less than $(2K)^p \delta$. To this end, we can write

$$
\begin{aligned}
\big|h(0)\big| &= \big|g(z_j)\big| = \big|g(z_j) - f(z_j)\big| = \Big|\sum_{i=1}^{p} (c'_i - c_i) z_j^{p-i}\Big| \\
&\leq \sum_{i=1}^{p} |c'_i - c_i| |z_j|^{p-i} \leq \sum_{i=1}^{p} K^i \delta |z_j|^{p-i} \\
&\leq \delta \sum_{i=1}^{p} K^i (2 \max_{1 \leq k \leq p} \sqrt[k]{|c_k|})^{p-i} < \delta \sum_{i=1}^{p} K^i (2K)^{p-i} \\
&= \delta K^p \sum_{i=1}^{p} 2^{p-i} = \delta K^p (2^p - 1) < (2K)^p \delta,
\end{aligned}
$$

yielding $\big|h(0)\big| < (2K)^p \delta$, as desired. Now, let $z'_1, \ldots, z'_p$ be the roots of $g(z) = 0$. It follows that $z'_1 - z_j, \ldots, z'_p - z_j$ are the roots of $h(z) = g(z + z_j) = 0$. Obviously, there is a $1 \leq k \leq p$ such that $|z'_k - z_j| = \min_{1 \leq i \leq p} |z'_i - z_j|$. Consequently,

$$|z'_k - z_j|^p \leq \prod_{1 \leq i \leq p} |z'_i - z_j| = \big|h(0)\big| < (2K)^p \delta,$$

implying that $|z'_k - z_j| < 2K \sqrt[p]{\delta}$, which is what we want.

(c) We have not been able to prove the assertion. However, we prove the following which can be thought of as a counterpart of the assertion at the local level.

*Under the hypothesis of the (c) part of the problem, prove that for every $t \in [t_1, t_2]$, there exists a $\delta = \delta(t) > 0$ such that if $s \in [t_1, t_2]$ and $|s - t| < \delta$,* *then*

$$|f(s) - f(t)| < 2K \sqrt[p]{|s - t|^\alpha}.$$

For a given $t \in [t_1, t_2]$, let $z_1 = f(t), z_2, \ldots, z_k$ $(1 \leq k \leq p)$ be distinct roots of $z^p + b_1(t)z^{p-1} + \cdots + b_{p-1}(t)z + b_p(t) = 0$. If $k = 1$, in fact $f$ need not be continuous and yet the assertion is a quick consequence of (b). To see this, note that by (b), for the root $f(s)$ of $z^p + b_1(s)z^{p-1} + \cdots + b_{p-1}(s)z + b_p(s) = 0$, where $s \in [t_1, t_2]$, there is a root of $z^p + b_1(t)z^{p-1} + \cdots + b_{p-1}(t)z + b_p(t) = 0$, which must be the only root of the equation, namely $f(t)$, such that

$$|f(s) - f(t)| < 2K \sqrt[p]{|s - t|^\alpha},$$

as desired. So, we may without loss of generality assume that $k > 1$. Now, set $d_0 = \frac{1}{2} \min_{1 \leq i < j \leq k} |z_i - z_j|$. Choose a $\delta = \delta(t) > 0$ such that $2K \sqrt[p]{|s - t|^\alpha} < d_0$ and $|f(s) - f(t)| < d_0$ whenever $|s - t| < \delta$ and $s \in [t_1, t_2]$. It follows from (b) that for every $s \in [t_1, t_2]$ with $|s - t| < \delta$, there is an $i_s \in \{1, \ldots, k\}$ such that

$$|f(s) - z_{i_s}| < 2K \sqrt[p]{|s - t|^\alpha} < d_0.$$

On the other hand,

$$|f(s) - f(t)| < d_0.$$

Consequently,

$$|z_1 - z_{i_s}| = |f(t) - z_{i_s}| \leq |f(t) - f(s)| + |f(s) - z_{i_s}| < 2d_0,$$

implying that $z_1 = f(t) = z_{i_s}$. Thus,

$$|f(s) - f(t)| < 2K \sqrt[p]{|s - t|^\alpha},$$

whenever $s \in [t_1, t_2]$ and $|s - t| < \delta$, which is what we want.  ∎

**2.3.2. Algebra. 1.** (a) Just apply the First Isomorphism Theorem for modules to the mapping $\varphi : M_2 \to \frac{M_1 + M_2}{M_1}$ defined by $\varphi(x) = x + M_1$, where $x \in M_2$.

(b) Suppose that $\{\alpha_i\}_{i=1}^m$ is a basis for $M_1 \cap M_2$. Enlarge $\{\alpha_i\}_{i=1}^m$ to bases $\{\alpha_i\}_{i=1}^m \cup \{\beta_i\}_{i=1}^n$ and $\{\alpha_i\}_{i=1}^m \cup \{\gamma_i\}_{i=1}^p$ for $M_1$ and $M_2$, respectively. It is easily seen that $\{\alpha_i\}_{i=1}^m \cup \{\beta_i\}_{i=1}^n \cup \{\gamma_i\}_{i=1}^p$ is a basis for $M_1 + M_2$. We can write

$$\begin{aligned} \dim M_1 + \dim M_2 &= (m + n) + (m + p) = (m + n + p) + m \\ &= \dim(M_1 + M_2) + \dim(M_1 \cap M_2) \\ &= \dim M_3 + \dim M_4, \end{aligned}$$

which is what we want.  ∎

**2.** If $G$ is abelian, we see that the function $f : G \to G$ defined by $f(x) = x^2$ is a homomorphism. Conversely, if $f$, defined by $f(x) = x^2$ for all $x \in G$, is a homomorphism, then $(xy)^2 = x^2y^2$ for all $x, y \in G$. Multiplying both sides by $x^{-1}$ and $y^{-1}$ on the left and right, respectively, we obtain $xy = yx$ for all $x, y \in G$. That is, $G$ is abelian, which is what we want.  ∎

**3.** (a) It is easily verified that the semigroup $S$ defined by

$$S := \{(a_{ij}) \in M_2(\mathbb{R}) : a_{11} = a_{21} = 0, a_{12}, a_{22} \in \mathbb{R} \setminus \{0\}\}$$

has a right identity element, namely $\begin{pmatrix} 0 & 1 \\ 0 & 1 \end{pmatrix}$, with respect to which every element of $S$ has a left inverse. And yet $S$ is not a group, for its right identity element is not a left identity element.

(b) Let $x \in S$ be an arbitrary element with a left inverse $\bar{x} \in S$ so that $\bar{x}x = e$. To prove the assertion, it suffices to show that $x\bar{x} = e$. To this end, let $\bar{\bar{x}}$ be a left inverse of $\bar{x}$. We have

$$ex = (\bar{\bar{x}}.\bar{x})x = \bar{\bar{x}}(\bar{x}x) = \bar{\bar{x}}e = \bar{\bar{x}}.$$

By showing that $x\bar{x}$ is a right identity element of $S$, we conclude that $x\bar{x} = e$, finishing the proof. Suppose that $y \in S$ is arbitrary. We can write

$$y.(x\bar{x}) = (ye)(x\bar{x}) = y(ex)\bar{x} = y(\bar{\bar{x}}.\bar{x}) = ye = y.$$

Thus, $x\bar{x} = e$, which is what we want. ∎

### 2.3.3. General.

**1.** Letting $z = x$ in the second condition, we see that

$$F(y, x) \leq F(x, x) + F(x, y) = F(x, y),$$

for all $x, y \in \mathbb{R}$. Therefore, $F(y, x) \leq F(x, y)$ for all $x, y \in \mathbb{R}$, implying that $F(x, y) \leq F(y, x)$ for all $x, y \in \mathbb{R}$. That is, $F(x, y) = F(y, x)$ for all $x, y \in \mathbb{R}$. This proves (b). Letting $y = x$ in the second condition, we get

$$0 = F(x, x) \leq F(x, z) + F(z, x),$$

from which, in view of (b), we obtain $2F(x, z) \geq 0$, implying that $F(x, z) \geq 0$ for all $x, z \in \mathbb{R}$. This proves (a), completing the proof. ∎

**2.** The identity $(1+x)^{2n} = (1+x)^n(1+x)^n$ in $\mathbb{R}[x]$ together with the Binomial Theorem implies that

$$\sum_{k=0}^{2n} C_{2n}^k x^k = \left(\sum_{k=0}^{n} C_n^k x^k\right)\left(\sum_{k=0}^{n} C_n^k x^k\right),$$

for all $n \in \mathbb{N}$, where $C_n^k = \dfrac{n!}{k!(n-k)!}$ is the binomial coefficient. Using the formula for Cauchy product of two polynomials, noting that $C_n^k = C_n^{n-k}$, and finally equating the coefficients of $x^n$ of both sides of the above identity, we see that

$$\frac{(2n)!}{(n!)^2} = C_{2n}^n = \sum_{k=0}^{n} \left(C_n^k\right)^2 = \sum_{k=0}^{n} \left(\frac{n!}{k!(n-k)!}\right)^2,$$

proving the assertion. ∎

**3.** It is plain that if from the point $M$, we shine a beam of light onto a point $R$ of the mirror $\Delta$ so that the reflected light beam hits the mirror $\Delta'$ at a point $S$, then the three points $S$, $R$, and $M'$ are collinear, where $M'$ is the image

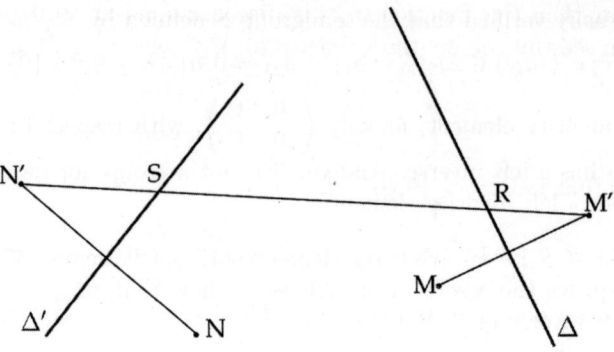

Figure 5

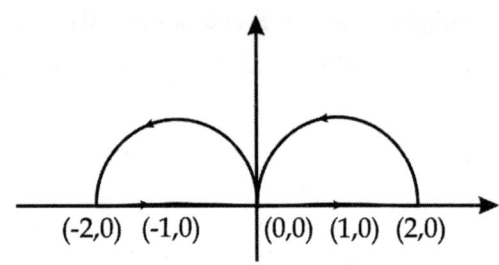

Figure 6

of $M$ in the mirror $\Delta$. Hence the problem is solved as follows. Find the image of $M$ and $N$ in the mirrors $\Delta$ and $\Delta'$, respectively, to obtain the points $M'$ and $N'$ of the plane $P$. The line joining $M'$ and $N'$ intersects $\Delta$ and $\Delta'$ at the points $R$ and $S$, respectively. It is now obvious that if we shine a beam of light onto the point $R$ on the mirror $\Delta$, then the reflected light beam after hitting the mirror $\Delta'$ at $S$ would pass through the point $N$ (see Figure 5).   ∎

## 2.4. Fourth Competition

**2.4.1. Analysis. 1.** Define the function $f : \mathbb{R} \to \mathbb{R}^2$ by

$$f(t) = \begin{cases} g(t) & t < -2, \\ (t, 0) & -2 \le t < 2, \\ h(t) & 2 \le t, \end{cases}$$

where $g(t) = \left( \frac{-2}{1+(t+2)^2}, \frac{2(t+2)}{1+(t+2)^2} \right)$ and $h(t) = \left( \frac{2}{1+(t-2)^2}, \frac{2(t-2)}{1+(t-2)^2} \right)$. It is straightforward to see that $f$ is continuous and one-to-one. In fact, $f$ is a one-to-one parametrization of the curve consisting of the upper semicircle with radius one centered at $(-1, 0)$ going from $(0, 0)$ to $(-2, 0)$, the line segment joining $(-2, 0)$ and $(2, 0)$, and the upper semicircle with radius one centered at $(1, 0)$ going from $(2, 0)$ to $(0, 0)$ (see Figure 6).

We claim that the function $f^{-1}$ is discontinuous at $(0,0)$, proving the assertion. To see this, define the sequence $(a_n)_{n=1}^{+\infty}$ by

$$a_n = \begin{cases} \left(\frac{1}{n}, 0\right) & n \text{ even,} \\ \left(\frac{2}{1+n^2}, \frac{2n}{1+n^2}\right) & n \text{ odd.} \end{cases}$$

It is obvious that $\lim_n a_n = (0,0) = f(0)$ but

$$f^{-1}(a_n) = \begin{cases} \frac{1}{n} & n \text{ even,} \\ n+2 & n \text{ odd,} \end{cases}$$

whose limit does not exist as $n \to +\infty$. Therefore, $f^{-1}$ is not continuous at $(0,0)$, proving the claim. ∎

**2.** Set $\mathbb{S} := \{n \in \mathbb{N} : \alpha_n > 0\}$. If the set $\mathbb{S}$ is finite, there is nothing to prove. So assume that $\mathbb{S}$ is an infinite set. For each $k \in \mathbb{S}$ with $k \geq 2$, define

$$\mathbb{T}_k := \left\{ N \in \mathbb{S} \mid N \geq k \ni \alpha_n \leq \alpha_k \ \forall n \geq N \right\}.$$

As $\lim_n \alpha_n = 0$, letting $\varepsilon = a_k$ in the definition of the limit for $\lim_n \alpha_n = 0$, we see that $\mathbb{T}_k \neq \emptyset$. Let $M_k$ be the initial element of $\mathbb{T}_k$. If $M_k = k$, we will have $\alpha_n \leq \alpha_k$ for all $n \geq k$, in which case set $N_k := k$. If $M_k > k$, set $N_k$ to be the greatest integer subject to $k \leq N_k < M_k$ and $N_k \in \mathbb{S}$. It follows that $N_k \notin \mathbb{T}_k$ and that $\alpha_n = 0$ for all $N_k < n < M_k$. Consequently, $\alpha_{N_k} > \alpha_k > 0$, for $M_k$ is the initial element of $\mathbb{T}_k$. Now, if $n \geq N_k$, then either $n = N_k$, in which case $\alpha_n = \alpha_{N_k} \leq \alpha_{N_k}$, or $N_k < n < M_k$, which yields $\alpha_n = 0 < \alpha_{N_k}$, or $n \geq M_k$, in which case $\alpha_n \leq \alpha_k < \alpha_{N_k}$. In other words, $\alpha_n \leq \alpha_{N_k}$ for all $n \geq N_k$. On the other hand, $\mathbb{S}$ is an infinite set and $N_k \geq k$ for all $k \in \mathbb{S}$. This shows that there are an infinite number of $N_k$'s such that $\alpha_n \leq \alpha_{N_k}$ for all $n \geq N_k$. ∎

### 2.4.2. Algebra.

**1.** (a) Let $x, y \in R(B)$ and $a \in A$ be arbitrary. It follows that there are $M, N \in \mathbb{N}$ such that $x^M \in B$ and $x^N \in B$. Noting that in a commutative ring the Binomial Theorem holds, $(ab)^n = a^n b^n$ for all $a, b \in A$ and $n \in \mathbb{N}$, and that $B$ is an ideal of $A$, we see that

$$(x-y)^{M+N} = \sum_{i=0}^{M+N} C_{M+N}^i x^i (-y)^{M+N-i} \in B, \ (ax)^M = a^M x^M \in B,$$

yielding $x - y \in R(B)$ and $ax \in R(B)$. That implies $R(B)$ is an ideal of $A$, which is what we want.

(b) **First proof.** Let $B$ be a prime ideal of $A$. To prove that $R(B)$ is prime, suppose for given $x, y \in A$, we have $xy \in R(B)$. We need to show that $x \in R(B)$ or $y \in R(B)$. As $xy \in R(B)$, there is an $n \in \mathbb{N}$ such that $x^n y^n = (xy)^n \in B$. But $B$ is prime. So we must have $x^n \in B$ or $y^n \in B$. In other words, $x \in\in R(B)$ or $y \in R(B)$, as desired.

**Second proof.** We prove the assertion by showing that $R(B) = B$. Evidently, $B \subseteq R(B)$. To show that $R(B) \subseteq B$, let $x \in R(B)$ be arbitrary. It follows that there is an $n \in \mathbb{N}$ such that $x^n \in B$. Now, a straightforward induction on $n$ together with the hypothesis that $B$ is prime reveals that $x \in B$. This proves that $R(B) \subseteq B$, and hence $R(B) = B$, as desired.

(c) **Remark.** We must have $a \neq 0, \pm 1$.

Without loss of generality, assume that $a > 1$. Write $a = p_1^{m_1} \cdots p_k^{m_k}$, where $m_i \in \mathbb{N}$ and $p_i$'s are distinct prime numbers $(1 \leq i \leq k)$. We show that $R(B) = \langle p_1 p_2 \cdots p_k \rangle$. First, let $x \in \langle p_1 p_2 \cdots p_k \rangle$ be arbitrary. It follows that $x = p_1 p_2 \cdots p_k r$ for some $r \in \mathbb{Z}$. Obviously, we can write $x^{m_1 + \cdots + m_k} = as \in B$ for some $s \in \mathbb{Z}$, implying that $x \in R(B)$. This yields $\langle p_1 p_2 \cdots p_k \rangle \subseteq R(B)$. Next, let $x \in R(B)$ be arbitrary. It follows that $x^n \in B = \langle p_1^{m_1} \cdots p_k^{m_k} \rangle$, form which, we obtain $x^n = p_1^{m_1} \cdots p_k^{m_k} r$ for some $r \in \mathbb{Z}$. Thus, $p_i | x^n$ for all $1 \leq i \leq k$, yielding $p_i | x$ because $p_i$ is prime. This implies $p_1 \cdots p_k | x$, for $p_i$'s are distinct primes. That is, $x \in \langle p_1 p_2 \cdots p_k \rangle$, and hence $R(B) \subseteq \langle p_1 p_2 \cdots p_k \rangle$. Therefore, $R(B) = \langle p_1 p_2 \cdots p_k \rangle$, finishing the proof. ∎

**2.** Let $\{x, y\}$ be a linearly independent set in $E$, which is a vector space over a field $F$. Use Zorn's Lemma to enlarge the independent sets $\{x, y\}$ and $\{x, x+y\}$ to $\{x, y\} \cup \{\alpha_i\}_{i \in I}$ and $\{x, x+y\} \cup \{\beta_i\}_{i \in I}$, respectively, each of which is a Hamel basis for $E$. It is well-known that there exists a linear transformation $u : E \to E$ satisfying $u(x) = x$, $u(y) = x + y$, and $u(\alpha_i) = \beta_i$ for all $i \in I$. The linear transformation $u$ is invertible because it maps the elements of the Hamel basis $\{x, y\} \cup \{\alpha_i\}_{i \in I}$ onto those of the Hamel basis $\{x, x+y\} \cup \{\beta_i\}_{i \in I}$. This means $u$ is an automorphism of the vector space $E$. Now, suppose that a linear transformation $f : E \to E$, also called an endomorphism of the vector space $E$, commutes with any invertible linear transformation $v : E \to E$, which is also called an automorphism of $E$. We claim, first of all, that for any nonzero vector $x$, we have $f(x) = \lambda_x x$ for some $\lambda_x \in F$. To prove this by contradiction, letting $u_x : E \to E$ be an automorphism such that $u_x(x) = x$ and $u_x(f(x)) = x + f(x)$, we must have $f u_x = u_x f$. In particular, we have $f(u_x(x)) = u_x(f(x))$ which means $f(x) = x + f(x)$, yielding $x = 0$, which is impossible. Therefore, for any $x \in E$ there exists $\lambda_x \in F$ such that $f(x) = \lambda_x x$. Next, fix a nonzero $x_0 \in E$. By showing that $\lambda_x = \lambda_{x_0}$, we prove the assertion. Without loss of generality, assume that $\{x, x_0\}$ is linearly independent. Therefore, there is an automorphism $u_x : E \to E$ such that $u_x(x) = x$ and $u_x(x_0) = x + x_0$. But $f$ commuting with $u_x$ implies that

$$
\begin{aligned}
\lambda_x x + \lambda_{x_0} x_0 &= f(x + x_0) = f(u_x(x_0)) \\
&= u_x(f(x_0)) = \lambda_{x_0} u_x(x_0) \\
&= \lambda_{x_0} x + \lambda_{x_0} x_0,
\end{aligned}
$$

yielding $\lambda_x x = \lambda_{x_0} x$. From this, we obtain $\lambda_x = \lambda_{x_0}$ because $x \neq 0$. So we have shown that $f(x) = \lambda_{x_0} x$ for all $x \in E$, or equivalently $f = \lambda_{x_0} I_E$, which is what we want, finishing the proof. ∎

**3.** It suffices to show that any nonzero $x \in R$ has a multiplicative inverse. To this end, let a nonzero $x \in R$ be given. It follows from the hypothesis that the following descending chain of the ideals of $R$

$$xR \supseteq x^2R \supseteq x^3R \supseteq \cdots$$

must terminate. That is, there exists $N \in \mathbb{N}$ such that $x^nR = x^NR$ for all $n \geq N$. In particular, $x^{N+1}R = x^NR$, implying that $x^N \in x^{N+1}R$. Hence, there exists $y \in R$ such that $x^N = x^{N+1}y$, yielding $x^N(1 - xy) = 0$. This, in view of the fact that $R$ is an integral domain, implies that $1 - xy = 0$, yielding $xy = yx = 1$. In other words, $x$ is invertible, which is what we want. ∎

**2.4.3. General. 1.** Doing the substitution $x = \frac{\pi}{2} - t$, we can write

$$I := \int_0^{\frac{\pi}{2}} \frac{\sin^n x}{\cos^n x + \sin^n x} dx = \int_{\frac{\pi}{2}}^0 \frac{\sin^n(\frac{\pi}{2} - t)}{\cos^n(\frac{\pi}{2} - t) + \sin^n(\frac{\pi}{2} - t)}(-dt)$$

$$= \int_0^{\frac{\pi}{2}} \frac{\cos^n t}{\sin^n t + \cos^n t} dt.$$

We have

$$I = \frac{1}{2}(I + I) = \frac{1}{2}\left(\int_0^{\frac{\pi}{2}} \frac{\sin^n x + \cos^n x}{\cos^n x + \sin^n x} dx\right) = \frac{1}{2} \times \frac{\pi}{2} = \frac{\pi}{4},$$

for all $n \in \mathbb{R}$. ∎

**2.** (a) Define the events $a, b, c, a_1, b_1, c_1$ as follows:

$a, b, c$: The event that the prize is in box $A$, $B$, or $C$, respectively.

$a_1, b_1, c_1$: The event that the box $A$, $B$, or $C$ is opened, respectively.

It is obvious that $\{a, b, c\}$ is a partition of the probability space and that the desired probability is $P(b|a_1)$. Using the Inverse Probability Theorem or Bayes' Theorem, we can write

$$P(b|a_1) = \frac{P(b)P(a_1|b)}{P(a_1)} = \frac{P(b)P(a_1|b)}{P(a)P(a_1|a) + P(b)P(a_1|b) + P(c)P(a_1|c)}$$

$$= \frac{\frac{1}{3} \times \frac{1}{2}}{\frac{1}{3} \times 0 + \frac{1}{3} \times \frac{1}{2} + \frac{1}{3} \times 1} = \frac{1}{3}.$$

So the probability that the chosen box, i.e. $B$, contains the prize is $\frac{1}{3}$.

(b) Plainly, the desired probability is $P(c|a_1)$. Again, using Bayes' Theorem, we can write

$$P(c|a_1) = \frac{P(c)P(a_1|c)}{P(a_1)} = \frac{P(c)P(a_1|c)}{P(a)P(a_1|a) + P(b)P(a_1|b) + P(c)P(a_1|c)}$$

$$= \frac{\frac{1}{3} \times 1}{\frac{1}{3} \times 0 + \frac{1}{3} \times \frac{1}{2} + \frac{1}{3} \times 1} = \frac{2}{3}.$$

That is, the probability that the other box, i.e., $C$, contains the prize is $\frac{2}{3}$. ∎

## 2.5. Fifth Competition

**2.5.1. Analysis. 1.** No, it is not. We present a counterexample on the closed interval $[0, 1]$ which will work on the open interval $(0, 1)$ as well. Define the functions $f, g : [0, 1] \to [0, 1]$ as follows

$$f(x) = \begin{cases} \frac{1}{q} & x \in \mathbb{Q} \cap (0, 1], x = \frac{p}{q}, p, q \in \mathbb{N}, \gcd(p, q) = 1, \\ 0 & x \in [0, 1] \setminus (\mathbb{Q} \cap (0, 1]). \end{cases}$$

$$g(x) = \begin{cases} 1 & x \in \{\frac{1}{q} | q \in \mathbb{N}\}, \\ 0 & x \notin \{\frac{1}{q} | q \in \mathbb{N}\}. \end{cases}$$

Obviously,

$$g \circ f(x) = \begin{cases} 1 & x \in \mathbb{Q} \cap (0, 1], \\ 0 & x \in [0, 1] \setminus (\mathbb{Q} \cap (0, 1]). \end{cases}$$

We claim that $f, g$ are Riemann integrable on $[0, 1]$ but $g \circ f$ is not integrable on $[0, 1]$. In fact, it turns out that $g \circ f$ is not integrable on any closed interval $[\varepsilon, \delta]$, where $0 < \varepsilon < \delta < 1$, from which it follows that $g \circ f$ in not integrable on the open interval $(0, 1)$. In view of Solution 3 of 2.1.1, we see that

$$\forall a \in (0, 1) : \lim_{x \to a} f(x) = 0, \quad \lim_{x \to 0^+} f(x) = 0, \quad \lim_{x \to 1^-} f(x) = 0.$$

Thus, the function $f$ is continuous at the irrational points of $(0, 1)$ and the set of discontinuity points of $f$, i.e., $\mathbb{Q} \cap (0, 1)$, is countable. It is easily seen that $g$ is discontinuous at any point of the set $\{\frac{1}{q} | q \in \mathbb{N}\} \cup \{0\}$ and it is continuous on the set $(0, 1] \setminus \{\frac{1}{q} | q \in \mathbb{N}\}$. Finally, using the fact that $\mathbb{Q} \cap (0, 1]$ is dense in $[0, 1]$, we see that $g \circ f$ is discontinuous at any point of the interval $[0, 1]$. From this point on, we present two proofs for the claim we made in the above. The first one uses some standard theorem from analysis and the second proof is self-contained.

**First proof.** We will make use of Lebesgue's Integrability Criterion for integrals in the Riemann sense which asserts that *a bounded function* $f :$ $[a, b] \to \mathbb{R}$ *is integrable in the Riemann sense if and only if the set consisting of the points at which $f$ is discontinuous has measure zero.* Recall that a set $A \subseteq \mathbb{R}$ is said to be a *set of measure zero* if for any $\varepsilon > 0$, there exists a sequence $(I_n)_{n=1}^{+\infty}$ of intervals such that $A \subseteq \bigcup_{n=1}^{+\infty} I_n$ and $\sum_{n=1}^{+\infty} \ell(I_n) < \varepsilon$, where $\ell(I_n) := \beta_n - \alpha_n$ and $\alpha_n$ and $\beta_n$ are the end-points of the interval $I_n$. It is easy to see that any finite or countable subset of reals has measure zero and that a countable union of sets of measure zero has measure zero. In view of all that, we see that the functions $f$ and $g$ are Riemann integrable on $[0, 1]$ and hence on $(0, 1)$ but $g \circ f$ is not integrable $(0, 1)$ because it is not integrable on any interval $[\varepsilon, \delta]$, where $0 < \varepsilon < \delta < 1$, for $g \circ f$ is continuous nowhere.

**Second proof.** Define the sequences $(f_n)_{n=1}^{+\infty}$ and $(g_n)_{n=1}^{+\infty}$, with $f_n, g_n$ : $[0,1] \to [0,1]$, as follows

$$f_n(x) = \begin{cases} \frac{1}{q} & x = \frac{p}{q}, p, q \in \mathbb{N}, \gcd(p,q) = 1, q \leq n, \\ 0 & \text{otherwise.} \end{cases}$$

$$g_n(x) = \begin{cases} 1 & x = \frac{1}{q}, q \in \mathbb{N}, q \leq n, \\ 0 & \text{otherwise.} \end{cases}$$

It is easily verified that the sequences $(f_n)_{n=1}^{+\infty}$ and $(g_n)_{n=1}^{+\infty}$ converge uniformly on $[0,1]$ to $f$ and $g$, respectively. As $f_n$ and $g_n$ $(n \in \mathbb{N})$ are discontinuous at only finitely many points of the interval $[0,1]$, it follows that $f$ and $g$ are integrable in the Riemann sense on $[0,1]$. Now, we see from the following lemma that $g \circ f$ is not integrable on any interval $[\varepsilon, \delta]$, where $0 < \varepsilon < \delta < 1$, which is what we want.

**Lemma.** *Let* $f : [a,b] \to \mathbb{R}$ *be bounded and integrable on* $[a,b]$ *in the Riemann sense. Then the set of points at which* $f$ *is continuous is infinite.*

**Proof.** Since $f$ integrable on any closed subinterval of $[a,b]$, it suffices to prove that $f$ is continuous at some point of the interval $[a,b]$, for it would then follow that any subinterval of $[a,b]$ contains a point at which $f$ is continuous. And hence, the set of points at which $f$ is continuous would be infinite, which is what we want. Taking $\varepsilon = b - a$ in Riemann's criterion for integrability on the interval $I_0 := [a,b]$, we obtain a partition $P_0 : a = t_{00} < \cdots < t_{0n_0} = b$ of $I_0$ such that

$$U(P_0, f) - L(P_0, f) < b - a,$$

where

$$U(P_0, f) = \sum_{i=1}^{n_0} M_{0i} \Delta t_{0i}, \; L(P_0, f) = \sum_{i=1}^{n_0} m_{0i} \Delta t_{0i}, \; \Delta t_{0i} = t_{0i} - t_{0(i-1)},$$

$$M_{0i} = \sup_{t_{0(i-1)} \leq x \leq t_{0i}} f(x), \; m_{0i} = \inf_{t_{0(i-1)} \leq x \leq t_{0i}} f(x).$$

Hence, there exists $1 \leq i_0 \leq n_0$ such that

$$M_{0i_0} - m_{0i_0} < 1,$$

for otherwise it would follow that $U(P_0, f) - L(P_0, f) \geq b - a$, which is impossible. Set $I_1 := [a_1, b_1] = [t_{0(i_0-1)}, t_{0i_0}]$. As $f$ is integrable on $I_1$, taking $\varepsilon = \frac{b_1 - a_1}{2}$, a similar argument shows that there exists a subinterval $I_2 := [a_2, b_2]$ of $I_1$ such that $\sup_{x \in I_2} f(x) - \inf_{x \in I_2} f(x) < \frac{1}{2}$. Continuing this way, we obtain a sequence of nested closed interval $I_1 \supset I_2 \supset \cdots$ such that $0 \leq \sup_{x \in I_n} f(x) - \inf_{x \in I_n} f(x) < \frac{1}{n}$. Since $\bigcap_{n=1}^{+\infty} I_n \neq \emptyset$ because $I_0$ is compact, in view of the above inequality, it follows that $f$ is continuous at any point of $\bigcap_{n=1}^{+\infty} I_n \neq \emptyset$, finishing the proof. ∎

**2.** We prove the assertion by presenting the following proposition, which characterizes all continuous real functions having the property that they do not assume any value more than twice.

**Proposition.** *Let $f : \mathbb{R} \longrightarrow \mathbb{R}$ be a continuous function that is neither increasing nor decreasing on $\mathbb{R}$ and that it assumes every value at most twice. Prove that one of the following statements holds.*

*(i) There exists an $a \in \mathbb{R}$ such that $f$ or $-f$ is strictly increasing on $(-\infty, a]$ and is strictly decreasing on $[a, +\infty)$.*

*(ii) There exist $a, b \in \mathbb{R}$ with $a < b$ such that $f$ or $-f$ is strictly increasing on $(-\infty, a]$, strictly decreasing on $[a, b]$, and strictly increasing on $[b, +\infty)$. Moreover, $\lim_{x \to -\infty} f(x) \geq \lim_{x \to +\infty} f(x)$ or $\lim_{x \to -\infty} -f(x) \geq \lim_{x \to +\infty} -f(x)$.*

To prove the proposition, we need the following two lemmas.

**Lemma 1.** *Let $R$ be a totally ordered set with at least three elements and $f : R \to R$ a function that is neither strictly increasing nor strictly decreasing. Then, there exist $x_1, x_2, x_3 \in R$ with $x_1 < x_2 < x_3$ such that $f(x_2) \geq \max\big(f(x_1), f(x_3)\big)$ or $f(x_2) \leq \min\big(f(x_1), f(x_3)\big)$.*

**Proof.** To prove the assertion by contradiction, suppose that for all $x_1, x_2, x_3 \in R$ with $x_1 < x_2 < x_3$, we have

$$f(x_2) < \max\big(f(x_1), f(x_3)\big) \text{ and } f(x_2) > \min\big(f(x_1), f(x_3)\big).$$

Now, since $\max\big(f(x_1), f(x_3)\big) = f(x_1)$ or $\max\big(f(x_1), f(x_3)\big) = f(x_3)$, we see that for all $x_1, x_2, x_3 \in R$ with $x_1 < x_2 < x_3$, we have

$$f(x_1) < f(x_2) < f(x_3) \text{ or } f(x_1) > f(x_2) > f(x_3). \qquad (*)$$

Pick $a, b, c \in R$ with $a < b < c$. It follows that either $f(a) < f(b) < f(c)$ or $f(a) > f(b) > f(c)$. Suppose $f(a) < f(b) < f(c)$. We show that $f$ is strictly increasing, a contradiction. To see this, let $x, y \in R$ with $x < y$ be arbitrary. There are five cases to consider. (i) $x < y < a$, (ii) $x < y = a$, (iii) $x < a < y$, (iv) $x = a < y$, and (v) $a < x < y$. In each case, in view of $(*)$, it is easy to see that $f(x) < f(y)$, from which, we conclude that $f$ is strictly increasing, which is a contradiction. Likewise, if $f(a) > f(b) > f(c)$, one can see, in a similar fashion, that $f$ is strictly decreasing which is again a contradiction. So the proof is complete by contradiction. $\square$

**Lemma 2.** *Let $f : \mathbb{R} \longrightarrow \mathbb{R}$ be a continuous function which assumes every value at most twice.*

*(i) Suppose that there are $x_1, x_2, x_3 \in \mathbb{R}$ with $x_1 < x_2 < x_3$ such that*

$$f(x_2) \geq \max\big(f(x_1), f(x_3)\big).$$

*Then, $f$ has a unique absolute maximum, on $\mathbb{R}$, which occurs in the interval $(x_1, x_3)$.*

*(ii) Suppose that there are $x_1, x_2, x_3 \in \mathbb{R}$ with $x_1 < x_2 < x_3$ such that*

$$f(x_2) \leq \min\big(f(x_1), f(x_3)\big).$$

*Then, $f$ has a unique absolute minimum, on $\mathbb{R}$, which occurs in the interval $(x_1, x_3)$.*

**Proof.** We only prove part (i). Part (ii) can be proven in a similar fashion or by replacing $f$ by $-f$ and applying (i). From $f(x_2) \geq \max\big(f(x_1), f(x_3)\big)$ and the continuity of $f$, we see that $f$ attains its absolute maximum on the interval

$[x_1, x_3]$ at a point $x_M \in (x_1, x_3)$. It suffices to prove that $f(x_M) > f(x)$ for all $x \in \mathbb{R} \setminus \{x_M\}$. First, we show that $f(x_M) > f(x)$ for all $x \in [x_1, x_3] \setminus \{x_M\}$. Proceed by way of contradiction, assuming that there is an $x_0 \in [x_1, x_3] \setminus \{x_M\}$ such that $f(x_0) = f(x_M)$. There are two cases to consider (a) $x_1 \leq x_0 < x_M$, and (b) $x_M < x_0 \leq x_3$. First, let $x_1 \leq x_0 < x_M$ and pick a $t \in (x_0, x_M)$. We see that $f(t) < f(x_0) = f(x_M)$ because $f(x_M)$ is the absolute maximum of $f$ on the interval $[x_1, x_3]$ and that $f$ assumes the value $f(x_M)$ at most twice. Set $\lambda = \frac{f(x_M) + \max\big(f(x_3), f(t)\big)}{2}$. It is obvious that $f(t) < \lambda < f(x_0)$, $f(t) < \lambda < f(x_M)$, and $f(x_3) < \lambda < f(x_M)$. Thus, it follows from the Intermediate Value Theorem that there exist $\xi_1, \xi_2, \xi_3$ with $x_0 < \xi_1 < t$, $t < \xi_2 < x_M$, and $x_M < \xi_3 < x_3$ such that $f(\xi_i) = \lambda$ for each $i = 1, 2, 3$, contradicting the hypothesis. Similarly, if $x_M < x_0 \leq x_3$, we obtain a contradiction. That is, we have shown that $f(x_M) > f(x)$ for all $x \in [x_1, x_3]$ with $x \neq x_M$. Next, we prove that $f(x) < f(x_M)$ for all $x \in \mathbb{R}$ with $x \neq x_M$. Again suppose to the contrary that there is a $t \in \mathbb{R} \setminus \{x_M\}$ such that $f(t) \geq f(x_M)$. From what we have proved so far, it follows that $t < x_1$ or $t > x_3$. If $t < x_1$, noting that $f(x_1) < f(x_M) \leq f(t)$, we see that there exists an $x_0 \in \mathbb{R}$ with $t \leq x_0 < x_1$ such that $f(x_0) = f(x_M)$. Now, set $\lambda = \frac{f(x_M) + \max\big(f(x_1), f(x_3)\big)}{2}$. Again, it is obvious that $f(x_1) < \lambda < f(x_0)$, $f(x_1) < \lambda < f(x_M)$, and $f(x_3) < \lambda < f(x_M)$. Hence, by the Intermediate Value Theorem, we obtain $\xi_1, \xi_2, \xi_3$ with $x_0 < \xi_1 < x_1$, $x_1 < \xi_2 < x_M$, and $x_M < \xi_3 < x_3$ such that $f(\xi_i) = \lambda$ for each $i = 1, 2, 3$, contradicting the hypothesis. Similarly, if $x_3 < t$, we obtain a contradiction. Therefore, $f(x) < f(x_M)$ for all $x \in \mathbb{R}$ with $x \neq x_M$, proving (i). So the proof is complete. $\qquad \square$

**Proof of Proposition.** Let $f : \mathbb{R} \longrightarrow \mathbb{R}$ be a continuous function such that it assumes every value at most twice. In view of Lemma 1, there are two cases to consider.

(a) There exist $x_1, x_2, x_3 \in \mathbb{R}$ with $x_1 < x_2 < x_3$ such that
$$f(x_2) \geq \max\big(f(x_1), f(x_3)\big).$$

(b) There exist $x_1, x_2, x_3 \in \mathbb{R}$ with $x_1 < x_2 < x_3$ such that
$$f(x_2) \leq \min\big(f(x_1), f(x_3)\big).$$

We prove the assertion in the case (a). Again, the case (b) can be proven in a similar fashion or by replacing $f$ by $-f$ and applying (a). It follows from Lemma 2 that there is an $x_M \in (x_1, x_3)$ such that $f(x_M) > f(x)$ for all $x \in \mathbb{R} \setminus \{x_M\}$. It is plain that $f$ is not strictly decreasing on $(-\infty, x_M]$. We now distinguish two cases. The function $f$ is either strictly increasing on $(-\infty, x_M]$ or it is not. First, assume that $f$ is not strictly increasing on $(-\infty, x_M]$. Since $f(x_M) > f(x)$ for all $x \in \mathbb{R} \setminus \{x_M\}$, from Lemmas 1 and 2, we see that there are $y_1, y_2, y_3 \in (-\infty, x_M]$ with $y_1 < y_2 < y_3$ such that
$$f(y_2) \leq \min\big(f(y_1), f(y_3)\big).$$

It thus follows from Lemma 2 that there is an $x_m \in (y_1, y_3)$ such that $f(x_m) < f(x)$ for all $x \in \mathbb{R} \setminus \{x_m\}$. Now, from Lemmas 1 and 2, we easily conclude that $f$ is strictly decreasing on $(-\infty, x_m]$, strictly increasing on $[x_m, x_M]$, and strictly decreasing on $[x_M, +\infty)$. Moreover, $\lim_{x\to-\infty} f(x) \leq \lim_{x\to+\infty} f(x)$, for $f$ assumes every value at most twice. That is, $-f$ is strictly increasing on $(-\infty, x_m]$, strictly decreasing on $[x_m, x_M]$, and strictly increasing on $[x_M, +\infty)$, and moreover, $\lim_{x\to-\infty} -f(x) \geq \lim_{x\to+\infty} -f(x)$, as desired. Next, assuming that $f$ is strictly increasing on $(-\infty, x_M]$, we look at the status of $f$ on $[x_M, +\infty)$. Note that $f$ cannot be strictly increasing on $[x_M, +\infty)$. If $f$ is strictly decreasing on $[x_M, +\infty)$, we will have nothing to prove. So suppose that $f$ is not strictly decreasing on $[x_M, +\infty)$ as well. Again, since $f(x_M) > f(x)$ for all $x \in \mathbb{R}\backslash\{x_M\}$, from Lemmas 1 and 2, we see that there are $y_1, y_2, y_3 \in [x_M, +\infty)$ with $y_1 < y_2 < y_3$ such that

$$f(y_2) \leq \min\left(f(y_1), f(y_3)\right).$$

It now follows from Lemma 2 that there is an $x_m \in (y_1, y_3)$ such that $f(x_m) < f(x)$ for all $x \in \mathbb{R} \setminus \{x_m\}$. Therefore, Lemmas 1 and 2 imply that $f$ is strictly increasing on $(-\infty, x_m]$, strictly decreasing on $[x_m, x_M]$, and strictly increasing on $[x_M, +\infty)$, and moreover, $\lim_{x\to-\infty} f(x) \geq \lim_{x\to+\infty} f(x)$, for $f$ assumes every value at most twice. This completes the proof. ∎

**3.** Let $\mathcal{B} = \{(p, q) : p, q \in \mathbb{Q}, p < q\}$, where $(p, q)$ denotes the open interval with end points $p$ and $q$ in $\mathbb{R}$. The set $\mathcal{B}$ is a countable basis for the ordinary topology of $\mathbb{R}$. Write $\mathcal{B} = \{I_n\}_{n\in\mathbb{N}}$, where $I_n = (p_n, q_n)$ for some $p_n, q_n \in \mathbb{Q}$, and set

$$T := \{n \in \mathbb{N} : \text{card}(I_n \cap A) \leq \aleph_0\}.$$

Obviously, $T \subsetneq \mathbb{N}$ because otherwise $A$ would be countable which is impossible. Now, set

$$A_1 = \left(\bigcup_{n\in T} I_n\right) \cap A = \bigcup_{n\in T} (I_n \cap A).$$

The set $A_1$ is countable because so is $I_n \cap A$ for all $n \in T$. By showing that any point $x \in A\backslash A_1$ is a congestion point of $A$, we finish the proof. To prove this by contradiction, suppose that there exists an $\varepsilon_0 > 0$ such that $(x - \varepsilon_0, x + \varepsilon_0) \cap A$ is countable. As $\mathcal{B} = \{I_n\}_{n\in\mathbb{N}}$ is a basis for the topology of $\mathbb{R}$, we see that there exists $n_0 \in \mathbb{N}$ such that $x \in I_{n_0} \subseteq (x - \varepsilon_0, x + \varepsilon_0)$, from which, we obtain

$$I_{n_0} \cap A \subseteq (x - \varepsilon_0, x + \varepsilon_0) \cap A,$$

implying that $I_{n_0} \cap A$ is countable because so is $(x - \varepsilon_0, x + \varepsilon_0) \cap A$. That is, $\text{card}(I_{n_0} \cap A) \leq \aleph_0$, yielding $n_0 \in T$, and hence $x \in I_{n_0} \cap A \subseteq \bigcup_{n\in T}(I_n \cap A) = A_1$. Consequently, $x \in A_1$, which contradicts the hypothesis that $x \in A \setminus A_1$. Thus, $x$ is a congestion point of $A$, which is what we want. ∎

      **2.5.2. Algebra. 1.** Let $A : V \to V$ be such that $A \neq I$, $A^2 \neq I$, $A^3 = I$. If $x \neq 0$ is an eigenvector corresponding to the eigenvalue $\lambda \in \mathbb{C}$, we can write

$$Ax = \lambda x \implies A^2 x = \lambda^2 x \implies A^3 x = \lambda^3 x,$$

which, in view of $A^3 = I$, implies that $\lambda^3 = 1$. That is, the set of the eigenvalues of $A$ is a subset of the third roots of unity. Likewise, if $A \neq I, \ldots, A^{k-1} \neq I, A^k = I$, then the set of the eigenvalues of $A$ is a subset of the $k$th roots of unity. The eigenvalues of $A$ do not necessarily form a group under multiplication. To see this, let $\omega_k = \cos\frac{2\pi}{k} + i\sin\frac{2\pi}{k}$, where $k - 1 \in \mathbb{N}$, and just note that $\omega_k \neq 1$ is the only eigenvalue of the linear transformation $\omega_k I_n$, where $I_n : \mathbb{C}^n \to \mathbb{C}^n$ is the identity transformation. Obviously, $\{\omega_k\}$ does not form a group under multiplication. As for a necessary and sufficient condition for the set of eigenvalues of $A$ to form a group, nothing interesting can be said. Having said that, it is easily seen that the set of eigenvalues of $A$ forms a group under multiplication, in fact a cyclic subgroup of the multiplicative group of the $k$th roots of unity, if and only if the set of the eigenvalues of $A$ is closed under multiplication. ∎

**2.** We have $\mathrm{ord}(a) = 7$ because $a^7 = 1$, $a \neq 1$, and 7 is prime. Induction on $k$ together with the relations $ab = b^2 a$ and $a = b^2 ab^{-1}$ yields

$$a^k = b^{2^k} a^k b^{-1},$$

for all $k \in \mathbb{N}$. In particular, $a^7 = b^{2^7} a^7 b^{-1}$ together with $a^7 = 1$ implies that $b^{2^7 - 1} = 1$. That is, $b^{127} = 1$. But 127 is prime and $b \neq 1$. Hence, $\mathrm{ord}(b) = 127$. Therefore, $\mathrm{ord}(a) = 7$ and $\mathrm{ord}(b) = 127$, which is what we want. ∎

**3.** (a) We can write

$$(-x)^3 + 2(-x)^2 + (-x) = 0 \implies -x^3 + 2x^2 - x = 0,$$

for all $x \in R$. On the other hand, $x^3 + 2x^2 + x = 0$ for all $x \in R$. Adding up the two equalities, we obtain $4x^2 = 0$ for all $x \in R$. We can also write

$$(2x)^3 + 2(2x)^2 + (2x) = 0 \implies 8x^3 + 8x^2 + 2x = 0,$$

for all $x \in R$. But $8x^3 = 8x^2 = 0$ because $4x^2 = 0$ for all $x \in R$. This together with the above yields $2x = 0$ for all $x \in R$, which is what we want.

(b) Using (a), we can write

$$(x+1)^3 + 2(x+1)^2 + (x+1) = 0 \implies x^2 + x = 0,$$

for all $x \in R$. Consequently, $x^2 = -x = x$ because $2x = 0$. Hence, $x^2 = x$ for all $x \in R$. Therefore, $R$ is a Boolean ring, which is commutative by the following argument.

$$(x+y)^2 = x + y \implies x^2 + y^2 + xy + yx = x^2 + y^2$$
$$\implies xy = -yx = yx,$$

for all $x, y \in R$, which is what we want. ∎

### 2.5.3. General. 1. We need the following lemma

**Lemma.** *The product of $k$ consecutive integers is divisible by $k!$.*
**Proof.** It suffices to show that the product of $k$ natural numbers is divisible by $k!$. To this end, let $k, n \in \mathbb{N}$ with $k \leq n$ be given. To show that $n(n -$

1)$\cdots(n-k+1)$ is divisible by $k!$, it is enough to prove that $C_n^k := \dfrac{n!}{k!(n-k)!}$
is an integer because

$$n(n-1)\cdots(n-k+1) = k!C_n^k.$$

By proving the polynomial identity below, which is also known as the Binomial Theorem

$$(1+x)^n = \sum_{k=0}^{n} C_n^k x^k, \qquad (*)$$

we conclude that $C_n^k$ is an integer, finishing the proof. Recall that, by Taylor's Formula, for two polynomial functions $p, q : \mathbb{R} \to \mathbb{R}$ of degree $n$, we have $p = q$ if and only if $p^{(k)}(0) = q^{(k)}(0)$ for each $k = 0, 1, \ldots, n$, where $p^{(k)}, q^{(k)}$ denote the $k$th derivative of $p$ and $q$, respectively, and the zeroth derivative of a function, by definition, is just the function. Also, it is readily verified that

$$\left((a+x)^k\right)^{(i)} = \begin{cases} k(k-1)\cdots(k-(i-1))(a+x)^{k-i} & 1 \le i < k, \\ k! & i = k, \\ 0 & i > k, \end{cases}$$

where $a \in \mathbb{R}$ is a constant. Use $p$ and $q$ to denote the left and right hand side of $(*)$. Obviously, $p(0) = q(0)$. We can write

$$p^{(k)}(0) = \left((1+x)^n\right)^{(k)}(0) = n(n-1)\cdots(n-k+1)(1+0)^{n-k} = k!C_n^k,$$

$$q^{(k)}(0) = \left(\sum_{i=0}^{n} C_n^i x^i\right)^{(k)} = C_n^k(x^k)^{(k)} = C_n^k k!,$$

for each $k = 1, \ldots, n$. So we have shown that $p^{(k)}(0) = q^{(k)}(0)$ for each $k = 0, 1, \ldots, n$, implying that $p = q$, which is what we want.      □

Now to prove the assertion, note first that

$$(mn)! = \prod_{k=0}^{m-1}\left(\prod_{j=1}^{n}(kn+j)\right).$$

But

$$\prod_{j=1}^{n}(kn+j) = \prod_{j=1}^{n-1}(kn+j) \times n(k+1).$$

It follows from the lemma that $\prod_{j=1}^{n-1}(kn+j)$ is divisible by $(n-1)!$, whence $\prod_{j=1}^{n}(kn+j)$ is divisible by $(n-1)! \times n(k+1) = n!(k+1)$. This implies $(mn)! = \prod_{k=0}^{m-1}\left(\prod_{j=1}^{n}(kn+j)\right)$ is divisible by $\prod_{k=0}^{m-1} n!(k+1) = (n!)^m m!$.
Thus, the number $\dfrac{(mn)!}{m!(n!)^m}$ is an integer, which is what we want.      ∎

**2.** The assertion is known as the Gauss-Lucas Theorem . As is shown in what follows, it suffices to prove that if the roots of a polynomial $p$ lie in a closed half plane, then so do those of $p'$. For a subset $S$ of $\mathbb{C}$, the *closed convex hull*

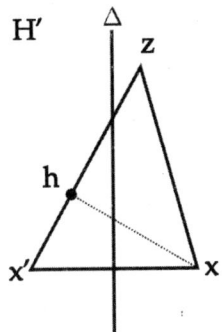

Figure 7

of $S$, denoted by the symbol $\overline{C(S)}$, is said to be the intersection of all closed convex subsets of $\mathbb{C}$ which include $S$, i.e.,

$$\overline{C(S)} := \bigcap \{C : S \subseteq C, C \text{ is closed and convex}\}.$$

It is straightforward to see that if $S = \{z_i\}_{i=1}^{n}$, where $n \in \mathbb{N}$, then $\overline{C(S)} = C(S)$. We claim that

$$\overline{C(S)} = C(S) = \bigcap \{H : S \subseteq H, H \text{ is a closed half plane}\}.$$

Set $C_1 := \bigcap \{H : S \subseteq H, H \text{ is a closed half-plane }\}$. As $C_1$ is closed and convex, it follows that $\overline{C(S)} = C(S) \subseteq C_1$. We establish our claim by showing that $\overline{C(S)}^c = C(S)^c \subseteq C_1^c$. To see this, let $x \notin \overline{C(S)} = C(S)$ be arbitrary. As $C(S)$ is closed and $\{x\}$ is compact, it follows that there exists an $x' \in C(S)$ such that

$$|x - x'| = \inf_{y \in C(S)} |x - y|.$$

Now, consider the perpendicular bisector $\Delta$ of the line segment joining $x$ and $x'$. Let $H'$ be the closed half-plane containing the point $x' \in S$. The half plane $H'$ includes $S$ because if, contrary to this, there exists a point $z \in S$ in the other half plane, then in the triangle $x'xz$ the perpendicular foot $h$ of the altitude passing through the vertex $x$ belongs $S$, for $x', z \in S$, and moreover $|x - h| < |x - x'|$, which is a contradiction. Therefore, the closed half plane $H'$ includes $S$, implying that $C_1 \subseteq C(S)$, which is what we want.

To complete the proof we need the following lemma.

**Lemma.** *If the roots of a polynomial with complex coefficients lie in a closed half plane, then so do those of the derivative of the polynomial.*

**Proof.** Suppose $z_1, \ldots, z_n$ are the roots of a polynomial $p$ so that $p(z) = c(z - z_1) \cdots (z - z_n)$, where $c \in \mathbb{C}$. Also suppose that the closed half plane $H = \{z \in \mathbb{C} : \operatorname{Im} \frac{z-a}{b} \leq 0\}$ includes the roots of $p$. We show that $H$ includes the roots of $p'$, the derivative of $p$. If $z = z_i$, for some $1 \leq i \leq n$, is a root of

$p$ of multiplicity greater than one, then $z_i$ is a root of $p'$ as well, in which case there is nothing to prove. So, without loss of generality, we may assume that

$$\frac{p'(z)}{p(z)} = \frac{1}{z - z_1} + \cdots + \frac{1}{z - z_n}.$$

Now, suppose $z \notin H$. We complete the proof by showing that $z$ cannot be a root of $p'$. To this end, as $z \notin H$ and $z_i \in H$, we obtain

$$\operatorname{Im}\frac{z - z_i}{b} = \operatorname{Im}\frac{z - a}{b} - \operatorname{Im}\frac{z_i - a}{b} > 0,$$

for all $1 \leq i \leq n$. It follows that $\operatorname{Im}\frac{b}{z - z_i} < 0$ for all $1 \leq i \leq n$. From this, we obtain

$$\operatorname{Im}\frac{bp'(z)}{p(z)} = \sum_{i=1}^{n} \operatorname{Im}\frac{b}{z - z_i} < 0,$$

implying that $p'(z) \neq 0$, proving the lemma.                                     □

Now, assume that $f$ is a polynomial of degree greater than or equal to 2 and $A$ and $A'$ denote the sets of the roots of the equations $f(z) = 0$ and $f'(z) = 0$, respectively. Suppose that $H$ is an arbitrary closed half plane including $A$. It follows from the lemma that $H$ includes $A'$. This yields

$$A' \subseteq \bigcap \{H : A \subseteq H, H \text{ is a closed half plane}\} = C(A).$$

But since $C(A)$ is a closed convex set which includes $A'$, we see that

$$\overline{C(A')} = C(A') \subseteq C(A),$$

which is what we want.                                                            ∎

**3. Remark.** The lemma below together with the proof following the lemma shows that if the initial ball trajectory intersects the line segment joining the two foci, then there exists a hyperbola confocal with the original ellipse to which all trajectory segments are tangent.

We need the following lemma.

**Lemma.** *(i) Let $F$ and $F'$ be two points and $l$ a line in a plane not intersecting the line segment $F'F$ (resp. intersecting the line segment $FF'$ between its end points except the perpendicular bisector of $F'F$). Then, there exists a unique ellipse (resp. hyperbola) whose foci are $F'$ and $F$ and to which the line $l$ is tangent.*

*(ii) Let $F$ and $F'$ be two points and $Pm$ and $Pn$ two half-lines through a point $P$ in the plane such that the half-lines do not intersect the line segment $F'F$ (resp. intersect $FF'$ between its end points, i.e., between $F$ and $F'$). Then, there exists a unique ellipse (resp. hyperbola) whose foci are $F'$ and $F$ and to which the half-lines $Pm$ and $Pn$ are tangent if and only if $\angle FPm = \angle F'Pn$.*

**Proof.** (i) Since the foci, i.e., $F$ and $F'$, of the desired ellipse (resp. hyperbola) are fixed, it suffices to show that the sum (resp. the difference) of the focal radii of any point of the ellipse (resp. hyperbola) is uniquely determined by the given line $l$. To this end, suppose that the line $l$ is tangent to an ellipse (resp. a hyperbola) whose foci are $F$ and $F'$. By the optical property of ellipses (resp. hyperbolas), the tangent point is obtained as follows. Reflect the focus

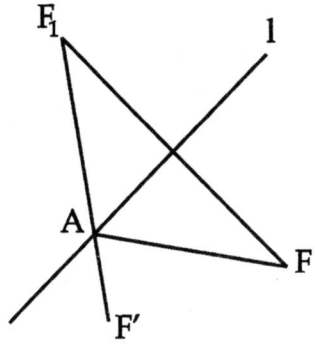
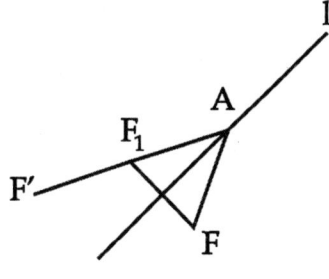

Figure 8

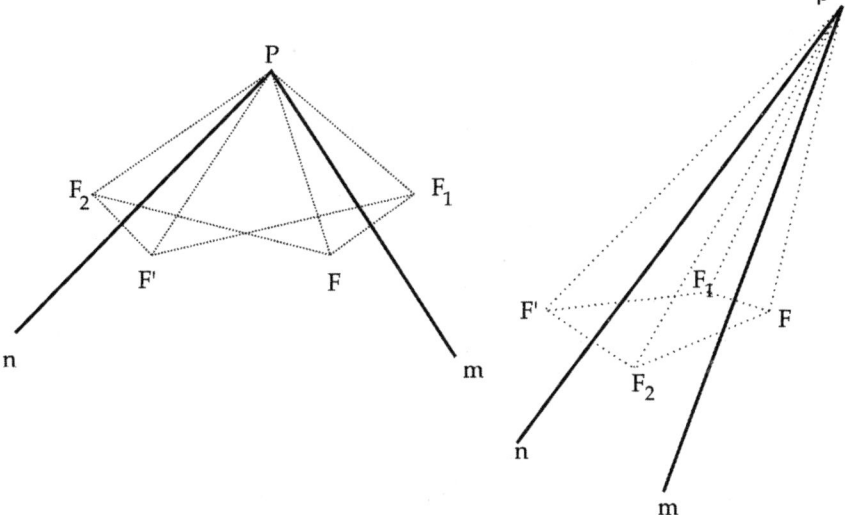

Figure 9

$F$ in the line $l$ to get $F_1$. Then, join $F_1$ and $F'$ to intersect the line $l$ at the tangent point, say, $A \in l$. It is plain that $AF + AF' = F_1F' = F_2F$ (resp. $|AF - AF'| = F_1F' = F_2F$), where $F_2$ is the reflection of $F'$ across the line $l$. In other words, the sum (resp. the difference) of the focal radii of the point $A$, and hence any point, of the desired ellipse (resp. hyperbola) is $F_1F' = F_2F$, which is uniquely determined by the given line $l$. This proves (i).

(ii) The uniqueness follows from (i). Let $F_1$ and $F_2$ be the reflection of $F$ and $F'$ across the half-lines $Pm$ and $Pn$, respectively.

First, suppose there exists an ellipse (resp. a hyperbola) whose foci are $F'$ and $F$ and to which the half-lines $Pm$ and $Pn$ are tangent. It is easily seen that the sum (resp. the difference) of the focal radii of any point of the ellipse (resp. hyperbola) is equal to $F_2F = F_1F'$. Now, the triangles $\triangle F_2PF$ and $\triangle F'PF_1$ are congruent because $F_2F = F_1F'$, $F_2P = F'P$, and $PF = PF_1$.

This, in particular, implies $\angle F_2 PF = \angle F'PF_1$, yielding

$$\angle F'PF + 2\angle F'Pn = \angle F'PF + 2\angle FPm$$

(resp.

$$\angle F'PF - 2\angle F'Pn = \angle F'PF - 2\angle FPm).$$

Thus, $\angle F'Pn = \angle FPm$, which, in turn, yields

$$\angle FPn = \angle F'PF + \angle F'Pn = \angle F'PF + \angle FPm = \angle F'Pm,$$

(resp.

$$\angle FPn = \angle F'PF - \angle F'Pn = \angle F'PF - \angle FPm = \angle F'Pm,)$$

as desired.

Next, suppose $\angle FPn = \angle F'Pm$. We can write

$$\angle F_2 PF = \angle F'PF + 2\angle FPn = \angle F'PF + 2\angle F'Pm = \angle F_1 PF',$$

(resp.

$$\angle F_2 PF = \angle F'PF - 2\angle FPn = \angle F'PF - 2\angle F'Pm = \angle F_1 PF').$$

That is, $\angle F_2 PF = \angle F_1 PF'$. It thus follows that the triangles $\triangle F_2 PF$ and $\triangle F'PF_1$ are congruent because $F_2 P = F'P$, $\angle F_2 PF = \angle F_1 PF'$, and $PF = PF_1$. In particular, we must have $F_2 F = F_1 F'$. In other words, the two ellipses (resp. hyperbolas) whose foci are $F$ and $F'$ and are tangent to the half-lines $Pm$ and $Pn$ coincide, proving the "if part" of the assertion. $\square$

Let $F$ and $F'$ be the foci of the ellipse $\xi$ and $PQ, QR, RS$ be three consecutive segments of the ball trajectory. It is well known that in any ellipse the focal radii of any point on the ellipse form equal angles with the tangent at the point. In other words, the normal and tangent lines at any point on an ellipse bisect the angles between the focal radii of the point. From this, it follows that if a billiard segment intersects the interior of the line segment $F'F$, so does all other billiard segments; if a billiard segment passes through one of the foci, then so does all other billiard segments; and finally, if a billiard segment does not intersect the line segment $F'F$, then neither does any other billiard segment. Now, with $P, Q, R, S$ as in the above, it follows that $\angle F'QP = \angle FQR$ and $\angle F'RQ = \angle FRS$. This together with the lemma implies that there are two ellipses $\eta_Q$ and $\eta_R$ confocal with $\xi$ each of which is tangent to $QR$. We see from part (i) of the lemma that $\eta_Q = \eta_R$ for any billiard segments $QR$. That is, there is an ellipse $\eta$ confocal with the ellipse $\xi$ such that the ball trajectory is always tangent to $\eta$, which is what we want. ∎

## 2.6. Sixth Competition

**2.6.1. Analysis. 1.** To prove the assertion by contradiction, suppose that there are three rational points $A = (x_1, y_1)$, $B = (x_2, y_2)$, and $C = (x_3, y_3)$ on the circle. Since $A, B, C \in \mathbb{Q}^2$, it follows that the equations of the perpendicular bisectors of the line segments $AB$ and $BC$ are of the form $a_1 x + b_1 y = c_1$ and

$a_2 x + b_2 y = c_2$, respectively, where $a_i, b_i, c_i \in \mathbb{Q}$ $(i = 1, 2)$ and $a_1 b_2 - b_1 a_2 \neq 0$. It is obvious that the center $(x, y)$ of the circle is the solution of

$$\begin{cases} a_1 x + b_1 y = c_1 \\ a_2 x + b_2 y = c_2 \end{cases},$$

which must be rational because $a_i, b_i, c_i \in \mathbb{Q}$ $(i = 1, 2)$. This is a contradiction, proving the assertion, which is what we want. ∎

**2.** As the sequence $(f_n)_{n=1}^{+\infty}$ is uniformly convergent to $f$ on $M$, for a given $\varepsilon > 0$, there exists a natural number $N_1$ such that for all $n \geq N_1$ and $x \in M$ we have

$$d\big(f_n(x), f(x)\big) < \frac{\varepsilon}{6}.$$

Now, since the function $f_{N_1}$ is continuous at the point $x \in M$, for the given $\varepsilon > 0$, there exists a $\delta > 0$ such that for all $y \in M$ with $d(x, y) < \delta$ we have

$$d\big(f_{N_1}(y), f_{N_1}(x)\big) < \frac{\varepsilon}{3}.$$

From $\lim_n x_n = x$, we see that for $\delta > 0$ obtained from the above, there exists a natural number $N_2$ such that for all $n \geq N_2$

$$d(x_n, x) < \delta.$$

Let $N = \max(N_1, N_2)$. For the given $\varepsilon > 0$ and for all $n \geq N$, we can write

$$\begin{aligned} d\big(f_n(x_n), f(x)\big) &\leq d\big(f_n(x_n), f(x_n)\big) + d\big(f(x_n), f_{N_1}(x_n)\big) + \\ &\quad d\big(f_{N_1}(x_n), f_{N_1}(x)\big) + d\big(f_{N_1}(x), f(x)\big) \\ &< \frac{\varepsilon}{6} + \frac{\varepsilon}{6} + \frac{\varepsilon}{3} + \frac{\varepsilon}{3} = \varepsilon. \end{aligned}$$

That is, $\lim_{n \to +\infty} f_n(x_n) = f(x)$, which is what we want. ∎

**3.** Since $\int_0^{+\infty} |f(x)| dx < +\infty$, we see that for given $\varepsilon > 0$, there exists a natural number $N_1$ such that for all $M \geq N_1$

$$\int_M^{+\infty} |f(x)| dx < \frac{\varepsilon}{3}.$$

The function $f$ is continuous on the compact interval $[0, N_1 + 1]$. Hence, it is uniformly continuous on $[0, N_1 + 1]$. Therefore, for given $\varepsilon > 0$, there exists a $\delta > 0$ such that for all $x, y \in [0, N_1 + 1]$ with $|x - y| < \delta$, we have

$$|f(x) - f(y)| < \frac{\varepsilon}{3N_1}.$$

Now, for a given $\varepsilon > 0$, find $N_1 > 0$ and $\delta > 0$ as in the above and pick $N \in \mathbb{N}$ such that $\frac{1}{N} < \delta$. For all $n \geq N$, we can write

$$
\begin{aligned}
\int_0^{+\infty} \left| f(x + \tfrac{1}{n}) - f(x) \right| dx &= \int_0^{N_1} \left| f(x + \tfrac{1}{n}) - f(x) \right| dx + \\
&\quad \int_{N_1}^{+\infty} \left| f(x + \tfrac{1}{n}) - f(x) \right| dx \\
&< \frac{\varepsilon}{3N_1}(N_1 - 0) + \int_{N_1}^{+\infty} \left| f(x + \tfrac{1}{n}) \right| dx + \\
&\quad \int_{N_1}^{+\infty} \left| f(x) \right| dx \\
&< \frac{\varepsilon}{3} + \frac{\varepsilon}{3} + \frac{\varepsilon}{3} = \varepsilon.
\end{aligned}
$$

That is,

$$
\left| \int_0^{+\infty} \left| f(x + \tfrac{1}{n}) - f(x) \right| dx - 0 \right| < \varepsilon,
$$

for all $n \geq N$. Therefore, $\lim_{n \to +\infty} \int_0^{+\infty} \left| f(x + \tfrac{1}{n}) - f(x) \right| dx = 0$, which is what we want. ∎

### 2.6.2. Algebra.

**1.** Proceed by contradiction. Suppose that $AB$ is a subgroup of $G$. Using Lagrange's Theorem , we see that $|AB| = p^n b$, where $b \in \mathbb{N}$ and $b|a$. Since $\gcd(p, a) = 1$ and $b|a$, $\gcd(p, b) = 1$. On the other hand, we can write

$$
|AB| = \frac{|A| \times |B|}{|A \cap B|}.
$$

But $A \cap B \subsetneq B$ because $B \not\subset A$. Hence, $|A \cap B| < |B|$. Now, in view of Lagrange's Theorem, let $|A \cap B| = p^i$ for some $0 \leq i < m$. We can write

$$
p^n b = \frac{p^n \cdot p^m}{p^i} \implies b = p^{m-i},
$$

implying that $\gcd(p, b) = p$, which is a contradiction. Therefore, $AB$ cannot be a subgroup of $G$, finishing the proof. ∎

**2.** We prove the assertion for matrices over a general division ring $\Delta$. Induct on $m$. If $m = 1$, the assertion is easily seen to hold. Assuming that the assertion holds for matrices of size less than $m$, we prove the assertion for matrices of size $m$. To this end, note first that $r := \text{rank}(A_1) \leq \dim \ker A_1$ because $A_1^2 = 0$. Now, from the Rank-Nullity Theorem, we see that $r \leq \frac{m}{2}$. Let $\mathcal{K} := \ker A_1$. Since $A_i$'s commute, $\mathcal{K}$ is invariant under all $A_i$'s $(1 \leq i \leq n)$. Therefore, after a similarity, one can write

$$
A_i = \begin{pmatrix} B_i & C_i \\ 0 & D_i \end{pmatrix},
$$

where $B_i \in M_{m-r}(\Delta)$ with $B_1 = 0$, $C_i \in M_{(m-r) \times r}(\Delta)$, and $D_i \in M_r(\Delta)$ for all $1 \leq i \leq n$. We have $m < 2^n$, yielding $r \leq \frac{m}{2} < 2^{n-1}$. As $A_i$'s commute, so do $D_i$'s in $M_r(\Delta)$ $(1 \leq i \leq n)$. Hence, from the induction hypothesis, we see that $D_2 \cdots D_n = 0$, which, in turn, implies that

$$A_2 \cdots A_n = \begin{pmatrix} B & C \\ 0 & 0 \end{pmatrix}, \qquad (*)$$

for some appropriate matrices $B \in M_{m-r}(\Delta)$ and $C \in M_{(m-r) \times r}(\Delta)$. On the other hand,

$$A_1 = \begin{pmatrix} 0 & C_1 \\ 0 & D_1 \end{pmatrix}. \qquad (**)$$

It now easily follows from $(*)$ and $(**)$ that

$$A_1 A_2 \cdots A_n = A_1(A_2 \cdots A_{n-1} A_n) = 0,$$

which is what we want.  ∎

**3.** Since $e^2 = e$, $f^2 = f$, and $ef = fe$, using the Binomial Theorem, we see that for all $n \in \mathbb{N}$, there exists a $K_n \in \mathbb{N}$ such that $(e - f)^n = e + (-1)^n f + K_n ef$. Suppose $(e - f)^n = 0$. Without loss of generality, we may assume that $n$ is even. It follows that

$$(e - f)^n = e + f + K_n ef = 0.$$

Multiplying both sides of the above by $e$ and $f$, respectively, and noting that $e^2 = e$, $f^2 = f$, and $ef = fe$, we obtain

$$e + ef + K_n ef = 0 = f + ef + K_n ef,$$

yielding $e = f = -(1 + K_n)ef$, which proves the assertion.  ∎

**2.6.3.  General. 1.** Noting that $(-1)^{|n|} = (-1)^n$ for all $n \in \mathbb{N}$, we can write

$$\begin{aligned}
(-1)^{|n_1 - m_1| + \cdots + |n_k - m_k|} &= (-1)^{|n_1 - m_1|} \times \cdots \times (-1)^{|n_k - m_k|} \\
&= (-1)^{n_1 - m_1} \times \cdots \times (-1)^{n_k - m_k} \\
&= (-1)^{(n_1 + \cdots + n_k) - (m_1 + \cdots + m_k)} = (-1)^0 = 1,
\end{aligned}$$

implying that $|n_1 - m_1| + \cdots + |n_k - m_k|$ is an even number.  ∎

**2.** Proceed by induction on $n$. If $n = 1$, the assertion is easy. Assuming that the assertion holds for $n \times n$ matrices, we prove it for $(n+1) \times (n+1)$ matrices. Suppose that $A$ is a $(n+1) \times (n+1)$ matrix such that the sum of the elements on any row and any column corresponding to any nonzero element of $A$ is at least $n + 1$. As the entries of $A$ are all nonnegative integers, if the entries are all nonzero, then the sum of the entries of the matrix $A$ is at least $(n + 1)^2$ which is no less than $\frac{(n+1)^2}{2}$. Now, assume that for some $1 \leq i, j \leq n + 1$, $a_{ij} = 0$. Let $\hat{A}_{ij}$ be the $n \times n$ matrix which is obtained from eliminating row $i$ and column $j$ of the matrix $A$. Obviously, the sum of the elements on any row and any column corresponding to any nonzero element of the matrix $\hat{A}_{ij}$ is at least $n$. So it follows from the induction hypothesis that the sum of the entries of $\hat{A}_{ij}$ is no less than $\frac{n^2}{2}$. Noting that the sum of the entries of row $i$

and column $j$ of $A$ is at least $n+1$, we see that the sum of the entries of the matrix $A$ is at least

$$\frac{n^2}{2} + n + 1 = \frac{(n+1)^2 + 1}{2} > \frac{(n+1)^2}{2},$$

which is what we want, finishing the proof.  ∎

**3.** The assertion is a special case of the following: *Suppose $k+m$ points inside a circle are given. Then, there exists a straight line not passing through any of the points that divides the circle into two sections one of which containing $k$ points and the other the remaining $m$ points.* To see this, note that the number of lines passing through a pair of points from these $k+m$ points is finite; in fact it is less than or equal to $\binom{k+m}{2}$. As the number of directions determined by these lines is finite, there is a line $d$ whose direction is not parallel to any of the lines determined by any pair of points from the given points. Let $AA'$ be a diameter of the circle whose direction is perpendicular to the line $d$. Project the given points on the line segment $AA'$ to get $k+m$ distinct points. Going from $A$ to $A'$, if $P$ is the $k$th point and $Q$ the $(k+1)$st point, then any line perpendicular to $AA'$ at any point between $P$ and $Q$ does not pass through any of the given points inside the circle and yet it divides the circle into two sections one of which contains $k$ points, including the point whose image is $P$, and the other contains the remaining $m$ points.  ∎

## 2.7. Seventh Competition

**2.7.1. Analysis. 1.** First of all, the hypothesis that $S$ is closed is redundant because it follows from the other hypothesis of the problem that such an $S$ is closed. The assertion is a consequence of the following. *Let $n \in \mathbb{N}$. Then there is no proper subset $S$ of $\mathbb{R}^n$ with the property that for all $x \in \mathbb{R}^n \setminus S$ there exists $n_x \in \mathbb{N}$ with $n_x > 1$ such that there are exactly $n_x$ points in $S$ as the closest point of $S$ to $x$.* To prove this by contradiction, pick a point $x \in \mathbb{R}^n \setminus S$. It follows from the hypothesis that there exists an $n_x \in \mathbb{N}$ with $n_x > 1$ and $n_x$ points $s_1, \ldots, s_{n_x}$ in $S$ such that

$$\inf_{s \in S} ||x - s|| = ||x - s_i|| = \delta_x,$$

for all $1 \leq i \leq n_x$. It is obvious that there is no point of $S$ in $B_{\delta_x}(x)$, the open ball centered at $x$ with radius $\delta_x$. Let $y = \frac{x+s_1}{2}$ and $\delta = \frac{\delta_x}{2}$. Plainly, the only point of $S$ in $\overline{B}_\delta(y)$, the closed ball centered at $y$ with radius $\delta$, is $s_1$. As $B_\delta(y) \subseteq B_{\delta_x}(x)$, it follows that for the point $y \in \mathbb{R}^n \setminus S$ there exists exactly one point $s_1 \in S$ such that

$$\inf_{s \in S} ||x - s|| = ||x - s_1||,$$

which is a contradiction, finishing the proof.  ∎

**2.** We need the following lemma.

> **Lemma.** *Let $(I_r)_{r \in \mathbb{Q}^+}$ and $A$ and $f$ be as in the statement of the problem. Then,*

*(i) for all $x \in A$ with $x \in \overline{I}_r$, we have $f(x) \leq r$.*

*(ii) for all $x \in A$ with $x \notin I_r$, we have $f(x) \geq r$.*

**Proof.** Set $S_x = \{r : x \in I_r\}$. We have $f(x) = \inf S_x$. To prove (i), note that if $x \in \overline{I}_r$, then $\overline{I}_r \subseteq I_s$ whenever $r < s$ where $s \in \mathbb{Q}$, from which, we obtain $x \in I_s$ for all $s \in \mathbb{Q}$ with $r < s$. Consequently, $S_x$ includes all rational numbers greater than $r$, whence $f(x) = \inf S_x \leq r$.

To prove (ii), note that if $x \notin I_r$, then $\overline{I}_s \subseteq I_r$ whenever $s < r$ where $s \in \mathbb{Q}$. This implies $x \notin I_s$ for all $s \in \mathbb{Q}$ with $s < r$. Therefore, $S_x$ contains no rational number less than $r$, whence $f(x) = \inf S_x \geq r$, proving the lemma. $\square$

We now prove that $f$ is continuous. Let $x_0$ be an arbitrary point of $A$. It suffices to prove that for any open interval $(c, d)$ with $f(x_0) \in (c, d)$ there exists an open neighborhood $N$ around $x_0$ such that $f(N) \subseteq (c, d)$. To this end, as $c < f(x_0) < d$, pick rational numbers $p$ and $q$ such that $c < p < f(x_0) < q < d$. We claim that the open set $N = I_q \setminus \overline{I}_p$ contains $x_0$ and moreover $f(N) \subseteq [p, q] \subseteq (c, d)$, finishing the proof. Firstly, $x_0 \in N$ because, in view of the lemma, $x_0 \in I_q$ and $x_0 \notin \overline{I}_p$. Secondly, $f(N) \subseteq [p, q] \subseteq (c, d)$. To see this, let $x \in N = I_q \setminus \overline{I}_p$ be arbitrary. It follows that $x \in I_q \subseteq \overline{I}_q$, which, in view of the lemma, implies $f(x) \leq q$. On the other hand, since $x \notin \overline{I}_p$, we conclude from the lemma that $f(x) \geq p$. That is, we have proved $f(x) \in [p, q]$ for all $x \in N$. Consequently, $f(N) \subseteq [p, q] \subseteq (c, d)$, which is what we want. $\blacksquare$

**3. First solution:** To find the limit, we need the following lemma whose first and second parts are known as Stolz's First and Second Theorems on limits.

**Lemma.** *(i) Let $(p_n)_{n=1}^{+\infty}$ be a sequence of reals whose elements are eventually nonnegative, $s_n = p_1 + \cdots + p_n$, and $\lim_n s_n = +\infty$. If $(a_n)_{n=1}^{+\infty}$ is a sequence of real numbers such that $\lim_n a_n \in \mathbb{R} \cup \{\pm\infty\}$, then*

$$\lim_n \frac{p_1 a_1 + \cdots + p_n a_n}{p_1 + \cdots + p_n} = \lim_n a_n.$$

*(ii) Let $(x_n)_{n=1}^{+\infty}$ and $(y_n)_{n=1}^{+\infty}$ be sequences of real numbers with $\lim_n y_n = +\infty$ and such that $(y_n)_{n=1}^{+\infty}$ is eventually increasing. If $\lim_n \frac{x_n - x_{n-1}}{y_n - y_{n-1}} \in \mathbb{R} \cup \{\pm\infty\}$, then*

$$\lim_n \frac{x_n}{y_n} = \lim_n \frac{x_n - x_{n-1}}{y_n - y_{n-1}}.$$

**Proof.** (i) There are two cases to consider.

(a) $\lim_n a_n = \pm\infty$.

As $\frac{p_1(-a_1) + \cdots + p_n(-a_n)}{p_1 + \cdots + p_n} = -\frac{p_1 a_1 + \cdots + p_n a_n}{p_1 + \cdots + p_n}$, it suffices to prove the assertion for the case when $\lim_n a_n = +\infty$. To this end, it follows from the hypothesis that for given $M > 0$, there exists an $N_1 > 0$ such that $a_n > 2M$ whenever $n > N_1$. Let $A_{N_1} = p_1 a_1 + \cdots + p_{N_1} a_{N_1}$. We can write

$$\frac{p_1 a_1 + \cdots + p_n a_n}{p_1 + \cdots + p_n} = \frac{A_{N_1}}{s_n} + \frac{p_{N_1+1} a_{N_1+1} + \cdots + p_n a_n}{s_n}$$

$$> \frac{A_{N_1}}{s_n} + 2M \frac{s_n - s_{N_1}}{s_n}$$

But $\lim_n \frac{A_{N_1}}{s_n} = 0$ and $\lim_n \frac{s_n - s_{N_1}}{s_n} = 1$. Thus, there exists $N_2 > 0$ such that $\frac{A_{N_1}}{s_n} > \frac{-M}{3}$ and $\frac{s_n - s_{N_1}}{s_n} > \frac{2}{3}$ whenever $n > N_2$. Letting $N = \max(N_1, N_2)$, for all $n > N$, we have

$$\frac{p_1 a_1 + \cdots + p_n a_n}{p_1 + \cdots + p_n} > \frac{-M}{3} + \frac{4M}{3} = M.$$

As $M > 0$ is arbitrary, this means $\lim_n \frac{p_1 a_1 + \cdots + p_n a_n}{p_1 + \cdots + p_n} = +\infty$, proving the assertion in this case.

(b) $\lim_n a_n = a \in \mathbb{R}$.

First, let $\lim_n a_n = 0$. It follows that for given $\varepsilon > 0$ there exists $N_1 > 0$ such that $|a_n| < \frac{\varepsilon}{2}$ whenever $n > N_1$. Let $A_{N_1}$ be as in (a). For all $n > N_1$, we can write

$$\left| \frac{p_1 a_1 + \cdots + p_n a_n}{p_1 + \cdots + p_n} \right| \leq \frac{|A_{N_1}|}{s_n} + \frac{\varepsilon}{2} \times \frac{s_n - s_{N_1}}{s_n} \leq \frac{|A_{N_1}|}{s_n} + \frac{\varepsilon}{2}.$$

Now, as $\lim_n \frac{A_{N_1}}{s_n} = 0$, we see that there exists an $N_2 > 0$ such that $\frac{|A_{N_1}|}{s_n} < \frac{\varepsilon}{2}$ whenever $n > N_2$. Letting $N = \max(N_1, N_2)$, for all $n > N$, we have

$$\left| \frac{p_1 a_1 + \cdots + p_n a_n}{p_1 + \cdots + p_n} \right| < \frac{\varepsilon}{2} + \frac{\varepsilon}{2} = \varepsilon.$$

That is, $\lim_n \frac{p_1 a_1 + \cdots + p_n a_n}{p_1 + \cdots + p_n} = 0$, which is what we want.

Next, let $\lim_n a_n = a \neq 0$. As $\lim_n (a_n - a) = 0$, from what we just proved, it follows that

$$\lim_n \left( \frac{p_1 a_1 + \cdots + p_n a_n}{p_1 + \cdots + p_n} - a \right) = \lim_n \frac{p_1(a_1 - a) + \cdots + p_n(a_n - a)}{p_1 + \cdots + p_n}$$
$$= \lim_n (a_n - a) = 0.$$

This implies that $\lim_n \frac{p_1 a_1 + \cdots + p_n a_n}{p_1 + \cdots + p_n} = a$, finishing the proof.

(ii) Define the two sequences $(p_n)_{n=1}^{+\infty}$ and $(a_n)_{n=1}^{+\infty}$, inductively, as follows

$$p_n = y_n - y_{n-1}, \; p_1 = y_1, \; (n \geq 2),$$
$$x_n - x_{n-1} = p_n a_n, \; x_1 = p_1 a_1, \; (n \geq 2).$$

Firstly, $p_n$'s are eventually nonnegative. Secondly, $y_n = p_1 + \cdots + p_n \to +\infty$ as $n \to +\infty$. Thirdly, $a_n = \frac{x_n - x_{n-1}}{y_n - y_{n-1}}$ and $p_1 a_1 + \cdots + p_n a_n = x_n$, and $p_1 + \cdots + p_n = y_n$. Therefore, (i) applies, proving the assertion. $\square$

Now, note that

$$x_n = \frac{\frac{2^1}{1} + \cdots + \frac{2^n}{n}}{\frac{2^{n+1}}{n+1}} = \frac{X_n}{Y_n}.$$

First of all, it is easily seen that the sequence $(Y_n)_{n=1}^{+\infty}$, where $Y_n = \frac{2^{n+1}}{n+1}$ for all $n \in \mathbb{N}$, is increasing. Secondly, $\lim_n Y_n = +\infty$. Thirdly, we can write

$$\lim_n \frac{X_n - X_{n-1}}{Y_n - Y_{n-1}} = \lim_n \frac{\frac{2^n}{n}}{\frac{2^{n+1}}{n+1} - \frac{2^n}{n}} = \lim_n \frac{1}{\frac{2n}{n+1} - 1} = 1.$$

Therefore, it follows from the lemma that $\lim_n x_n = \lim_n \frac{X_n}{Y_n}$ exists, and moreover

$$\lim_n x_n = \lim_n \frac{X_n}{Y_n} = \lim_n \frac{X_n - X_{n-1}}{Y_n - Y_{n-1}} = 1,$$

which is what we want.

**Second solution:** It is readily checked that

$$x_{n+1} = \frac{n+2}{2^{n+2}} \sum_{k=1}^{n+1} \frac{2^k}{k} = \frac{n+2}{2(n+1)}(x_n + 1),$$

from which, we obtain

$$x_{n+2} - x_{n+1} = \frac{(n+2)^2(x_{n+1} - x_n) - x_{n+1} - 1}{2(n+1)(n+2)},$$

for all $n \in \mathbb{N}$. As $x_n > 0$ for all $n \in \mathbb{N}$, in view of the above equality, we see that $x_{n+2} - x_{n+1} < 0$ whenever $x_{n+1} - x_n \leq 0$. But a straightforward calculation reveals that $x_3 = x_4 = \frac{5}{3}$. Thus, by an easy induction on $n \geq 3$, we see that the sequence $(x_n)_{n=1}^{+\infty}$ is decreasing for $n \geq 3$. From this, together with the fact that $x_n > 0$ for all $n \in \mathbb{N}$, we conclude that $\lim_{n \to +\infty} x_n$ exists, and hence $\lim_{n \to +\infty} x_n = L$ for some $L \in \mathbb{R}$. Consequently,

$$L = \lim_{n \to +\infty} x_{n+1} = \lim_{n \to +\infty} \frac{n+2}{2(n+1)}(x_n + 1) = \frac{1}{2}(L + 1) \implies L = 1.$$

Therefore, $\lim_{n \to +\infty} x_n = 1$, which is what we want. ∎

**2.7.2. Algebra. 1.** (a) Let $x_0 \in R$ be arbitrary. We show that $x_0 e = x_0$. To this end, note that for all $x \in R$, we can write

$$(x_0 + e - x_0 e)x = x_0 x + ex - x_0(ex) = x_0 x + x - x_0 x = x.$$

That is, $(x_0 + e - x_0 e)x = x$ for all $x \in R$. As the left identity element is unique, we obtain $x_0 + e - x_0 e = e$, implying $x_0 e = x_0$. This means $e$ a right identity element as well. Consequently, $e$ is the identity element of $R$.

(b) Let $a$ be an idempotent of $R$, i.e. $a^2 = a$, and $x \in R$ be arbitrary. It is straightforward to see that $(axa - ax)^2 = 0$. That is, $axa - ax$ is a nilpotent element of $R$, and hence $axa - ax = 0$, by the hypothesis. Likewise, $(axa - xa)^2 = 0$, implying $axa - xa = 0$ for all $x \in R$. So we have shown that $ax = xa = axa$ for all $x \in R$, which is what we want. ∎

**2. First solution:** Let $H$ be an arbitrary subgroup of $G$. Set $H_1 = H \cap G_1$. We show that $H_1$ is commutative, $H_1 \trianglelefteq H$, and that $\frac{H}{H_1}$ is commutative. First, $H_1$ is commutative because $H_1 \subseteq G_1$ and $G_1$ is commutative. Next, $H_1 \leq H$ because $H_1 = H \cap G_1$ and $H$ and $G_1$ are subgroups of $G$. That $H_1 \trianglelefteq H$ follows from $G_1 \trianglelefteq G$ and $H_1 = H \cap G_1$. Finally, to show that $\frac{H}{H_1}$ is commutative, using the Second Isomorphism Theorem for groups, we can write

$$\frac{H}{H_1} = \frac{H}{H \cap G_1} \cong \frac{HG_1}{G_1}.$$

But $HG_1 \leq G$. Hence, $\frac{HG_1}{G_1} \leq \frac{G}{G_1}$. Consequently, $\frac{HG_1}{G_1}$ is commutative because so is $\frac{G}{G_1}$. This, in view of the above isomorphism, completes the proof.

**Second solution:** Use $G'$ to denote the derived subgroup of $G$. Since $\frac{G}{G_1}$ is abelian, it follows that $G' \leq G_1$. Now for a given subgroup $H$ of $G$, set

$$H_1 \;=\; H' := \; \langle \{ h_1 h_2 h_1^{-1} h_2^{-1} : h_1, h_2 \in H \} \rangle.$$

It is plain that $H_1$ is a normal subgroup of $H$ and that $\frac{H}{H_1}$ is abelian because $H_1$ is the derived subgroup of $H$. To see that $H_1$ is abelian, just note that $H_1 \leq G' \leq G_1$ and that $G_1$ is abelian by the hypothesis. This completes the proof.  ∎

**3.** For the given $\theta : V \to V$, it is obvious that

$$\ker \theta \subseteq \ker \theta^2 \subseteq \cdots$$

is an ascending chain of the subspaces of $V$. As $V$ is finite-dimensional, it follows that there exists an $s \in \mathbb{N}$ such that $\ker \theta^s = \ker \theta^{s+1}$.

We claim that if $\ker \theta^s = \ker \theta^{s+1}$, then $\ker \theta^s = \ker \theta^{s+k}$ for all $k \in \mathbb{N}$. We prove this by induction on $k$. If $k = 1$, there is nothing to prove. Suppose the assertion holds for $k$. To prove the assertion for $k + 1$, note first that $\ker \theta^s \subseteq \ker \theta^{s+k+1}$. Next, let $x \in \ker \theta^{s+k+1}$ be arbitrary. We obtain $\theta^{s+k}(\theta x) = 0$. That is, $\theta x \in \ker \theta^{s+k}$, which, in view of the induction hypothesis, yields $\theta x \in \ker \theta^s$. This implies $\theta^{s+1} x = 0$ which means $x \in \ker \theta^{s+1}$. But $\ker \theta^{s+1} = \ker \theta^s$. Thus, $x \in \ker \theta^s$. As $x \in \ker \theta^{s+k+1}$ was arbitrary, we see that $\ker \theta^{s+k+1} \subseteq \ker \theta^s$. Therefore, $\ker \theta^{s+k+1} = \ker \theta^s$, proving the claim by induction on $k$.

We now show that $V = \operatorname{im}(\theta^s) \oplus \ker(\theta^s)$, where $s$ is as in the above. Note that by the Rank-Nullity Theorem

$$\dim \operatorname{im}(\theta^s) + \dim \ker(\theta^s) \;=\; \dim V.$$

On the other hand, by Problem 1(a) of 1.3.2, we can write

$$\dim(\operatorname{im}(\theta^s) + \ker(\theta^s)) \;=\; \dim \operatorname{im}(\theta^s) + \dim \ker(\theta^s) - \dim \left( \operatorname{im}(\theta^s) \cap \ker(\theta^s) \right),$$

which, in view of the preceding equality, implies $V = \operatorname{im}(\theta^s) \oplus \ker(\theta^s)$ as soon as we show that $\operatorname{im}(\theta^s) \cap \ker(\theta^s) = \{0\}$. To see this, let $x \in \operatorname{im}(\theta^s) \cap \ker(\theta^s)$ be arbitrary. It follows that there exists $x_1 \in V$ such that $x = \theta^s x_1$, from which, we obtain $\theta^{s+s} x_1 = 0$ because $x \in \ker \theta^s$. On the other hand, by the claim we made in the above, we have $\ker \theta^{s+s} = \ker \theta^s$, yielding $x_1 \in \ker \theta^s$. In other words, $x = \theta^s x_1 = 0$. Since $x \in \operatorname{im}(\theta^s) \cap \ker(\theta^s)$ was arbitrary, we conclude that $\operatorname{im}(\theta^s) \cap \ker(\theta^s) = \{0\}$, which is what we want. Therefore, $V = \operatorname{im}(\theta^s) \oplus \ker(\theta^s)$, completing the proof.  ∎

## 2.8. Eighth Competition

**2.8.1. Analysis. 1.** (a) Define the function $h : F \cup G \to \mathbb{R}$ by

$$h(x) = \begin{cases} f(x) & x \in F, \\ g(x) & x \in G. \end{cases}$$

As $f(x) = g(x)$ on $F \cap G$, the function $h$ is well-defined. It is obvious that $h$ is an extension of $f$ and $g$ to $F \cup G$. To prove that $h$ is continuous, it suffices to show that $h^{-1}(C)$ is a closed subset of $F \cup G$ whenever $C$ is a closed subset of $\mathbb{R}$. To this end, note that for any closed subset $C$ of $\mathbb{R}$, we can write

$$h^{-1}(C) = f^{-1}(C) \cup g^{-1}(C).$$

Now, as $f$ and $g$ are continuous and $C$ is a closed subset of $\mathbb{R}$, we see that $f^{-1}(C)$ and $g^{-1}(C)$ are closed subsets of $F$ and $G$, respectively. On the other hand, $F$ and $G$ are closed in $F \cup G$. Hence, $f^{-1}(C)$ and $g^{-1}(C)$ are closed subsets of $F \cup G$. This together with the above equality implies that $h^{-1}(C)$ is a closed subset of $F \cup G$, which is what we want.

(b) For given $n \in \mathbb{N}$, let $F = B_1(0) := \{x \in \mathbb{R}^n : ||x|| < 1\}$ and $G = F^c = \{x \in \mathbb{R}^n : ||x|| \geq 1\}$. Obviously, $F \cap G = \emptyset$ and $F \cup G = \mathbb{R}^n$. Now, define the functions $f : F \to \mathbb{R}$ and $g : G \to \mathbb{R}$ by $f(x) = 0$ and $g(x) = 1$ whenever $x \in F$ and $x \in G$, respectively. For this $f$ and $g$, we show that the conclusion of (a) does not hold. Suppose to the contrary that $h : \mathbb{R}^n = F \cup G \to \mathbb{R}$ is an extension of $f$ and $g$ to $\mathbb{R}^n = F \cup G$. We must have

$$h(x) = \begin{cases} 0 & x \in F, \\ 1 & x \in G, \end{cases}$$

which is not continuous on $\mathbb{R}^n = F \cup G$, for otherwise $h(\mathbb{R}^n) = \{0,1\}$, which is disconnected whereas $\mathbb{R}^n$ is connected. This leads to a contradiction, proving the assertion. ∎

**2. Remark.** It is worth mentioning that using a proof almost identical to the proof below one can show that *if $(a_n)_{n=1}^{+\infty}$ is a decreasing sequence of real numbers with $\lim_n a_n = 0$, then $\sum_{n=1}^{+\infty} a_n$ is convergent if and only if $\sum_{n=1}^{+\infty} n(a_n - a_{n+1})$ is convergent.* Also, see Problem 2 of 1.12.1.

We have $\lim_n a_n = 0$ because $\sum_{n=1}^{+\infty} a_n$ is convergent. Since $(a_n)_{n=1}^{+\infty}$ is decreasing, it follows that $a_n \geq a_{n+k}$ for all $n, k \in \mathbb{N}$. Letting $k \to +\infty$, we obtain $a_n \geq 0$ for all $n \in \mathbb{N}$. Now, as $\sum_{n=1}^{+\infty} a_n$ is convergent, it follows that it is Cauchy. Thus, for given $\varepsilon > 0$, there exists an $N > 0$ such that

$$\left| \sum_{k=m}^{n} a_k \right| < \frac{\varepsilon}{2},$$

for all $n \geq m \geq N$. Let $m = [\frac{n}{2}] > N$. We obtain $n > m \geq N$. This together with the fact that $a_n$'s are nonnegative implies

$$\sum_{k=[\frac{n}{2}]}^{n} a_k < \frac{\varepsilon}{2}.$$

As $a_n$'s are decreasing, we have $(n - [\frac{n}{2}] + 1)a_n \leq \sum_{k=[\frac{n}{2}]}^{n} a_k$, yielding

$$\frac{na_n}{2} \leq \frac{na_n}{2} + \left(\frac{n}{2} - [\frac{n}{2}] + 1\right)a_n$$

$$= \left(n - [\frac{n}{2}] + 1\right)a_n \leq \sum_{k=[\frac{n}{2}]}^{n} a_k$$

$$< \frac{\varepsilon}{2},$$

for all $n > 2N + 2$. Consequently, $na_n < \varepsilon$ for all $n > 2(N + 1)$. That is, $\lim_n na_n = 0$. We can write

$$\sum_{n=1}^{+\infty} n(a_n - a_{n+1}) = \sum_{n=1}^{+\infty} \left((na_n - (n+1)a_{n+1}) + a_{n+1}\right).$$

Since $\lim_n na_n = 0$, we see that the telescopic series $\sum_{n=1}^{+\infty} (na_n - (n+1)a_{n+1})$ converges to $a_1$. On the other hand, $\sum_{n=1}^{+\infty} a_{n+1}$ is convergent. Therefore, the series $\sum_{n=1}^{+\infty} n(a_n - a_{n+1})$ converges and we have

$$\sum_{n=1}^{+\infty} n(a_n - a_{n+1}) = \sum_{n=1}^{+\infty} \left((na_n - (n+1)a_{n+1}) + a_{n+1}\right)$$

$$= \sum_{n=1}^{+\infty} (na_n - (n+1)a_{n+1}) + \sum_{n=1}^{+\infty} a_{n+1}$$

$$= a_1 + \sum_{n=1}^{+\infty} a_{n+1},$$

finishing the proof. ∎

**3. First solution:** As the function $g$ is continuously differentiable on $[0, 1]$, it follows that $g$ is of bounded variation on $[0, 1]$. We show that $g$ can be written as the difference of two increasing functions $g_1$ and $g_2$ each of which is continuously differentiable on $[0, 1]$. To see this, let $g' = f$, where $f$ is continuous on $[0, 1]$. We can write $f = f^+ - f^-$, where $f^+ = \max(f, 0) = \frac{f}{2} + \frac{|f|}{2}$ and $f^- = \max(-f, 0) = \frac{-f}{2} + \frac{|f|}{2}$. Obviously, $f^+$ and $f^-$ are nonnegative continuous functions on $[0, 1]$. Using the Second Fundamental Theorem of Calculus, we can write

$$g(x) = g(0) + \int_0^x f = \left(g(0) + \int_0^x f^+\right) - \int_0^x f^- = g_1(x) - g_2(x),$$

where $g_1(x) = g(0) + \int_0^x f^+$ and $g_2(x) = \int_0^x f^-$. Applying the First Fundamental Theorem of Calculus, we obtain

$$g_1'(x) = f^+(x) \geq 0, \quad g_2'(x) = f^-(x) \geq 0,$$

for all $x \in [0, 1]$. Therefore, $g = g_1 - g_2$, where $g_1$ and $g_2$ are continuously differentiable and increasing on $[0, 1]$. We can write

$$\int_0^1 x^n dg = \int_0^1 x^n d(g_1 - g_2) = \int_0^1 x^n dg_1 - \int_0^1 x^n dg_2.$$

Now, as $g_1$ and $g_2$ are continuously differentiable on $[0, 1]$, we have

$$\int_0^1 x^n dg = \int_0^1 x^n g_1'(x) dx - \int_0^1 x^n g_2'(x) dx = \int_0^1 x^n \big(g_1'(x) - g_2'(x)\big) dx.$$

Hence,

$$\int_0^1 x^n dg = \int_0^1 x^n g'(x) dx.$$

Now, as $g'$ is continuous on $[0, 1]$, there is an $M > 0$ such that $|g'| \le M$ on $[0, 1]$, from which, we see that

$$\left| \int_0^1 x^n dg \right| = \left| \int_0^1 x^n g'(x) dx \right| \le \int_0^1 x^n |g'(x)| dx$$

$$\le M \int_0^1 x^n dx = \frac{M}{n+1},$$

implying that $\lim_n \int_0^1 x^n dg = 0$ because $\lim_n \frac{M}{n+1} = 0$.

**Second solution:** As $g$ is continuously differentiable on $[0, 1]$, it is of bounded variation on $[0, 1]$. Hence, there exists increasing functions $g_1$ and $g_2$ on $[0, 1]$ such that $g = g_1 - g_2$. By the definition of the Riemann-Stieltjes integrals with respect to integrands of bounded variations, we have

$$\int_0^1 x^n dg = \int_0^1 x^n dg_1 - \int_0^1 x^n dg_2.$$

Using integration by parts, we obtain

$$\int_0^1 x^n dg_i = x^n g_i(x) \Big|_0^1 - \int_0^1 g_i(x) dx^n = g_i(1) - \int_0^1 n g_i(x) x^{n-1} dx,$$

for each $i = 1, 2$. So we have

$$\int_0^1 x^n dg = g(1) - \int_0^1 n g(x) x^{n-1} dx = \int_0^1 n\big(g(1) - g(x)\big) x^{n-1} dx.$$

Now, for given $\varepsilon > 0$, as $g$ is continuous at 1, there exists a $0 < \delta < 1$ such that $|g(x) - g(1)| < \frac{\varepsilon}{2}$ whenever $1 - \delta < x < 1$. We can write

$$\left| \int_0^1 x^n dg \right| = \left| \int_0^1 n\big(g(1) - g(x)\big) x^{n-1} dx \right| \le \int_0^1 n |g(1) - g(x)| x^{n-1} dx$$

$$= \int_0^{1-\delta} n |g(1) - g(x)| x^{n-1} dx + \int_{1-\delta}^1 n |g(1) - g(x)| x^{n-1} dx.$$

By the continuity of $g$ on $[0, 1]$, there exists $M > 0$ such that $|g| \le M$ on $[0, 1]$. With this in mind, we have

$$\left| \int_0^1 x^n dg \right| \le 2Mn \int_0^{1-\delta} x^{n-1} dx + \frac{n\varepsilon}{2} \int_{1-\delta}^1 x^{n-1} dx$$

$$= 2M(1 - \delta)^n + \frac{\varepsilon}{2}\big(1 - (1 - \delta)^n\big) < 2M(1 - \delta)^n + \frac{\varepsilon}{2}.$$

Consequently,

$$\left| \int_0^1 x^n dg \right| < 2M(1 - \delta)^n + \frac{\varepsilon}{2}.$$

Now, $\lim_n (1 - \delta)^n = 0$ because $0 < 1 - \delta < 1$. Hence, there exists $N > 0$ such that $(1 - \delta)^n < \frac{\varepsilon}{4M}$ for all $n \geq N$. Therefore,

$$\left| \int_0^1 x^n \, dg \right| < 2M \times \frac{\varepsilon}{4M} + \frac{\varepsilon}{2} = \varepsilon,$$

for all $n \geq N$. This means $\lim_n \int_0^1 x^n dg(x) = 0$, which is what we want.    ■

**2.8.2. Algebra. 1.** Note that $b = I + N$, where $I$ is the identity matrix, $N = b - I$, and $N^2 = 0$. From this, using the Binomial Theorem and noting that $b^{-1} = I - N$, it is easily verified that

$$b^n = \begin{pmatrix} 1 & n \\ 0 & 1 \end{pmatrix},$$

for all $n \in \mathbb{Z}$. Therefore,

$$H = \{b^n : n \in \mathbb{Z}\} = \left\{ \begin{pmatrix} 1 & n \\ 0 & 1 \end{pmatrix} : n \in \mathbb{Z} \right\}.$$

To find $aHa^{-1}$, note that $a^{-1} = \begin{pmatrix} \frac{1}{2} & 0 \\ 0 & 1 \end{pmatrix}$. We can write

$$ab^n a^{-1} = \begin{pmatrix} 2 & 0 \\ 0 & 1 \end{pmatrix} \begin{pmatrix} 1 & n \\ 0 & 1 \end{pmatrix} \begin{pmatrix} \frac{1}{2} & 0 \\ 0 & 1 \end{pmatrix} = \begin{pmatrix} 1 & 2n \\ 0 & 1 \end{pmatrix}.$$

Therefore,

$$aHa^{-1} = \left\{ \begin{pmatrix} 1 & 2n \\ 0 & 1 \end{pmatrix} : n \in \mathbb{Z} \right\}.$$

This implies that $aHa^{-1} \subsetneqq H$. That is, $aHa^{-1}$ is a proper subgroup of $H$, which is what we want.    ■

**2.** We need the "only if part" of the following standard lemma.

**Lemma.** *Let $R$ be a commutative ring with identity. Then, $x \in J(R) := \bigcap \{m \triangleleft R : m \text{ is maximal in } R\}$ if and only if $1 - xy$ is a unit in $R$ for all $y \in R$.*

**Proof.** "$\Longleftarrow$" Let $x \in J(R)$. Suppose to the contrary that there exists a $y_0 \in R$ such that $1 - xy_0$ is not a unit. As $1 - xy_0$ is not invertible, a standard argument using Zorn's Lemma shows that there exists a maximal ideal $m_0$ such that $1 - xy_0 \in m_0$. On the other hand, since $x \in J(R)$, $x \in m_0$, and hence $xy_0 \in m_0$. So, we obtain $1 = (1 - xy_0) + xy_0 \in m_0$, yielding $m_0 = R$, which is a contradiction. Thus the assertion follows by contradiction.

"$\Longrightarrow$" Again, we proceed by contradiction. Suppose that there exists a maximal ideal $m_0$ such that $x \notin m_0$. As $m_0 \subsetneqq m_0 + xR$ and $m_0$ is maximal, we see that $m_0 + xR = R$. Hence, there exist $a \in m_0$ and $y_0 \in R$ such that $1 = a + xy_0$. This together with the hypothesis implies that $a = 1 - xy_0 \in m_0$ is a unit in $R$, which, in turn, implies $m_0 = R$, a contradiction. Therefore, the assertion follows by contradiction.    □

To prove the assertion by contradiction, suppose that $R$ has finitely many maximal ideals $m_1, \ldots, m_n$ for some $n \in \mathbb{N}$. Firstly, $\{0\} \subsetneqq m_i$ for each $i =$

$1,\ldots,n$ because otherwise $\{0\}$ and $R$ are the only ideal of $R$, implying that $R$ is a field, which is a contradiction, for $R$ would then have infinitely many units. On the other hand, we have

$$\bigcap_{i=1}^{n} m_i \supseteq m_1 \cdots m_n.$$

We also have $m_1 \cdots m_n \neq \{0\}$ because $\{0\} \subsetneq m_i$ for each $i = 1,\ldots,n$ and $R$ is an integral domain. Now, pick $0 \neq a \in m_1 \cdots m_n \subseteq \bigcap_{i=1}^{n} m_i = \bigcap\{m \lhd R : m \text{ is maximal in } R\}$. It follows from the lemma that $1 - ax$ is a unit in $R$ for $x \in R$. But, as $R$ is an integral domain, $1 - ax \neq 1 - ay$ whenever $x \neq y$. This together with the fact that $R$ is an infinite integral domain implies that $R$ has infinitely many units, which is a contradiction. So the assertion follows by contradiction. ∎

**3.** Let $A = (a_{ij}) \in M_n(\mathbb{R})$. Use the symbol $(A^2)_{ij}$ to denote the $ij$ entry of the matrix $A^2$. Using $A^2 = I$, we can write

$$\sum_{i=1}^{n}\sum_{j=1}^{n}(A^2)_{ij} = \sum_{i=1}^{n}\sum_{j=1}^{n}(I)_{ij} = n.$$

On the other hand,

$$\sum_{i=1}^{n}\sum_{j=1}^{n}(A^2)_{ij} = \sum_{i=1}^{n}\sum_{j=1}^{n}\sum_{k=1}^{n}a_{ik}a_{kj} = \sum_{i=1}^{n}\sum_{k=1}^{n}a_{ik}\sum_{j=1}^{n}a_{kj}$$
$$= \sum_{i=1}^{n}\sum_{k=1}^{n}a_{ik}a = a\sum_{i=1}^{n}\sum_{k=1}^{n}a_{ik} = a\sum_{i=1}^{n}a = na^2.$$

Therefore, $na^2 = n$, yielding $a = \pm 1$. Note that both cases $a = 1$ or $a = -1$ can happen because $A = \pm I$ yields $a = \pm 1$, respectively. ∎

**2.8.3. General. 1.** Note that $f(x,y,z) = (x-3)^2 + 4(y+\frac{1}{2}) + z^2 - 10$. This yields $f(x,y,z) \geq -10$ for all $x,y,z \in \mathbb{R}$ and that $f(x,y,z) = -10$ if and only if $x = 3$, $y = -\frac{1}{2}$, and $z = 0$. Therefore, the function $f$ assumes it absolute minimum at $(3, -\frac{1}{2}, 0)$, which is what we want. ∎

**2.** Let $x_1 < \cdots < x_5$ be the amount of money owned by these five people in increasing order. We have $x_1|x_2$, $x_2|x_3$, $x_3|x_4$, and $x_4|x_5$. Hence, there exist natural numbers $k_i$ $(1 \leq i \leq 4)$ such that $x_{i+1} = k_i x_i$. Using the hypothesis, we can write

$$x_1(1 + k_1 + k_1k_2 + k_1k_2k_3 + k_1k_2k_3k_4) = 719.$$

But 719 is prime. So, we obtain

$$x_1 = 1, 1 + k_1 + k_1k_2 + k_1k_2k_3 + k_1k_2k_3k_4 = 719.$$

Consequently, $k_1(1 + k_2 + k_2k_3 + k_2k_3k_4) = 718 = 2 \times 359$. As 2 and 359 are primes, we must have

$$k_1 = 2, \ 1 + k_2 + k_2k_3 + k_2k_3k_4 = 359, \ \text{or,}$$
$$k_1 = 1, \ 1 + k_2 + k_2k_3 + k_2k_3k_4 = 718, \ \text{or,}$$
$$k_1 = 359, \ 1 + k_2 + k_2k_3 + k_2k_3k_4 = 2, \ \text{or,}$$
$$k_1 = 718, \ 1 + k_2 + k_2k_3 + k_2k_3k_4 = 1.$$

The last three cases are easily seen to be refuted. Thus,

$$k_1 = 2, \ 1 + k_2 + k_2k_3 + k_2k_3k_4 = 359.$$

Simplifying, we obtain $k_2(1 + k_3 + k_3k_4) = 358 = 2 \times 179$. Since 2 and 179 are primes, we see that

$$k_2 = 2, \ 1 + k_3 + k_3k_4 = 179, \ \text{or,}$$
$$k_2 = 1, \ 1 + k_3 + k_3k_4 = 358, \ \text{or,}$$
$$k_2 = 179, \ 1 + k_3 + k_3k_4 = 2, \ \text{or,}$$
$$k_2 = 358, \ 1 + k_3 + k_3k_4 = 1.$$

Once again, refuting the last three cases, we obtain $k_2 = 2, \ 1 + k_3 + k_3k_4 = 179$, which yields $k_3(1 + k_4) = 178 = 2 \times 89$. Likewise, noting that 2 and 89 are primes, we get

$$k_3 = 2, \ 1 + k_4 = 89, \ \text{or,}$$
$$k_3 = 1, \ 1 + k_4 = 178, \ \text{or,}$$
$$k_3 = 89, \ 1 + k_4 = 2, \ \text{or,}$$
$$k_3 = 178, \ 1 + k_4 = 1.$$

Again, refuting the last three cases, we obtain $k_3 = 2, \ 1 + k_4 = 89$, which yields $k_4 = 88$. Therefore, the only solution of $x_1 + \cdots + x_5 = 719$ subject to the imposed conditions is $x_1 = 1$, $x_2 = k_1x_1 = 2$, $x_3 = k_2x_2 = 4$, $x_4 = k_3x_3 = 8$, and $x_5 = k_4x_4 = 704$. That is, these five people in increasing order have

$$x_1 = 1, x_2 = 2, x_3 = 4, x_4 = 8, x_5 = 704,$$

amounts of money, which is what we want.                                    ∎

**3.** Without loss of generality, we may assume that starting with every minute, the traffic-light of the crossroad $B$ is 30 seconds red and 30 seconds green, respectively. If $t$ denotes the $t$th second of a particular minute at the crossroad $B$, then the stop time function of the bus at the crossroad, denoted by $f$, on that particular minute, is

$$f(t) = \begin{cases} 30 - t & 0 \leq t \leq 30, \\ 0 & 30 \leq t \leq 60. \end{cases}$$

Let $\mu_f$ denote the average value of the function $f$, which means the average of the stop time at the crossroad $B$. By definition, we have

$$\mu_f = \frac{1}{60} \int_0^{60} f(t)dt = \frac{1}{60} \int_0^{30} (30 - t)dt = 7.5.$$

But the bus passes through the crossroad $B$ of the city with the probability of $\frac{1}{3}$. Therefore, the average of the stop time of the bus at the crossroad is equal to $\frac{1}{3}\mu_f = \frac{7.5}{3} = 2.5$, which is what we want.                                    ∎

## 2.9. Ninth Competition

**2.9.1. Analysis. 1.** Suppose that $C$ is an equivalence class of the equivalence relation $\sim$. Pick an $x_0 \in C$. It is plain that

$$C = \{x_0 + r \mid r \in \mathbb{Q}\}.$$

It suffices to show that $C \cap (a, b) \neq \emptyset$ for all open intervals $(a, b)$ of $\mathbb{R}$. To this end, for a given open interval $(a, b)$, as $\mathbb{Q}$ is dense in $\mathbb{R}$, we see that there exists an $r \in \mathbb{Q} \cap (a - x_0, b - x_0)$. This yields $x_0 + r \in C \cap (a, b)$, implying $C \cap (a, b) \neq \emptyset$, which is what we want. ∎

**2.** First, we claim that the function $f$ is strictly increasing or strictly decreasing. To prove this by contradiction, suppose that $f$ is neither strictly increasing nor strictly decreasing. By the lemma presented in Solution 2 of 2.5.1, there exist $x_1, x_2, x_3 \in \mathbb{R}$ with $x_1 < x_2 < x_3$ such that $f(x_2) \geq \max\big(f(x_1), f(x_3)\big)$ or $f(x_2) \leq \min\big(f(x_1), f(x_3)\big)$. As $f$ is one-to-one and $x_1 < x_2 < x_3$, we see that $f(x_2) > \max\big(f(x_1), f(x_3)\big)$ or $f(x_2) < \min\big(f(x_1), f(x_3)\big)$. First, suppose $f(x_2) > \max\big(f(x_1), f(x_3)\big)$. Letting $\lambda = \frac{f(x_2) + \max\big(f(x_1), f(x_3)\big)}{2}$, we see that $f(x_1) < \lambda < f(x_2)$ and $f(x_3) < \lambda < f(x_2)$. Hence, by the Intermediate Value Theorem, there exist $x_1 < \xi_1 < x_2$ and $x_2 < \xi_2 < x_3$ such that $f(\xi_1) = \lambda = f(\xi_2)$. As $f$ is one-to-one, this yields $\xi_1 = \xi_2$, which is a contradiction.

Likewise, if $f(x_2) < \min\big(f(x_1), f(x_3)\big)$, considering $\lambda = \frac{f(x_2) + \min\big(f(x_1), f(x_3)\big)}{2}$, we obtain a contradiction, proving the claim. Thus, the function $f$ is either strictly increasing or strictly decreasing. We prove the assertion in the case when $f$ is strictly increasing. The case when $f$ is strictly decreasing can be done similarly or by replacing $f$ by $-f$ and repeating the above argument. To prove that $f^{-1}$ is continuous, it suffices to show that $f$ is an open map. To this end, note first that $f(-\infty) = \lim_{x \to -\infty} f(x)$ and $f(+\infty) = \lim_{x \to +\infty} f(x)$ exist, where $f(\pm\infty) \in \mathbb{R} \cup \{\pm\infty\}$, and that, in view of the Intermediate Value Theorem, $f(\mathbb{R}) = \big(f(-\infty), f(+\infty)\big)$. Analogously, if $(a, b)$ is an open interval in $\mathbb{R}$, we obtain $f(a, b) = \big(f(a), f(b)\big)$. Now, suppose that $G$ is an open subset of $\mathbb{R}$. It follows that there exists a family $\{(a_i, b_i)\}_{i \in I}$ of open intervals of $\mathbb{R}$ such that $G = \bigcup_{i \in I}(a_i, b_i)$. We can write

$$f(G) = f\Big(\bigcup_{i \in I}(a_i, b_i)\Big) = \bigcup_{i \in I} f(a_i, b_i) = \bigcup_{i \in I}\big(f(a_i), f(b_i)\big).$$

But $\big(f(a_i), f(b_i)\big)$ is an open interval, hence an open subset, of $\big(f(-\infty), f(+\infty)\big)$ $= f(\mathbb{R})$. This, in view of the above equality, implies that $f(G)$ is an open subset of $f(\mathbb{R})$. Therefore, $f$ is an open map, which is what we want. ∎

**3.** To show $f(E) = E$, assume first that $x \in E$ is arbitrary. We need to show that $x \in f(E)$. To see this, note that $f(x) \geq x$. If $f(x) = x$, there is nothing to prove. If $f(x) > x$, we have $f(a) = a \leq x < f(x)$. It thus follows from the Intermediate Value Theorem that there exists a $c \in [a, x]$ such that $f(c) = x$, implying $x \in f[a, b]$. By showing $c \in E$, we conclude that $x \in f(E)$, which is what we want. If $c \notin E$, then $f(c) < c$, from which, as $f$ is increasing, we see that $f\big(f(c)\big) < f(c)$. But $f(c) = x$, which obtains $f(x) < x$. This is a contradiction, yielding $x \in f(E)$, implying $E \subseteq f(E)$.

Next, suppose that $y = f(x) \in f(E)$, where $x \in E$, is arbitrary. As $x \in E$, we have $f(x) \geq x$, from which, as $f$ is increasing, we obtain $f\big(f(x)\big) \geq f(x)$. That is, $y \in [a, b]$ and $f(y) \geq y$, which means $y = f(x) \in E$. In other words, $f(E) \subseteq E$, proving the assertion. ∎

**4. Remark.** The function $f$, as defined, in not well-defined at 0. Since the value of $f$ at any point has no affect on its integrability, we redefine $f : [0, 1] \to \mathbb{R}$ by

$$f(x) = \begin{cases} \frac{1}{q} & x = \frac{p}{q}, p, q \in \mathbb{N}, \gcd(p, q) = 1, \\ 0 & \text{otherwise.} \end{cases}$$

**First solution:** Just as explained in Solution 1 of 2.5.1, the sequence $(f_n)_{n=1}^{+\infty}$, with $f_n : [0, 1] \to [0, 1]$, defined by

$$f_n(x) = \begin{cases} \frac{1}{q} & x = \frac{p}{q}, p, q \in \mathbb{N}, \gcd(p, q) = 1, q \leq n, \\ 0 & \text{otherwise,} \end{cases}$$

converges uniformly to $f$ on $[0, 1]$. As $f_n$ $(n \in \mathbb{N})$ is zero everywhere but at finitely many points of the interval $[0, 1]$, it follows that $f_n$'s are integrable on $[0, 1]$. Hence, $f$ is Riemann integrable on $[0, 1]$, and moreover

$$\int_0^1 f(x)dx = \lim_n \int_0^1 f_n(x)dx = 0.$$

**Second solution:** As shown in Solution 3 of 2.1.1, we have $\lim_{x \to a} f(x) = 0$ for all $a \in (0, 1)$, $\lim_{x \to 0^+} f(x) = 0$, and $\lim_{x \to 1^-} f(x) = 0$. Thus, the function $f$ is continuous on the set $[0, 1] \setminus \big(\mathbb{Q} \cap (0, 1]\big)$ and it is discontinuous on $\mathbb{Q} \cap (0, 1]$, which is a countable set. From this point on, we present two proofs for the integrability of $f$ on $[0, 1]$.

**First proof.** Just note that the set of points at which $f$ is discontinuous is countable and hence has measure zero. Therefore, using the Lebesgue's Integrability Criterion for Riemann integrals, we see that $f$ is Riemann integrable, which is what we want.

**Second proof.** Using Lebesgue's Number Lemma, we directly prove that $f$ is Riemann integrable on $[0, 1]$. Recall that Lebesgue's Number Lemma asserts that *if an open cover of a compact metric space $X$ is given, then there exists a number $\delta > 0$, called a Lebesgue number for the cover, such that every subset of $X$ of diameter less than $\delta$ is contained in some element of the cover.* We use Riemann's criterion for integrability to show that $f$ is integrable. We need to show that for given $\varepsilon > 0$, there exists a partition $P \in \mathcal{P}[0, 1]$, the set of all partitions of $[0, 1]$, such that

$$U(P, f) - L(P, f) = \sum_{i=1}^{n} (M_i - m_i)\Delta x_i < \varepsilon,$$

where $P : 0 = x_0 < \cdots < x_n = 1$, $\Delta x_i = x_i - x_{i-1}$, $M_i = \sup_{x_{i-1} \leq x \leq x_i} f(x)$, and $m_i = \inf_{x_{i-1} \leq x \leq x_i} f(x)$. To this end, for given $\varepsilon > 0$, choose $N \in \mathbb{N}$ such that $\frac{1}{2^N} < \frac{\varepsilon}{2}$ and let $\mathbb{Q} \cap (0, 1] = \{r_i\}_{i=1}^{+\infty}$. Set $N_i = \left(r_i - \frac{1}{2^{N+i}}, r_i + \frac{1}{2^{N+i}}\right) \cap [0, 1]$. Obviously, $\sum_{i=1}^{+\infty} l(N_i) \leq \frac{1}{2^N} < \frac{\varepsilon}{2}$, where $l(N_i)$ denotes the length of the interval $N_i$. On the other hand, $\mathbb{Q} \cap (0, 1] = \{r_i\}_{i=1}^{+\infty} \subseteq \bigcup_{i=1}^{+\infty} N_i$. Therefore, for any $x \in C := [0, 1] \setminus \bigcup_{i=1}^{+\infty} N_i$, the function $f$ is continuous at $x$. Consequently, for given $\varepsilon > 0$, there exists an open interval $N_x$ containing $x$ such that

$$M_{N_x}(f) - m_{N_x}(f) < \frac{\varepsilon}{2},$$

where $M_{N_x}(f) = \sup_{t \in N_x} f(t)$ and $m_{N_x}(f) = \inf_{t \in N_x} f(t)$. It is obvious that $\{N_i\}_{i=1}^{+\infty} \cup \{N_x\}_{x \in C}$ is an open cover for the compact interval $[0, 1]$. Let $\delta > 0$ be a Lebesgue number for this open cover. Choose $P \in \mathcal{P}[0, 1]$ such that the length of its subintervals are all less than $\delta$. That is, if

$$P : 0 = x_0 < \cdots < x_n = 1,$$

then $\Delta x_i = x_i - x_{i-1} < \delta$ for all $1 \leq i \leq n$. Therefore, for all $1 \leq i \leq n$, we have $[x_{i-1}, x_i] \subseteq N_j$ for some $j \in \mathbb{N}$ or $[x_{i-1}, x_i] \subseteq N_x$ for some $x \in C$. Letting

$$A = \{i \in \mathbb{N} : 1 \leq i \leq n, \exists j \in \mathbb{N} \ni [x_{i-1}, x_i] \subseteq N_j\},$$
$$B = \{i \in \mathbb{N} : 1 \leq i \leq n, \exists x \in C \ni [x_{i-1}, x_i] \subseteq N_x\},$$

we can write

$$
\begin{aligned}
U(P, f) - L(P, f) &= \sum_{i=1}^{n} (M_i - m_i) \Delta x_i \\
&\leq \sum_{i \in A} (M_i - m_i) \Delta x_i + \sum_{i \in B} (M_i - m_i) \Delta x_i \\
&\leq \sum_{i \in A} 1 \times \Delta x_i + \sum_{i \in B} \frac{\varepsilon}{2} \Delta x_i \leq \sum_{i=1}^{+\infty} l(N_i) + \frac{\varepsilon}{2} \sum_{i=1}^{n} \Delta x_i \\
&\leq \frac{1}{2^N} + \frac{\varepsilon}{2} < \varepsilon.
\end{aligned}
$$

This means, $U(P, f) - L(P, f) < \varepsilon$, which is what we want.

To find the value of the integral, it is obvious that

$$L(P, f) = \sum_{i=1}^{n} m_i \Delta x_i = 0,$$

for all $P \in \mathcal{P}[0, 1]$. Since $f$ is integrable, we have

$$\int_0^1 f(x)\,dx = \sup_{P \in \mathcal{P}[a,b]} L(P, f) = 0. \qquad \blacksquare$$

**5.** First, we show that no function $f : [0, 1] \rightarrow [0, 1] \times [0, 1]$ can have the three properties, namely, continuity, injectivity, and surjectivity. Suppose to the contrary that $f$ is a continuous and one-to-one function from $[0, 1]$ onto $[0, 1] \times [0, 1]$. As $f$ is invertible and $[0, 1]$ is compact, it follows that $f$ is a closed map, and hence $f^{-1} : [0, 1] \times [0, 1] \rightarrow [0, 1]$ is a continuous function.

Let $f(\frac{1}{2}) = (a, b)$ for some $a, b \in [a, b]$. Obviously, $f^{-1}(a, b) = \frac{1}{2}$. Letting $C = [0, 1] \times [0, 1] \setminus \{(a, b)\}$, we have

$$f^{-1}(C) = [0, 1] \setminus \{\frac{1}{2}\}.$$

Since $C$ is connected and $f^{-1}$ is continuous, it follows that $[0, 1] \setminus \{\frac{1}{2}\}$ is connected, which is a contradiction. Therefore, the assertion follows by contradiction.

We now prove that $f$ can have any two properties of the aforementioned properties. There are three cases to consider.

(i) $f : [0, 1] \to [0, 1] \times [0, 1]$ can be continuous and one-to-one.

The function $f : [0, 1] \to [0, 1] \times [0, 1]$ defined by $f(t) = (t, 0)$ is continuous and one-to-one.

(ii) $f : [0, 1] \to [0, 1] \times [0, 1]$ can be continuous and onto.

Such a function exists. We present the well-known example due to Schoenberg (1938). First, define the function $\phi$ on $[0, 2]$ by

$$\phi(t) = \begin{cases} 0 & 0 \leq t \leq \frac{1}{3} \text{ or } \frac{5}{3} \leq t \leq 2, \\ 3t - 1 & \frac{1}{3} \leq t \leq \frac{2}{3}, \\ 1 & \frac{2}{3} \leq t \leq \frac{4}{3}, \\ -3t + 5 & \frac{4}{3} \leq t \leq \frac{5}{3}. \end{cases}$$

Then, extend $\phi$ to $\mathbb{R}$ via $\phi(t+2) = \phi(t)$. It is plain that $\phi : \mathbb{R} \to \mathbb{R}$ is 2-periodic and that $0 \leq \phi(t) \leq 1$ for all $t \in \mathbb{R}$. Now, define $f_1, f_2 : \mathbb{R} \to \mathbb{R}$ by

$$f_1(t) = \sum_{n=1}^{+\infty} \frac{\phi(3^{2n-2}t)}{2^n}, \quad f_2(t) = \sum_{n=1}^{+\infty} \frac{\phi(3^{2n-1}t)}{2^n}.$$

As $0 \leq \phi \leq 1$ on $\mathbb{R}$, we see that the series defining $f_1$ and $f_2$ converge uniformly and absolutely on $\mathbb{R}$, and hence $f_1$ and $f_2$ are continuous on $\mathbb{R}$ and moreover, $0 \leq f_1, f_2 \leq 1$ on $\mathbb{R}$ because $0 \leq \phi \leq 1$ on $\mathbb{R}$. Thus, the function $f : [0, 1] \to [0, 1] \times [0, 1]$ defined by $f(t) = (f_1(t), f_2(t))$ is well-defined and continuous. To prove that $f$ is onto, let $(a, b) \in [0, 1] \times [0, 1]$ be arbitrary. We need to show that there exists a $c \in [0, 1]$ such that $f(c) = (a, b)$. To this end, write the binary expansions of $a, b \in [0, 1]$ as follows

$$a = \sum_{n=1}^{+\infty} \frac{a_n}{2^n}, \quad b = \sum_{n=1}^{+\infty} \frac{b_n}{2^n},$$

where $a_n, b_n = 0$ or $1$ for all $n \in \mathbb{N}$. Let

$$c = 2 \sum_{n=1}^{+\infty} \frac{c_n}{3^n},$$

where

$$c_n = \begin{cases} a_{\frac{n+1}{2}} & n \text{ odd}, \\ b_{\frac{n}{2}} & n \text{ even}. \end{cases}$$

Thus, $c_n = 0$ or $1$ for all $n \in \mathbb{N}$. Consequently, as $2\sum_{n=1}^{+\infty} \frac{1}{3^n} = 1$, we obtain $0 \le c \le 1$. We claim that

$$\phi(3^k c) = c_{k+1}, \tag{$*$}$$

for all $k \in \mathbb{Z}$ with $k \ge 0$. Suppose that $(*)$ holds. We would then have

$$\phi(3^{2n-2}c) = c_{2n-1} = a_n, \quad \phi(3^{2n-1}c) = c_{2n} = b_n,$$

for all $n \in \mathbb{N}$. This easily yields $f(c) = (f_1(c), f_2(c)) = (a, b)$, which is what we want. Now, to prove $(*)$, suppose $k \in \mathbb{Z}$ with $k \ge 0$. We can write

$$3^k c = 2\sum_{n=1}^{k} \frac{c_n}{3^{n-k}} + 2\sum_{n=k+1}^{+\infty} \frac{c_n}{3^{n-k}} = 2N_k + d_k,$$

where $N_k = \sum_{n=1}^{k} 3^{k-n}c_n \in \mathbb{N}$ and $d_k = 2\sum_{n=k+1}^{+\infty} \frac{c_n}{3^{n-k}}$. As $\phi$ is periodic and its period is 2, we have

$$\phi(3^k c) = \phi(d_k).$$

If $c_{k+1} = 0$, then

$$0 \le d_k \le 2\sum_{n=2}^{+\infty} \frac{1}{3^n} = \frac{1}{3},$$

in which case, $\phi(d_k) = 0$, whence $\phi(3^k c) = c_{k+1} = 0$. If $c_{k+1} = 1$, then $\frac{2}{3} \le d_k \le 1$, in which case, $\phi(d_k) = 1$, whence $\phi(3^k c) = c_{k+1} = 1$. That is,

$$\phi(3^k c) = c_{k+1},$$

for all $k \in \mathbb{Z}$ with $k \ge 0$, proving the claim and hence finishing the proof.

(iii) $f : [0, 1] \to [0, 1] \times [0, 1]$ can be one-to-one and onto.

Such a function exists. First, define $f : [0, 1] \to [0, 1] \times [0, 1]$ by

$$f(t) = (0.t_1 t_3 t_5 \ldots, 0.t_2 t_4 t_6 \ldots),$$

where $0.t_1 t_2 t_3 \ldots$, with $t_i = 0$ or $1$ for all $i \in \mathbb{N}$, denotes the binary expansion of the number $t$. We set the convention that in the binary expansion of a number $t$ the number 1 is not allowed to be repeated for all the digits after any digit except when $t = 1$, in which case its binary expansion, by definition, is set to be $0.111\ldots$. It is obvious that $f$ is onto. Next, as mentioned in (i), the function $f : [0, 1] \to [0, 1] \times [0, 1]$ defined by $f(t) = (t, 0)$ is (continuous and) one-to-one. It thus follows from the Schroder-Bernstein Theorem that there is a function $f : [0, 1] \to [0, 1] \times [0, 1]$, which is one-to-one and onto, as desired. ∎

**2.9.2. Algebra. 1. First solution:** Proceed by contradiction. Suppose that there exists a nonnormal subgroup $H$ of index $p$ in $G$ so that $|G| = p|H|$, where $p$ is the smallest prime dividing the order of $G$. It follows that there is a $g \in G$ such that

$$K = H^g := g^{-1}Hg = \{g^{-1}hg : h \in H\} \ne H.$$

It is obvious that $|K| = |H|$. Also, we have $|H \cap K| < |H| = |K|$, for otherwise $|H \cap K| = |H| = |K|$, yielding $H \cap K = H = K$, which is a contradiction. Now, we claim that $\frac{|K|}{|H \cap K|} \ge p$. To see this, let $q$ be a prime dividing $\frac{|K|}{|H \cap K|}$. It follows that $q$ divides $|K|$, and hence $|G|$. This obtains $p \leq q \leq \frac{|K|}{|H \cap K|}$ because $p$ is the smallest prime dividing $|G|$. On the other hand,

$$|G| \geq |HK| = |H| \frac{|K|}{|H \cap K|} \geq |H| p = |G|,$$

implying that $HK = G$. Now, let $g = hk$ for some $h \in H, k \in K$. We can write

$$H^g = H^{hk} = (H^h)^k = H^k.$$

Thus,

$$G = G^k = (HK)^k = H^k K^k = H^g K = KK = K,$$

which is a contradiction. So, the assertion follows by way of contradiction.

**Second solution:** Let $H$ be a subgroup of $G$ of index $p$ in $G$, where $p$ is the smallest prime dividing the order of $G$. First, we claim that if $x \notin H$, then $x^i \notin H$ for all $i \in \mathbb{N}$ with $1 \leq i \leq p-1$. By way of contradiction, suppose that there exists the smallest $j \in \mathbb{N}$ with $1 < j \leq p-1$ such that $x^j \in H$. It follows that $x^i \notin H$ for all $1 \leq i < j$. Set $n = |G|$ and $m = \text{ord}(x)$. We have $m | n$. As $j < p$ and $p$ is the smallest prime that divides $n$, it follows that $m$ is not divisible by $j$. Hence, there are numbers $r, q \in \mathbb{N}$ with $0 < r < j$ such that $m = jq + r$. Noting that $m = \text{ord}(x)$, we can write

$$e = x^m = x^{jq+r} = (x^j)^q x^r,$$

yielding $(x^j)^q x^r \in H$. But $x^j \in H$, from which, we obtain $x^r \in H$, a contradiction because $0 < r < j$. Consequently, $x^i \notin H$ for all $1 \leq i \leq p-1$, as we claimed.

Now, to prove the assertion, we proceed by contradiction again. Suppose that there exist $g \in G$ and $h \in H$ such that $ghg^{-1} \notin H$. This yields $g \notin H$, which, in turn together with the above claim, implies that $g^i \notin H$ for all $i \in \mathbb{N}$ with $1 \leq i \leq p-1$. On the other hand, $g^i H \neq g^j H$ whenever $1 \leq i < j \leq p-1$ because otherwise $g^j = g^i h'$ for $h' \in H$, implying $g^{j-i} = h' \in H$, which is a contradiction, for $1 \leq j - i \leq p-1$. Therefore, $H, gH, \ldots, g^{p-1}H$ are $p$ distinct left cosets of $H$. Consequently,

$$\frac{G}{H} = \{H, gH, \ldots, g^{p-1}H\}.$$

Let $g_1 = ghg^{-1}$. Likewise, as $g_1 \notin H$, we conclude that $H, g_1 H, \ldots, g_1^{p-1}H$ are $p$ distinct left cosets of $H$, implying

$$\frac{G}{H} = \{H, g_1 H, \ldots, g_1^{p-1}H\}.$$

Therefore, $gH = g_1^r H$ for some $1 \leq r \leq p-1$, and hence $g = g_1^r h_1$ for some $h_1 \in H$. On the other hand, $g_1^r = gh^r g^{-1}$, from which, we obtain $g = gh^r g^{-1} h_1$. This yields $g = h_1 h^r$, implying $g \in H$, which is a contradiction. Therefore, $H$ is a normal subgroup of $G$, finishing the proof.

**Third solution:** Let $H$ be a subgroup of $G$ of index $p$ in $G$, where $p$ is the smallest prime dividing the order of $G$. Use $L$ to denote the set of all left cosets of $H$ in $G$, which has $p$ elements by the hypothesis. For each $g \in G$ define the mapping $\tau_g : L \longrightarrow L$ by $\tau_g(xH) = gxH$. In other words, $G$ acts on $L$ by multiplication from the left. It is readily verified that $\tau_g$'s $(g \in G)$ give rise to a group homomorphism $\phi : G \longrightarrow S(L) \cong S_p$ which is defined by $\phi(g) = \tau_g$, where $S(L)$ denotes the group of all permutations of $L$ which is isomorphic to $S_p$ because $|L| = p$ (here, as is usual, $S_p$ denotes the symmetric group of degree $p$, whose order is $p!$). Obviously, $K := \ker \phi \subseteq H$. We prove the assertion by showing that $H = K$. By the First Isomorphism Theorem for groups, $\frac{G}{K}$ is isomorphic to a subgroup of $S_p$ implying that $|\frac{G}{K}| \, | \, p!$. On the other hand, every divisor of $|\frac{G}{K}|$ divides $|G|$. Also, of the prime divisors of $p!$ only $p$ divides $|G|$ because $p$ is the smallest prime dividing the order of $G$. Thus, $|\frac{G}{K}| = p$ or $|\frac{G}{K}| = 1$. The latter is impossible because $|\frac{G}{K}| \geq |\frac{G}{H}| = p$. Therefore, $|\frac{G}{K}| = p = |\frac{G}{H}|$, implying that $|H| = |K|$. This obtains $H = K$ because $K \subseteq H$, completing the proof.

**Fourth solution:** By a standard result from Galois theory, e.g., see Proposition V.2.16 of "Algebra" by T.W. Hungerford, there are fields $E, K$ such that $K$ is an extension of $E$ and that $\mathrm{Gal}(K/E) \cong G$. In view of this, and two other standard results from Galois theory, e.g., see Theorem V.2.5 and Corollary V.3.15 of "Algebra" by T.W. Hungerford, it suffices to prove the following.

*Under the hypothesis of the problem, let $K$ be an extension field of $E$ such that $\mathrm{Gal}(K/E) \cong G$. If $F \subseteq K$ is an extension field of $E$ such that $[F : E] = p$, then $F$ is normal over $E$.*

To prove this, note first that every element of $K$ is separable over $E$, i.e., the minimal polynomial of every element of $K$ over $E$ splits into distinct linear factors, for $K$ is Galois over $E$. Therefore, $F$ is a finite separable extension of $E$ and hence, by the Primitive Element Theorem (see Proposition V.6.15 of "Algebra" by T.W. Hungerford, there is an $\alpha \in F$ such that $F = E(\alpha)$. Let $g \in E[x]$ be an arbitrary irreducible polynomial that has a root, say $h(\alpha)$ for some $h \in E[x]$, in $F = E(\alpha)$. It follows that the splitting field of $g$ is contained in $K$, for $h(\alpha) \in F \subseteq K$ and $K$ is Galois, and hence normal, over $E$. Note that $\deg(g) = [E(h(\alpha)) : E] \leq [E(\alpha) : E] = p$. We need to show that $g$ splits into linear factors in $E(\alpha)$. Let $\beta \neq h(\alpha)$ be a root of $g$ in the splitting field of $g$, which, as we just saw, is contained in $K$. It suffices to prove that $\beta \in F = E(\alpha)$. To this end, we note that $[F(\beta) : F] < \deg(g) \leq p$ because $g(\beta) = 0$, $g(h(\alpha)) = 0$, and $h(\alpha) \in F$. This yields $[F(\beta) : F] = 1$ because $p$ is the smallest prime dividing $|G|$ and $[F(\beta) : F] \, | \, |G|$, for $\beta \in K$ and $[K : E] = |G|$ since $\mathrm{Gal}(K/E) \cong G$. Consequently, $\beta \in F$, and hence $F$ is normal over $E$, as desired. ∎

**2.** We determine all isomorphisms from the field $\dfrac{\mathbb{Z}_3[x]}{(x^2 + 1)}$ onto the field $\dfrac{\mathbb{Z}_3[x]}{(x^2 + x + 2)}$. Since $f : \dfrac{\mathbb{Z}_3[x]}{\langle x^2+1 \rangle} \to \dfrac{\mathbb{Z}_3[x]}{\langle x^2+x+2 \rangle}$ is an isomorphism of fields, we see that $f(a + \langle x^2 + 1 \rangle) = a + \langle x^2 + x + 2 \rangle$ for all $a \in \mathbb{Z}_3$. Thus, to determine $f$, we only need to determine $f(x + \langle x^2 + 1 \rangle)$. To this end, suppose that

$f(x + \langle x^2 + 1 \rangle) = ax + b + \langle x^2 + x + 2 \rangle$ for some $a, b \in \mathbb{Z}_3$. We must have

$$\left( f(x + \langle x^2 + 1 \rangle) \right)^2 \ = \ f(x^2 + \langle x^2 + 1 \rangle) = f(-1 + \langle x^2 + 1 \rangle),$$

implying that $(ax + b)^2 + 1 \in \langle x^2 + x + 2 \rangle$. But the polynomial $(ax + b)^2 + 1 = a^2 x^2 + 2abx + b^2 + 1$ is divisible by $x^2 + x + 2$ in $\mathbb{Z}_3[x]$ if and only if so is the polynomial $(2ab - a^2)x + (a^2 + b^2 + 1)$, which is obtained from $a^2 x^2 + 2abx + b^2 + 1$ by letting $x^2 = -x - 2 = -x + 1$, if and only if $a$ and $b$ satisfy the equations $a(2b - a) = 0$ and $a^2 + b^2 + 1 = 0$. It is readily seen that $a = 2, b = 1$ and $a = 1, b = 2$ are the only solutions of the above equations in $\mathbb{Z}_3$. Therefore, an isomorphism $f : \frac{\mathbb{Z}_3[x]}{\langle x^2 + 1 \rangle} \to \frac{\mathbb{Z}_3[x]}{\langle x^2 + x + 2 \rangle}$ must be given by

$$f(ax + b + \langle x^2 + 1 \rangle) \ = \ a(2x + 1) + b + \langle x^2 + x + 2 \rangle,$$

or by

$$f(ax + b + \langle x^2 + 1 \rangle) \ = \ a(x + 2) + b + \langle x^2 + x + 2 \rangle.$$

That is, either

$$f(ax + b + \langle x^2 + 1 \rangle) \ = \ 2ax + (a + b) + \langle x^2 + x + 2 \rangle,$$

or

$$f(ax + b + \langle x^2 + 1 \rangle) \ = \ ax + (2a + b) + \langle x^2 + x + 2 \rangle,$$

which are both easily seen to be monomorphisms of rings. It follows that such an $f$ is an isomorphism from the field $\dfrac{\mathbb{Z}_3[x]}{\langle x^2 + 1 \rangle}$ onto the field $\dfrac{\mathbb{Z}_3[x]}{\langle x^2 + x + 2 \rangle}$ because they both have 9 elements. ∎

**3.** Note that the ring $\mathbb{Z}_7[x]$ is a Euclidean ring because $\mathbb{Z}_7$ is a field. With that in mind, we can use the Euclidean algorithm to find the greatest common divisor of the two polynomials $4x^4 - 2x^2 + 1$ and $-3x^3 + 4x^2 + x + 1$ in $\mathbb{Z}_7[x]$. We have

$$4x^4 - 2x^2 + 1 \ = \ (-3x^3 + 4x^2 + x + 1)(x - 1) + x^2 + 2,$$
$$-3x^3 + 4x^2 + x + 1 \ = \ (x^2 + 2)(-3x + 4).$$

Therefore,

$$\gcd(4x^4 - 2x^2 + 1, -3x^3 + 4x^2 + x + 1) \ = \ x^2 + 2,$$

which is what we want. ∎

**4.** Let $\{1, y_1\}$, where $y_1 \in K$, be a basis for the vector space $K$ over $F$. If $y_1^2 \in F$, there is nothing to prove. Suppose $y_1^2 \notin F$. As $y_1^2 \in K$, it follows that there exist scalars $a, b \in F$ with $b \neq 0$ such that $y_1^2 = a + by_1$. We have

$$y_1^2 - by_1 \ = \ b.$$

On the other hand, since $\mathrm{ch}(F) \neq 2$, we have $2^{-1} = \frac{1}{2} \in F$, and hence we can write

$$y_1^2 - by_1 \ = \ (y_1 - \frac{b}{2})^2 - (\frac{b}{2})^2,$$

whence

$$(y_1 - \frac{b}{2})^2 \ = \ a + (\frac{b}{2})^2 \in F.$$

Let $y = y_1 - \frac{b}{2}$. It is easily verified that $\{1, y\}$ is a basis for $K$ over $F$ and $y^2 = a + (\frac{b}{2})^2 \in F$, which is what we want. ∎

**5.** Let $S = T - I$. Obviously, $W = \ker S$. On the other hand, by the Rank-Nullity Theorem, we have

$$\dim \ker S + \dim SV = \dim V = n.$$

We claim that $SV \subseteq \ker S$. To see this, let $y = Sx \in SV$, where $x \in V$, be arbitrary. We have

$$Sy = S^2 x = (T - I)^2 x = (T^2 - 2T + I)x = (2I - 2T)x = 0,$$

because $T^2 = I$ and $\operatorname{ch}(F) = 2$. This means, $y \in \ker S$, proving the claim. Now, since $SV \subseteq \ker S$, we can write

$$n = \dim \ker S + \dim SV \leq 2 \dim \ker S,$$

yielding $\dim W = \dim \ker S \geq \frac{n}{2}$, which is what we want. ∎

### 2.9.3. General. 1. Let

$$A = \begin{pmatrix} r & \lambda & \lambda & \cdots & \lambda & \lambda \\ \lambda & r & \lambda & \cdots & \lambda & \lambda \\ \vdots & \ddots & \ddots & \ddots & \vdots & \vdots \\ \vdots & \vdots & \ddots & \ddots & \ddots & \vdots \\ \lambda & \lambda & \cdots & \lambda & r & \lambda \\ \lambda & \lambda & \cdots & \cdots & \lambda & r \end{pmatrix}.$$

Subtracting the first row from all the other rows and then adding the $j$th column $(2 \leq j \leq n)$ to the first column, we obtain an upper triangular matrix whose determinant is the product of the entries on its main diagonal. Thus, we can write

$$\det A = \det \begin{pmatrix} r & \lambda & \lambda & \cdots & \lambda & \lambda \\ \lambda - r & r - \lambda & 0 & \cdots & 0 & 0 \\ \lambda - r & 0 & r - \lambda & 0 & \cdots & 0 \\ \vdots & \vdots & \ddots & \ddots & \ddots & \vdots \\ \lambda - r & 0 & \cdots & 0 & r - \lambda & 0 \\ \lambda - r & 0 & \cdots & \cdots & 0 & r - \lambda \end{pmatrix}$$

$$= \det \begin{pmatrix} r + (n-1)\lambda & \lambda & \lambda & \cdots & \lambda & \lambda \\ 0 & r - \lambda & 0 & \cdots & 0 & 0 \\ 0 & 0 & r - \lambda & 0 & \cdots & 0 \\ \vdots & \vdots & \ddots & \ddots & \ddots & \vdots \\ 0 & 0 & \cdots & 0 & r - \lambda & 0 \\ 0 & 0 & \cdots & \cdots & 0 & r - \lambda \end{pmatrix}$$

$$= (r + (n-1)\lambda)(r - \lambda)^{n-1}.$$

That is, $\det A = (r + (n-1)\lambda)(r - \lambda)^{n-1}$, which is what we want. ∎

**2.** Let $x$ denote the number of chickens, $a$ the amount of food a chicken consumes per day, and $t$ the number of the days before the farmers run out of chicken food. The amount of food in the chicken farm is equal to

$$(t+20)(x-75)a = (t-15)(x+100)a = txa.$$

Simplifying, we obtain

$$\begin{cases} (t+20)(x-75) = (t-15)(x+100) \\ (t-15)(x+100) = tx \end{cases} \implies \begin{cases} x = 5t. \\ 20t - 3x - 300 = 0. \end{cases}$$

Substituting $x = 5t$ into the second equation yields $x = 300$, which implies $t = 60$. Therefore, there are $x = 300$ chickens in the farm. ∎

**3.** To prove the assertion by contradiction, suppose that such a function $f$ exists. We can write

$$\int_0^1 (x-\alpha)^2 f(x)dx = \int_0^1 x^2 f(x)dx - 2\alpha \int_0^1 xf(x)dx + \alpha^2 \int_0^1 f(x)dx$$

$$= \alpha^2 - 2\alpha^2 + \alpha^2 = 0.$$

That is, $\int_0^1 (x-\alpha)^2 f(x) = 0$. As $(x-\alpha)^2 f(x)$ is continuous and nonnegative on $[0,1]$, it follows that $(x-\alpha)^2 f(x) = 0$ for all $x \in [0,1]$. This implies $f = 0$ on $[0,1]$ except possibly at $x = \alpha$. Now, from the continuity of $f$, we see that $f = 0$ on $[0,1]$, a contradiction. Therefore, no such function $f$ exists, finishing the proof. ∎

**4. First solution:** Consider the points with integer coordinates in the closed interval $[0,n]$ of the real line. To any solution $(x_1, \ldots, x_m)$ of the equation

$$x_1 + x_2 + \cdots + x_m = n$$

in $\mathbb{N}$, there corresponds $m-1$ points $p_1 < \cdots < p_{m-1}$ from the set $\{i\}_{i=1}^{n-1}$ as follows

$$p_1 = x_1, \quad p_2 = x_1 + x_2, \quad \ldots, \quad p_{m-1} = x_1 + \cdots + x_{m-1}.$$

Conversely, for any $m-1$ points $p_1 < \cdots < p_{m-1}$ from the set $\{i\}_{i=1}^{n-1}$, a solution $(x_1, \ldots, x_m)$ of the equation

$$x_1 + x_2 + \cdots + x_m = n$$

is determined as follows

$$x_1 = p_1, \quad x_2 = p_2 - p_1, \quad \ldots, \quad x_{m-1} = p_{m-1} - p_{m-2}, \quad x_m = n - p_{m-1}.$$

Therefore, the number of the solutions of the equation $x_1 + x_2 + \cdots + x_m = n$ in $\mathbb{N}$ is equal to the number of ways of choosing $m-1$ points $\{p_i\}_{i=1}^{m-1}$ from $n-1$ points $\{i\}_{i=1}^{n-1}$, which, as is well-known, is equal to $\binom{n-1}{m-1}$.

**Second solution:** We only briefly elaborate on this solution. It is not difficult to see that the number of the solutions of the equation

$$x_1 + x_2 + \cdots + x_m = n$$

in $\mathbb{N}$ is equal to the coefficient of $x^n$ in the expansion of

$$(x + x^2 + \cdots + x^{n-m})^m$$

or in that of

$$(x + x^2 + x^3 + \cdots)^m = (\frac{x}{1-x})^m = x^m \sum_{n=0}^{+\infty} \binom{m+n-1}{m-1} x^n,$$

which is easily seen to be $\binom{n-1}{m-1}$.  ■

**5.** Yes, just cut the paper along the heavy lines.

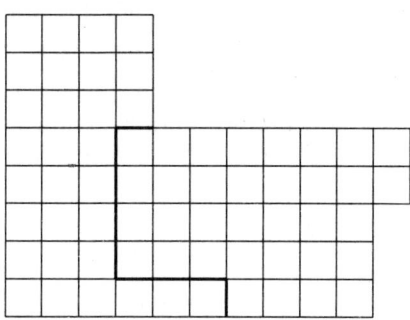

■

**6.** Since the time needed to get to work from home obeys the uniform distribution, it follows that the density probability function is equal to

$$f(x) = \begin{cases} 0 & x < 10 \\ \frac{1}{10} & 10 \le x \le 20 \\ 0 & x > 20 \end{cases}.$$

(a) As the worker needs 15 minutes to get to work on time, the desired probability is

$$P(x > 15) = \int_{15}^{20} \frac{dx}{10} = \frac{1}{2}.$$

(b) **First solution.** Noting that it takes the worker at least 10 minutes to get to work, let $x_0 > 10$ denote the time needed for the worker to get to work on time and with the probability of 75% to have time to eat breakfast. We must have

$$P(10 < x < x_0) = \int_{10}^{\min(x_0, 20)} \frac{dx}{10} = 0.75 \iff \min(x_0, 20) = 17.5$$

$$\iff x_0 = 17.5.$$

Therefore, the latest time that this worker can leave home to get to work on time and with the probability of 75% to have time to eat breakfast is equal to $7, 45' - 17.5' = 7, 27.5' = 7, 27', 30''$.

**Second solution.** Suppose the worker heads off for work $m$ minutes before 8 o'clock in the morning. In order for the worker to get to work on time and have time to eat breakfast, we must have $m - 15 \ge 10$, implying $m \ge 25$. Assuming that it takes the worker $x$ minutes to get to work, the worker has $m - x$ minutes to have breakfast. Thus, to make sure that the $m - x$ minutes, which is the remaining time before the start hour of the factory, is enough for the worker to have breakfast with the probability of 75%, we must have

$$P(m - x \geq 15) = 0.75 \iff P(x \leq m - 15) = 0.75$$

$$\iff \int_{10}^{\min(m-15,20)} \frac{dx}{10} = 0.75$$

$$\iff \min(m - 15, 20) = 17.5$$

$$\iff m = 32.5.$$

Therefore, if the worker heads off for work 32.5 minutes before 8 a.m., i.e., at $8,00' - 32.5' = 7,27.5' = 7,27',30''$, s/he will get to work on time and with the probability of 75% have time to eat breakfast.  ∎

**7. First solution:** We solve the problem on a general field $F$. Let $n \in \mathbb{N}$, $A = (c_i c_j) \in M_n(F)$, and $I_n$ the identity matrix of size $n$. To evaluate $\det(I_n + A)$, we may, without loss of generality, assume that $c_i \neq 0$ for all $1 \leq i \leq n$ because if $c_{i_0} = 0$ for some $i_0 \in \{1, \ldots, n\}$, then $\det(I_n + A) = \det(I_{n-1} + A')$ where $A' = (c_i' c_j') \in M_{n-1}(F)$ with $c_i' \in \{c_1, \ldots, c_n\} \setminus \{c_{i_0}\}$, which has the same general form of $A = (c_i c_j) \in M_n(F)$. Use the basic properties of the determinant function and perform the following operations: factor out $c_i$ of row $i$ and of column $i$ for each $i = 1, \ldots, n$; then subtract the first row from all the other rows; then multiply column $i$ by $c_i^2$ $(1 \leq i \leq n)$; and finally add row $i$ to row 1 for each $i = 2, \ldots, n$, to obtain an upper triangular whose determinant is equal to that of $I_n + A$. We can write

$$\det(I_n + A) = c_1^2 \cdots c_n^2 \det \begin{pmatrix} 1 + \frac{1}{c_1^2} & 1 & \cdots & 1 & 1 \\ 1 & 1 + \frac{1}{c_2^2} & \cdots & 1 & 1 \\ \vdots & \vdots & \ddots & \vdots & \vdots \\ 1 & 1 & \cdots & 1 + \frac{1}{c_{n-1}^2} & 1 \\ 1 & 1 & \cdots & 1 & 1 + \frac{1}{c_n^2} \end{pmatrix}$$

$$= c_1^2 \cdots c_n^2 \det \begin{pmatrix} 1 + \frac{1}{c_1^2} & 1 & \cdots & 1 & 1 \\ -\frac{1}{c_1^2} & \frac{1}{c_2^2} & \cdots & 0 & 0 \\ \vdots & \vdots & \ddots & \vdots & \vdots \\ -\frac{1}{c_1^2} & 0 & \cdots & \frac{1}{c_{n-1}^2} & 0 \\ -\frac{1}{c_1^2} & 0 & \cdots & 0 & \frac{1}{c_n^2} \end{pmatrix}$$

$$= \det \begin{pmatrix} 1 + c_1^2 & c_2^2 & \cdots & c_{n-1}^2 & c_n^2 \\ -1 & 1 & \cdots & 0 & 0 \\ \vdots & \vdots & \ddots & \vdots & \vdots \\ -1 & 0 & \cdots & 1 & 0 \\ -1 & 0 & \cdots & 0 & 1 \end{pmatrix}$$

$$= \det \begin{pmatrix} 1 + c_1^2 + \cdots + c_n^2 & c_2^2 & \cdots & c_{n-1}^2 & c_n^2 \\ 0 & 1 & \cdots & 0 & 0 \\ \vdots & & \ddots & \ddots & \vdots & \vdots \\ 0 & 0 & \cdots & 1 & 0 \\ 0 & 0 & \cdots & 0 & 1 \end{pmatrix}$$

$$= (1 + c_1^2 + \cdots + c_n^2) \times 1 \times \cdots \times 1 = 1 + c_1^2 + \cdots + c_n^2.$$

Therefore, $\det(I_n + A) = 1 + c_1^2 + \cdots + c_n^2$, which is what we want.

**Second solution:** Recall that if $A \in M_n(F)$, $\lambda_i$'s $(1 \le i \le n)$ are the eigenvalues of $A$ in the algebraic closure of $F$, and $f(x) = \det(xI_n - A)$ is the characteristic polynomial of $A$ so that

$$f(x) = x^n + f_{n-1}x^{n-1} + \cdots + f_0,$$

then, $f_{n-1} = -\sum_{i=1}^{n} \lambda_i$, $f_0 = (-1)^n \lambda_1 \cdots \lambda_n$, $\sum_{i=1}^{n} \lambda_i = \mathrm{tr}(A)$, and $\lambda_1 \cdots \lambda_n = \det(A)$. Also, if $\alpha \in F$ and $g(x) = \det\big(xI_n - (A + \alpha I_n)\big)$ is the characteristic polynomial of $A + \alpha I_n$, then

$$g(x) = \det\big((x - \alpha)I_n - A\big) = f(x - \alpha).$$

With all that in mind, note first that the rank of $A$ is 1 because row $i$ is $c_i(c_1, \ldots, c_n)$ which is a multiple of the fixed vector $(c_1, \ldots, c_n)$. It follows from the Rank-Nullity Theorem that $\dim \ker A = n - 1$, and hence 0 is an eigenvalue of $A$ of multiplicity $n - 1$. Let $\lambda$ be the only other eigenvalue of $A$. We must have

$$f(x) = \det(xI_n - A) = \prod_{i=1}^{n}(x - \lambda_i) = x^{n-1}(x - \lambda) = x^n - \lambda x^{n-1}.$$

On the other hand,

$$\sum_{i=1}^{n} \lambda_i = \lambda = \mathrm{tr}(A) = \sum_{i=1}^{n} c_i^2,$$

implying $\lambda = \sum_{i=1}^{n} c_i^2$. If $g$ denotes the characteristic polynomial of $I_n + A$, we have

$$\begin{aligned} g(x) &= f(x - 1) \\ &= (x - 1)^n - \lambda(x - 1)^{n-1} \\ &= x^n + g_{n-1}x^{n-1} + \cdots + g_1 x + (-1)^n(1 + \lambda). \end{aligned}$$

Recall that

$$(-1)^n(1 + \lambda) = (-1)^n \det(I_n + A),$$

whence

$$\det(I_n + A) = 1 + \lambda = 1 + \mathrm{tr}(A) = 1 + \sum_{i=1}^{n} c_i^2,$$

which is what we want. ∎

## 2.10. Tenth Competition

**2.10.1. Analysis. 1.** From (b), we see that $f'(x) > 0$ for all $x \geq 1$, implying that $f$ is strictly increasing on $[1, +\infty)$ and hence $f(t) > f(1) = 1$ for all $t > 1$. From this, we obtain

$$f'(t) = \frac{1}{t^2 + (f(t))^2} < \frac{1}{1 + t^2},$$

for all $t > 1$. It thus follows that $\int_1^{+\infty} f'(t)dt < \int_1^{+\infty} \frac{dt}{1+t^2} = \frac{\pi}{4}$ because $f'$ is continuous and $f'(t) < \frac{1}{1+t^2}$ on $[1, +\infty)$. Using the Second Fundamental Theorem of Calculus , we can write

$$f(x) = 1 + \int_1^x f'(t)dt < 1 + \int_1^x \frac{dt}{1+t^2} < 1 + \int_1^{+\infty} \frac{dt}{1+t^2} = 1 + \frac{\pi}{4}.$$

Therefore, $f$ is bounded from the above. This together with the hypothesis that $f$ is increasing implies that $\lim_{x \to +\infty} f(x)$ exists and that

$$\lim_{x \to +\infty} f(x) = 1 + \int_1^{+\infty} f'(t)dt < 1 + \int_1^{+\infty} \frac{1}{1+t^2} = 1 + \frac{\pi}{4},$$

implying $\lim_{x \to +\infty} f(x) < 1 + \frac{\pi}{4}$, which is what we want. ∎

**2.** For an $x \in (a, b)$, define $g : [a, b] \to \mathbb{R}$ by

$$g(t) = (t - a)(t - b)f(x) - (x - a)(x - b)f(t).$$

Plainly, $g(a) = g(x) = g(b) = 0$. From this, applying Rolle's Theorem twice to $g$ and $g'$, respectively, we obtain a $c \in (a, b)$ such that $g''(c) = 0$. We can write

$$g''(c) = 2f(x) - (x - a)(x - b)f''(c) = 0,$$

which implies

$$|f(x)| \leq \frac{M(x-a)(b-x)}{2},$$

for all $x \in [a, b]$. Taking integrals of both sides of the above yields

$$\int_a^b |f(x)|dx \leq \frac{M}{2} \int_a^b (-x^2 + (a+b)x - ab)dx = M\frac{(b-a)^3}{12}.$$

That is, $\int_a^b |f(x)|dx \leq M\frac{(b-a)^3}{12}$, which is what we want. ∎

**3. First solution:** We can write

$$\left| \int_0^3 \frac{x^2(1-x)x^n}{1+x^{2n}}dx \right| \leq \int_0^3 \frac{x^2|1-x|x^n}{1+x^{2n}}dx$$

$$= \int_0^1 \frac{x^2|1-x|x^n}{1+x^{2n}}dx + \int_1^3 \frac{x^2|1-x|x^n}{1+x^{2n}}dx$$

$$\leq \int_0^1 x^n dx + 18\int_1^3 \frac{x^n}{x^{2n}}dx$$

$$= \frac{1}{n+1} + \frac{18}{n-1} - \frac{18}{3^{n-1}(n-1)}.$$

It thus follows from the Squeeze Lemma that

$$\lim_{n \to +\infty} \int_0^3 \frac{x^2(1-x)x^n}{1+x^{2n}} dx = 0,$$

which is what we want.

**Second solution:** In view of a standard theorem from classical analysis, it suffices to show that the sequence $\left(\frac{x^2(1-x)x^n}{1+x^{2n}}\right)_{n=1}^{+\infty}$ uniformly converges to the zero function on $[0,3]$. Note first that

$$\left| \frac{x^2(1-x)x^n}{1+x^{2n}} \right| \leq |1-x|, \qquad (*)$$

for all $x \in [0,3]$. To see this, we have $x^{n+2} \leq 1$ on $[0,1]$, implying $x^{n+2} \leq 1+x^{2n}$ on $[0,1]$. Also if $n \geq 2$, then $x^{n+2} \leq x^{2n}$ on $[1,3]$, which implies $x^{n+2} \leq 1+x^{2n}$ on $[1,3]$. Thus, $x^{n+2} \leq 1+x^{2n}$ on $[0,3]$ of which the above inequality is a quick consequence. Now, let $0 < \varepsilon < 1$ be given. It follows from $(*)$ that

$$\left| \frac{x^2(1-x)x^n}{1+x^{2n}} \right| < \varepsilon,$$

for all $n \geq 2$ and $x \in (1-\varepsilon, 1+\varepsilon)$. On the interval $[0, 1-\varepsilon]$, for all $n > 2$ we can write

$$\left| \frac{x^2(1-x)x^n}{1+x^{2n}} \right| \leq \frac{x^{n+2}}{1+x^{2n}} \leq x^{n+2} \leq (1-\varepsilon)^{n+2}.$$

As $0 < 1-\varepsilon < 1$, we have $\lim_n (1-\varepsilon)^{n+2} = 0$. Hence, there exists an $N_1 > 1$ such that $(1-\varepsilon)^{n+2} < \varepsilon$ for all $n \geq N_1$. On the interval $[1+\varepsilon, 3]$, we have

$$\left| \frac{x^2(1-x)x^n}{1+x^{2n}} \right| < \frac{2x^{n+2}}{1+x^{2n}} = \frac{2x^2}{x^{-n}+x^n} < \frac{2}{x^{n-2}} \leq \frac{2}{(1+\varepsilon)^{n-2}},$$

for all $n \geq 2$. As $\lim_n \frac{2}{(1+\varepsilon)^{n-2}} = 0$, there exists an $N_2 > 1$ such that $\frac{2}{(1+\varepsilon)^{n-2}} < \varepsilon$ for all $n \geq N_2$. Letting $N = \max(N_1, N_2)$, for all $x \in [0,3]$ and $n \geq N$ we have

$$\left| \frac{x^2(1-x)x^n}{1+x^{2n}} \right| < \varepsilon.$$

That is, the sequence $\left(\frac{x^2(1-x)x^n}{1+x^{2n}}\right)_{n=1}^{+\infty}$ uniformly converges to the zero function on $[0,3]$, yielding

$$\lim_{n \to +\infty} \int_0^3 \frac{x^2(1-x)x^n}{1+x^{2n}} dx = \int_0^3 \lim_{n \to +\infty} \frac{x^2(1-x)x^n}{1+x^{2n}} dx = 0,$$

which is what we want. ∎

**2.10.2. Algebra. 1.** We prove the following lemma of which the assertion is a quick consequence.

**Lemma.** *Let $V$ be a left (resp. right) vector space over a division ring $D$ whose characteristic is zero. Then the additive group of $V$ has no maximal subgroup.*

**Proof.** To prove the assertion by contradiction, suppose that $M$ is a maximal subgroup of the additive group of $V$. As $M \le V$, $V$ is abelian, and $M$ is maximal, it follows that $\frac{V}{M}$ is a simple abelian group. Consequently, $\frac{V}{M}$ is a finite cyclic group, implying that there exists a prime number $p$ such that $\frac{V}{M} = \mathbb{Z}_p$, and hence $M$ is of finite index in $V$. By proving that $(V, +)$ has no proper subgroup of finite index, we obtain a contradiction, proving the assertion. Suppose to the contrary that there exists a proper subgroup $H$ of the additive group of $V$ such that $[V : H] = n \in \mathbb{N}$. As $|\frac{V}{H}| = n$, it follows from Lagrange's Theorem that $nx \in H$ for all $x \in V$. On the other hand, $x = n(\frac{x}{n})$ and $\frac{x}{n} \in V$ for all $x \in V$, implying $x \in H$ for all $x \in V$. This yields $V = H$, a contradiction. Thus, $V$ has no proper subgroup of finite index, which is what we want, finishing the proof. $\qquad\square$

To prove the assertion, just let $V = D = \mathbb{R}$ in the lemma. $\qquad\blacksquare$

**2.** Since the ring $R$ is unital, it follows from $(*)$ that there exist elements $a, a', a'' \in A$, $b \in B$, $c \in C$, and $d \in D$ such that

$$a + b = 1, \quad a' + c = 1, \quad a'' + d = 1.$$

Multiplying the above equalities and simplifying, we see that there is a $u \in A$ such that

$$1 \;=\; (a + b)(a' + c)(a'' + d) = u + bcd,$$

implying $1 = u + bcd \in A + M$ because $bcd \in M$. This obviously yields $A + M = R$, which is what we want.

Let $R = \mathbb{Z}$, $A = 2\mathbb{Z}$, $B = 3\mathbb{Z}$, $C = 5\mathbb{Z}$, and $D = 7\mathbb{Z}$. It is easy to verify that $M = B \cap C \cap D = 105\mathbb{Z}$ and that $R$ and its ideals $A, B, C, D$ satisfy $(*)$.

$\qquad\blacksquare$

**3.** We solve the problem under the weaker hypothesis that $\gcd(n, c) = 1$, where $c = \operatorname{lcm}(1, 2, \ldots, m - 1)$. It is obvious that $F \subseteq F(\alpha^m) \subseteq F(\alpha)$. So we can write

$$[F(\alpha) : F] \;=\; [F(\alpha) : F(\alpha^m)][F(\alpha^m) : F]. \tag{$*$}$$

As $[F(\alpha) : F] = n$, it follows that there exists a polynomial $P \in F[x]$ of degree $n$ such that

$$P(\alpha) \;=\; p_0 + p_1\alpha + \cdots + p_n\alpha^n = 0. \tag{$*'$}$$

Using the division algorithm, dividing $k$ by $m$ for each $k = 0, \ldots, n$, and collecting terms appropriately in $(*')$, we see that there exists a polynomial $Q \in F(\alpha^m)[x]$ with $\deg(Q) \le m - 1$ such that

$$P(\alpha) \;=\; Q(\alpha) = q_0 + q_1\alpha + \cdots + q_{m-1}\alpha^{m-1} = 0.$$

This clearly shows that $[F(\alpha) : F(\alpha^m)] \leq m - 1$, implying that $[F(\alpha) : F(\alpha^m)]$ divides $\text{lcm}(1, 2, \ldots, m - 1) = c$. But as, in view of $(*)$, $[F(\alpha) : F(\alpha^m)]$ divides $n$ and $\gcd(n, c) = 1$, we conclude that $[F(\alpha) : F(\alpha^m)] = 1$, whence $F(\alpha) = F(\alpha^m)$, yielding $[F(\alpha^m) : F] = [F(\alpha) : F]$. This, in turn, implies $F(\alpha) = F(\alpha^m)$, which is what we want. ∎

**4.** Recall that the trace of any nilpotent matrix $N \in M_n(F)$ is zero. With that in mind, suppose by way of contradiction that there exist nilpotent matrices $N_1, \ldots, N_k \in M_n(F)$ ($k \in \mathbb{N}$) that span $M_n(F)$. In particular, for the matrix $E_{11} \in M_n(F)$, where $E_{11}$ denote the matrix with 1 in the 11 place and zero elsewhere, there are scalars $c_1, \ldots, c_k \in F$ such that

$$E_{11} = c_1 N_1 + \cdots + c_k N_k,$$

which implies

$$1 = \text{tr}(E_{11}) = c_1 \text{tr}(N_1) + \cdots + c_k \text{tr}(N_k) = 0,$$

a contradiction, proving the assertion. ∎

### 2.10.3. General. 1. As the tests are independent, we can write

$$
\begin{aligned}
P\big(X = 0\big) &= P(FS) + P(SSFS) + P(FFFS) + P(SSSSFS) \\
&\quad + P(SSFFFS) + P(FFSSFS) + P(FFFFFS) + \cdots \\
&= qp + p^2 qp + q^2 qp + p^4 qp + p^2 q^2 qp + p^2 p^2 qp + q^4 qp + \cdots \\
&= qp\big(1 + (p^2 + q^2) + (p^2 + q^2)^2 + \cdots\big) \\
&= \frac{qp}{1 - p^2 - q^2} = \frac{\frac{1}{4}}{\frac{1}{2}} = \frac{1}{2}.
\end{aligned}
$$

That is, $P\big(X = 0\big) = \frac{1}{2}$. Similarly, one can show that $P\big(X = 1\big) = \frac{1}{2}$. ∎

**2.** We prove the assertion for the more general case where $A$ is a subset of a topological space. With that in mind, let $A$ be a closed subset of a topological space $X$ such that $A^\circ = \emptyset$. We prove the assertion by showing that $A = \partial A = \partial A^c$, where $\partial A$ denotes the boundary of $A$. Recall that by definition $\partial A = \overline{A} \cap \overline{A^c}$. So, we can write

$$\partial A^c = \overline{A^c} \cap \overline{A^{cc}} = \overline{A^c} \cap \overline{A} = \partial A.$$

That is, $\partial A = \partial A^c$. On the other hand,

$$\partial A = \overline{A} \setminus A^\circ = \overline{A} \setminus \emptyset = \overline{A} = A.$$

So we have $A = \partial A = \partial A^c$, where $A^c$ is an open set because $A$ is a closed set, which is what we want. ∎

**3.** Let $x(t)$ be a solution of $(*)$. We have

$$x'' = f(t, x), \quad x(0) = x_0, \quad x'(0) = y_0.$$

As $f$ is continuous, using the First Fundamental Theorem of Calculus and changing the order of integration, we can write

$$x'(t) = x'(0) + \int_0^t f(s, x(s)) ds \implies x(t) = x(0) + y_0 t +$$

$$\int_0^t \left( \int_0^\tau f(s, x(s)) ds \right) d\tau,$$

$$\implies x(t) = x_0 + y_0 t +$$

$$\int_0^t \left( \int_s^t d\tau \right) f(s, x(s)) ds,$$

$$\implies x(t) = x_0 + y_0 t +$$

$$\int_0^t (t - s) f(s, x(s)) ds,$$

which means $x(t)$ is a solution of $(**)$.

Now, let $x(t)$ be a solution of $(**)$. We can write

$$x(t) = x_0 + y_0 t + \int_0^t (t - s) f(s, x(s)) ds,$$

implying that

$$x(t) = x_0 + y_0 t + t \int_0^t f(s, x(s)) ds - \int_0^t s f(s, x(s)) ds.$$

Again, using the First Fundamental Theorem of Calculus, we can write

$$x'(t) = y_0 + \int_0^t f(s, x(s)) ds + t f(t, x(t)) - t f(t, x(t)),$$

which obtains

$$x'(t) = y_0 + \int_0^t f(s, x(s)) ds.$$

Taking derivative of both sides of the above yields

$$x''(t) = f(t, x(t)).$$

It is obvious that $x(0) = x_0$ and $x'(0) = y_0$. Therefore, $x(t)$ satisfies $(*)$, which is what we want.                                                           ∎

**4. First Solution:** See Solution 1 of 2.5.3.

**Second solution:** Just as in the first solution, we settle the problem by proving the polynomial identity

$$(1 + x)^n = \sum_{k=0}^n C_n^k x^k, \qquad (*)$$

where $C_n^k = \dfrac{n!}{k!(n - k)!}$. Proceed by induction on $n$. If $n = 1$, the assertion is easy. Suppose that $(*)$ holds for $n$. To prove that $(*)$ holds for $n + 1$, we can write

$$(1+x)^{n+1} = (1+x)^n(1+x) = (1+x)\sum_{k=0}^{n} C_n^k x^k$$

$$= \sum_{k=0}^{n} C_n^k x^k + \sum_{k=0}^{n} C_n^k x^{k+1}$$

$$= 1 + \sum_{k=1}^{n} C_n^k x^k + \sum_{k=1}^{n} C_n^{k-1} x^k + x^{n+1}$$

$$= 1 + \sum_{k=1}^{n} (C_n^k + C_n^{k-1}) x^k + x^{n+1}.$$

A straightforward calculation shows that $C_n^k + C_n^{k-1} = C_{n+1}^k$, from which, we easily obtain

$$(1+x)^{n+1} = \sum_{k=0}^{n} C_{n+1}^k x^k,$$

proving the induction assertion, which is what we want.

**Third Solution:** For $n, k \in \mathbb{N}$ with $k \leq n$, use $P_n^k$ to denote $n(n-1)\cdots(n-k+1)$. We have $P_n^k = k! C_n^k$ for each $n, k \in \mathbb{N}$ with $k \leq n$, where $C_n^k = \dfrac{n!}{k!(n-k)!}$. Thus, to prove the assertion, we need to show that $P_n^k$ is divisible by $k!$ for all $n, k \in \mathbb{N}$ with $k \leq n$. We proceed by induction on $n$. If $n = 1$, there is nothing to prove. Suppose that $P_n^k$ is divisible by $k!$ for all $k \leq n$. We need to show that $P_{n+1}^k$ is divisible by $k!$ for all $k \leq n+1$. If $k = n+1$, then $P_{n+1}^k = k!$, in which case, the assertion is trivial. If $k \leq n$, we can write

$$P_{n+1}^k = (n+1)n(n-1)\cdots(n+1-k+1)$$

$$= (n-k+1+k)n(n-1)\cdots(n-k+2)$$

$$= kP_n^{k-1} + P_n^k.$$

By the induction hypothesis, $P_n^{k-1}$ and $P_n^k$ are divisible by $(k-1)!$ and $k!$, respectively. Thus, $P_{n+1}^k = kP_n^{k-1} + P_n^k$ is divisible by $k!$, which is what we want.

**Fourth solution:** As we know, it suffices to prove that $C_n^k$ is an integer. To this end, we need, first of all, the following lemma, which is due to Legendre (1808).

**Lemma.** *Let $p$ a prime number and $n \in \mathbb{N}$. Then, the largest power of $p$ dividing $n!$ is equal to*

$$p^{\sum_{i=1}^{+\infty}\left[\frac{n}{p^i}\right]},$$

*where $[.]$ denotes the integer part function.*

**Proof.** As $n! = 1 \times 2 \times \cdots \times n$, without loss of generality, we may assume that $p \leq n$. Among the consecutive integers $1, \ldots, n$ those that are divisible by $p$ are

$$p, \ 2p, \ \ldots, \ m_1 p,$$

where $m_1 = \left[\frac{n}{p}\right]$. So, there are $\left[\frac{n}{p}\right]$ such integers. Likewise, there are $\left[\frac{n}{p^i}\right]$ consecutive integers from 1 up to $n$ that are divisible by $p^i$ (note that if $p^i > n$, then the number of integers between and including 1 and $n$ that are divisible by $p^i$ is zero). Note that any multiple of $p^k$ which is not a multiple of $p^{k+1}$ contributes as much as $k$ to the largest power of $p$ that divides $n!$. Therefore, the largest power of $p$ that divides $n!$ is equal to

$$p^{\sum_{i=1}^{+\infty} \left[\frac{n}{p^i}\right]},$$

proving the lemma. $\qquad\square$

Now to prove the assertion, note first that

$$[a+b] \geq [a] + [b],$$

for all $a, b \in \mathbb{R}$. Thus, we can write

$$\left[\frac{n}{p^i}\right] = \left[\frac{n-k+k}{p^i}\right] \geq \left[\frac{n-k}{p^i}\right] + \left[\frac{k}{p^i}\right],$$

for all $i \in \mathbb{N}$. Adding up these inequalities, we obtain

$$\sum_{i=1}^{+\infty} \left[\frac{n}{p^i}\right] \geq \sum_{i=1}^{+\infty} \left[\frac{n-k}{p^i}\right] + \sum_{i=1}^{+\infty} \left[\frac{k}{p^i}\right].$$

This means that the largest power of any prime $p$ that divides $n!$ is divisible by the largest power of $p$ that divides $k!(n-k)!$, proving the assertion. $\qquad\blacksquare$

## 2.11. Eleventh Competition

**2.11.1. Analysis. 1.** To prove the assertion by contradiction, suppose that the set of zeros of $f$, i.e.,

$$Z := \{x \in [0,1] | f(x) = 0\} = f^{-1}(\{0\})$$

is infinite. As the interval $[0,1]$ is compact, it follows that there exists a sequence $(x_n)_{n=1}^{+\infty}$ of distinct points in $Z$ such that $\lim_n x_n = x_0$ for some $x_0 \in [0,1]$. This implies $f(x_0) = 0$ because $f$ is continuous. But the function $f$ is differentiable at $x_0$. So we can write

$$f'(x_0) = \lim_n \frac{f(x_n) - f(x_0)}{x_n - x_0} = \lim_n \frac{0}{x_n - x_0} = 0.$$

That is, $f'(x_0) = 0$. On the other hand, $f(x_0) = 0$, implying that $x_0$ is a common zero of $f$ and $f'$, which is in contradiction with the hypothesis of the problem. Thus, $Z$ is finite, which is what we want. $\qquad\blacksquare$

**2.** (a) To show that the sequence $(f_n)_{n=1}^{+\infty}$ is pointwise convergent to a function $f$ on $[1, e^{\frac{1}{e}}]$, it suffices to prove that the sequence $(f_n(a))_{n=1}^{+\infty}$ is increasing and bounded from the above for all $a \in [1, e^{\frac{1}{e}}]$. To this end, letting $a \in [1, e^{\frac{1}{e}}]$ be arbitrary, we prove these assertions by induction on $n$. We have

$$f_1(a) = a, \quad f_{n+1}(a) = a^{f_n(a)} \ \forall n \in \mathbb{N}.$$

First, by induction on $n$, we show that $1 \leq f_n(a) < e$ for all $n \in \mathbb{N}$. If $n = 1$, then $1 \leq f_1(a) = a \leq e^{\frac{1}{e}} < e$. Assuming that $1 \leq f_n(a) < e$, we can write

$$1 \leq f_{n+1}(a) = a^{f_n(a)} < (e^{\frac{1}{e}})^e = e.$$

This proves $1 \leq f_n(a) < e$ for all $n \in \mathbb{N}$ by induction. We note that the function $g : \mathbb{R} \to \mathbb{R}$ defined by $g(x) = a^x$ is increasing whenever $a > 1$. To prove that the sequence $(f_n(a))_{n=1}^{+\infty}$ is increasing, again we use induction on $n$. If $n = 1$, we can write

$$f_2(a) = g(a) > g(1) = f_1(a).$$

Now assuming that $f_{n+1}(a) > f_n(a)$, we have

$$f_{n+2}(a) = g(f_{n+1}(a)) > g(f_n(a)) = f_{n+1}(a).$$

This proves that the sequence $(f_n(a))_{n=1}^{+\infty}$ is increasing. Thus, the sequence $(f_n)_{n=1}^{+\infty}$ is pointwise convergent to a function $f : [1, e^{\frac{1}{e}}] \to \mathbb{R}$. In view of the continuity of $g$, we can write

$$f(a) = \lim_n f_{n+1}(a) = \lim_n g(f_n(a)) = g(\lim_n f_n(a)) = g(f(a)) = a^{f(a)},$$

implying that $f(a) = a^{f(a)}$ and $1 \leq f(a) \leq e$ for all $a \in [1, e^{\frac{1}{e}}]$, as desired.

(b) Define the function $g : [1, e] \to [1, e^{\frac{1}{e}}]$ by $g(x) = e^{\frac{\ln x}{x}}$. By inspecting the derivative of $g$, one can easily verify that the function $g : [1, e] \to [1, e^{\frac{1}{e}}]$ is one-to-one and onto. Moreover, $g$ is continuous and so is its inverse $g^{-1} :$ $[1, e^{\frac{1}{e}}] \to [1, e]$; in fact $g$ and, hence its inverse, $g^{-1}$ are differentiable. From $f(x) = x^{f(x)}$, we easily obtain $g(f(x)) = x$ for all $x \in [1, e^{\frac{1}{e}}]$. This shows that $f = g^{-1}$. Thus, $f$ is one-to-one and continuous because so is $g^{-1}$.

Now, the functional sequence $(f_n)_{n=1}^{+\infty}$ of continuous functions converges pointwise to the continuous function $f$ on the compact interval $[1, e^{\frac{1}{e}}]$ and moreover $f_n(x) < f_{n+1}(x)$ for all $x \in [1, e^{\frac{1}{e}}]$. It thus follows from Dini's Theorem that the sequence $(f_n)_{n=1}^{+\infty}$ converges uniformly to the function $f$ on $[1, e^{\frac{1}{e}}]$, which is what we want. ∎

**3.** Set

$$M = \sup\{|\phi_n''(x)| : n \in \mathbb{N}, x \in [-1, 1]\}.$$

It follows from the Mean Value Theorem that there exists a $c$ between $0$ and $x$ such that

$$\phi_n'(x) - \phi_n'(0) = x\phi_n''(c),$$

from which, we obtain

$$|\phi_n'(x)| = |1 + x\phi_n''(c)| \leq M + 1,$$

for all $n \in \mathbb{N}$ and $x \in [-1, 1]$. Using integration by parts, we get

$$\left| \int_{-1}^1 \phi_n(t) \cos n\pi t \, dt \right| \leq \frac{1}{n\pi} \int_{-1}^1 |\phi_n'(t) \sin n\pi t| \, dt$$

$$\leq \frac{M+1}{n\pi} \int_{-1}^1 |\sin n\pi t| \, dt = \frac{M+1}{n\pi^2},$$

implying that

$$a_n = \frac{1}{n}\left|\int_{-1}^{1}\phi_n(t)\cos n\pi t dt\right| \le \frac{M+1}{n^2\pi^2},$$

for all $n \in \mathbb{N}$. This, in view of the Limit Comparison Test for series, implies that $\sum_{n=1}^{+\infty} a_n$ is convergent, which is what we want. ∎

### 2.11.2. Algebra. 1. Set

$$I = \{xu - x : x \in R\}.$$

It is obvious that $I$ is a two-sided ideal of $R$. It thus follows from the hypothesis that $I = \{0\}$ or $I = R$. If $I = \{0\}$, we see that $xu = x$, yielding $xu = ux = x$ for all $x \in R$, which means $u = 1_R$. By showing that the case $I = R$ is impossible, we finish the proof. Suppose to the contrary that $I = R$. Thus, there exists an $x_0 \in R$ such that $u = x_0u - x_0$. As $u^2 = u$, we can write

$$u = u^2 = (x_0u - x_0)u = x_0u - x_0u = 0,$$

and hence

$$x = ux = 0x = 0,$$

for all $x \in R$. This means $R = \{0\}$, implying that $R$ has only one ideal, which is a contradiction. Therefore, $I = \{0\}$, finishing the proof. ∎

**2.** To prove the assertion by contradiction, suppose that $\mathbb{A}$ is a finite extension of $\mathbb{Q}$ so that $[\mathbb{A} : \mathbb{Q}] = n$ for some $n \in \mathbb{N}$. Pick a prime number $p$, e.g., $p = 2$, and note that, by Eisenstein's Criterion, the monic polynomial $x^{n+1} - p$ is irreducible over $\mathbb{Z}$ and hence over $\mathbb{Q}$. If $\alpha$ is one of the roots of the equation $x^{n+1} - p = 0$, we have

$$\mathbb{Q} \subseteq \mathbb{Q}(\alpha) \subseteq \mathbb{A}.$$

On the other hand, $[\mathbb{A} : \mathbb{Q}] = n < n + 1 = [\mathbb{Q}(\alpha) : \mathbb{Q}]$, which is obviously impossible. Thus, $\mathbb{A}$ is not a finite extension of $\mathbb{Q}$, which is what we want. ∎

**3.** First we need to recall that *if $G$ is a finite p-group, where $p$ is a prime number, then the center of $G$, denoted by $Z(G)$, is nontrivial, i.e., $|Z(G)| > 1$.* To prove this, first recall that the class equation of $G$ is

$$|G| = |Z(G)| + \sum_{i=1}^{k}[G : C_G(x_i)],$$

where $\{x_1, \ldots, x_k\}$ is a maximal set of nonconjugate elements of $G \setminus Z(G)$ and $C_G(x_i)$ is the centralizer of $x_i$ in $G$ $(1 \le i \le k)$. Note that since $x_i$'s are in $G \setminus Z(G)$, $C_G(x_i)$'s are proper subgroups of $G$, implying that whose orders are all powers of $p$, and hence so are $[G : C_G(x_i)]$'s because $G$ is a finite $p$-group. Consequently, from the class equation of $G$, in view of $|G| = p^n$ for some $n \in \mathbb{N}$, we conclude that $p$ divides $|Z(G)|$, yielding $|Z(G)| > 1$, as desired.

We now prove the assertion. It follows from the hypothesis that any two elements of $G$ which are different from the identity share the same orders. Now suppose that $p$ is a prime dividing $|G|$. It follows from Cauchy's Theorem that there exists an $x \in G$ whose order is $p$. Hence, all of the elements of $G$ but $e$ have order $p$, whence the order of $G$ is a power of $p$. That is, $G$

is a $p$-group. Consequently, the center of $G$ is nontrivial, i.e., $Z(G) \neq \{e\}$. Since $\alpha(Z(G)) = Z(G)$ for all $\alpha \in \text{Aut}(G)$, from the hypothesis, we see that $Z(G) = G$. That is, $G$ is abelian. Therefore, by the Fundamental Theorem of finite abelian groups, we obtain

$$G \cong \mathbb{Z}_p \oplus \cdots \oplus \mathbb{Z}_p,$$

which is what we want. ∎

**2.11.3. General. 1.** Note that for an $A \in M_2(\mathbb{Z})$, the evenness of $\det(A)$ is equivalent to saying that $\det(A') = 0$, where $A' \in M_2(\mathbb{Z}_2)$ is the matrix whose entries are obtained from those of $A$ viewed as elements of $\mathbb{Z}_2$. With that in mind, the desired probability is equal to the probability that a randomly chosen matrix $A' \in M_2(\mathbb{Z}_2)$ has determinant zero. Obviously, $|M_2(\mathbb{Z}_2)| = 2^4$. On the other hand, the matrices in $M_2(\mathbb{Z}_2)$ with zero determinants are as follows

$$\begin{pmatrix} 0 & 0 \\ 0 & 0 \end{pmatrix}, \begin{pmatrix} 0 & 1 \\ 0 & 1 \end{pmatrix}, \begin{pmatrix} 1 & 0 \\ 1 & 0 \end{pmatrix}, \begin{pmatrix} 0 & 0 \\ 1 & 1 \end{pmatrix}, \begin{pmatrix} 1 & 1 \\ 0 & 0 \end{pmatrix},$$

$$\begin{pmatrix} 1 & 1 \\ 1 & 1 \end{pmatrix}, \begin{pmatrix} 0 & 0 \\ 1 & 0 \end{pmatrix}, \begin{pmatrix} 1 & 0 \\ 0 & 0 \end{pmatrix}, \begin{pmatrix} 0 & 1 \\ 0 & 0 \end{pmatrix}, \begin{pmatrix} 0 & 0 \\ 0 & 1 \end{pmatrix}.$$

Therefore, the desired probability is equal to $\frac{10}{16} = \frac{5}{8}$. ∎

**2.** Plainly, the equation is equivalent to the following

$$4x^3 - 3x = p,$$

where $p = -4c$. Now, letting $x = \cos z$, where $z \in \mathbb{C}$, the above equation becomes equivalent to the following equation

$$\cos 3z = p. \qquad (*)$$

As the original equation has a root $x_0$ in the closed interval $[-1, 1]$, it follows that $p \in [-1, 1]$, whence there exists $z_0 \in \mathbb{R}$ such that $x_0 = \cos z_0$. It is now obvious that $z_0$, $z_0 + \frac{2\pi}{3}$, and $z_0 + \frac{4\pi}{3}$ satisfy $(*)$. Thus, the corresponding $x$'s which are $\cos z_0, \cos(z_0 + \frac{2\pi}{3}), \cos(z_0 + \frac{4\pi}{3}) \in [-1, 1]$ are all of the roots of the original equation, proving the assertion. ∎

**3.** In view of the hypotheses of the problem, we see that $C$, endowed with the symmetric difference of sets, denoted by $\Delta$, which is defined by

$$A \Delta B = (A \setminus B) \cup (B \setminus A)$$

forms a subgroup of $\mathcal{P}(X)$, where $\mathcal{P}(X)$ denotes the power set of $X$ which obviously forms a group under $\Delta$ itself. As is well-known, $|\mathcal{P}(X)| = 2^n$, where $n = |X|$. Thus, it follows from Lagrange's Theorem that $|C|$ divides $|\mathcal{P}(X)| = 2^n$, yielding $|C| = 2^k$ for some $k \leq n$, which is what we want. ∎

## 2.12. Twelfth Competition

**2.12.1. Analysis. 1. First Solution:** Suppose that $f(x) = f(y)$ for some $x, y \in \mathbb{R}$. It follows that $M|x - y| \leq 0$, implying $x = y$. That is, $f$ is one-to-one. Now, as $f$ is continuous and one-to-one, we see from Problem 2 of 1.9.1 that $f$ is strictly monotonic. Without loss of generality, we may assume that $f$ is strictly increasing. It now easily follows from the hypothesis that $\lim_{x \to -\infty} f(x) = -\infty$ and $\lim_{x \to +\infty} f(x) = +\infty$. This, in view of the Intermediate Value Theorem, implies that $f$ is onto, which is what we want.

**Second Solution:** Just as we saw in the first solution, the function $f$ is one-to-one and strictly monotonic, and hence the inverse of $f$, i.e., $f^{-1} : f(\mathbb{R}) \to \mathbb{R}$ is continuous. This implies that $f$ is both an open and closed map. Now, since $\mathbb{R}$ is both open and closed in $\mathbb{R}$, we see that so is $f(\mathbb{R})$, yielding $f(\mathbb{R}) = \mathbb{R}$ because $\mathbb{R}$ is connected. This proves the assertion. $\blacksquare$

**2.** "$\Longrightarrow$" Since $\sum_{n=1}^{+\infty} f_n$ converges uniformly on $S$, it is uniformly Cauchy on $S$, and hence for given $\varepsilon > 0$, there exists an $N > 0$ such that

$$\left| \sum_{k=m}^{n} f_k \right| < \frac{\varepsilon}{2},$$

for all $m, n \geq N$. Now, letting $m = \left[ \frac{n}{2} \right] > N$, it follows that $n > m \geq N$, from which, in view of the fact that $(f_n)_{n=1}^{+\infty}$ is a decreasing sequence of nonnegative functions on $S$, we see that

$$\sum_{k=\left[\frac{n}{2}\right]}^{n} f_k < \frac{\varepsilon}{2}$$

$$\Longrightarrow \frac{n f_n}{2} \leq \left( n - \left[ \frac{n}{2} \right] + 1 \right) f_n \leq \sum_{k=\left[\frac{n}{2}\right]}^{n} f_k < \frac{\varepsilon}{2},$$

for all $n \geq 2N + 1$. This obviously implies $n f_n < \varepsilon$ for all $n \geq 2N + 1$. That is, the sequence $(n f_n)_{n=1}^{+\infty}$ converges uniformly to zero on $S$. On the other hand,

$$\sum_{n=1}^{+\infty} n(f_n - f_{n+1}) = \sum_{n=1}^{+\infty} \Big( \big( n f_n - (n+1) f_{n+1} \big) + f_{n+1} \Big). \qquad (*)$$

But $(n+1) f_{n+1}$ converges uniformly to zero on $S$. Hence, the telescopic series $\sum_{n=1}^{+\infty} \big( n f_n - (n+1) f_{n+1} \big)$ converges uniformly to $f_1$ on $S$. This together with the hypothesis and $(*)$ implies that $\sum_{n=1}^{+\infty} n(f_n - f_{n+1})$ converges uniformly to $\sum_{n=1}^{+\infty} f_n$ on $S$, which is what we want.

"$\Longleftarrow$" Note that as pointed out in the footnote of the problem in Section 1.12.1, for this implication we need to assume further that the sequence $(f_n)_{n=1}^{+\infty}$ converges uniformly to zero on the set $S$. That is because for the numerical sequence $(f_n)_{n=1}^{+\infty}$, where $f_n = 2 + \frac{1}{n^2}$, the series $\sum_{n=1}^{+\infty} n(f_n - f_{n+1})$ converges but $\sum_{n=1}^{+\infty} f_n$ is divergent.

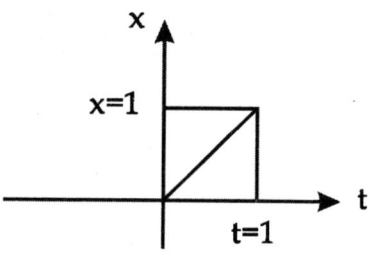

Figure 10

Now, since $(f_n)_{n=1}^{+\infty}$ converges uniformly to zero on $S$, it follows that the telescopic series $\sum_{n=1}^{+\infty}(f_n - f_{n+1})$ converges uniformly to $f_1$ on $S$. We can write

$$nf_n = n\sum_{k=n}^{+\infty}(f_k - f_{k+1}) = \sum_{k=n}^{+\infty} n(f_k - f_{k+1}) \leq \sum_{k=n}^{+\infty} k(f_k - f_{k+1}), \qquad (**)$$

for all $n \in \mathbb{N}$. By the hypothesis, the series $\sum_{n=1}^{+\infty} n(f_n - f_{n+1})$ is uniformly Cauchy on $S$. This, in view of $(**)$, easily implies that the sequence $(nf_n)_{n=1}^{+\infty}$ converges uniformly to zero on $S$. Thus, the telescopic series $\sum_{n=1}^{+\infty} \left(nf_n - (n+1)f_{n+1}\right)$ converges uniformly to $f_1$ on $S$. We can write

$$\sum_{n=1}^{+\infty} f_{n+1} = \sum_{n=1}^{+\infty} \left(n(f_n - f_{n+1}) - \left(nf_n - (n+1)f_{n+1}\right)\right),$$

implying that $\sum_{n=1}^{+\infty} f_{n+1}$ converges uniformly on $S$, for so do the series $\sum_{n=1}^{+\infty} n(f_n - f_{n+1})$ and $\sum_{n=1}^{+\infty} \left(nf_n - (n+1)f_{n+1}\right)$ on $S$. Therefore, $\sum_{n=1}^{+\infty} f_{n+1}$, and hence $\sum_{n=1}^{+\infty} f_n$, converges uniformly on $S$. Moreover,

$$\sum_{n=1}^{+\infty} f_n = \sum_{n=1}^{+\infty} n(f_n - f_{n+1}).$$

So the proof is complete. ■

**3. First solution:** Using integration by parts, we can write

$$\int_0^1 \left(\int_x^1 g(t)dt\right) dx = x\int_x^1 g(t)dt\Big|_{x=0}^{1} - \int_0^1 xd\left(\int_x^1 g(t)dt\right)$$

$$= 0 + \int_0^1 xg(x)dx$$

$$= \int_0^1 xg(x)dx,$$

which is what we want.

**Second solution:** Changing the order of integration, we can write

$$\int_{x=0}^{1} \left( \int_{t=x}^{1} g(t)dt \right) dx \;=\; \int_{t=0}^{1} \left( \int_{x=0}^{t} g(t)dx \right) dt$$

$$=\; \int_{t=0}^{1} g(t) \left( \int_{x=0}^{t} dx \right) dt$$

$$=\; \int_{0}^{1} tg(t)dt,$$

which is what we want.                                                  ∎


**2.12.2. Algebra. 1.** By Sylow's Third Theorem, the number of subgroups of order $p$ is equal to $1 + kp$ for some $k \in \mathbb{N} \cup \{0\}$ and that $1 + kp$ divides $|G|$. That is, $1 + kp | 2p$. This yields $1 + kp = 1$, implying $k = 0$. Thus, $G$ has only one Sylow $p$-subgroup. Therefore, $G$ has only one subgroup of order $p$.

Using Sylow's Third Theorem again, we see that the number of subgroups of order 2 is equal to $1 + 2k$ for some $k \in \mathbb{N} \cup \{0\}$ and that $1 + 2k$ divides $|G|$. That is, $1 + 2k | 2p$, from which we obtain $1 + 2k = 1$ or $1 + 2k = p$. Therefore, $G$ has either one and only one subgroup of order 2 or it has $p$ subgroups of order 2.

Now, suppose that $G$ has only one subgroup of order 2. Let $H$ and $K$ be the only Sylow 2-subgroup and Sylow $p$-subgroup, respectively. We have $|H| = 2$ and $|K| = p$. This implies $|H \cap K| = 1$. In other words, $H \cap K = e$. Pick $h \in H$ and $k \in K$ such that $\text{ord}(h) = 2$ and $\text{ord}(k) = p$. As $H \cap K = e$, we see that $hk = kh$ because $hkh^{-1}k^{-1} \in H \cap K$, and hence $\text{ord}(hk) = 2p$, for $\text{ord}(h) = 2$, $\text{ord}(k) = p$, and $\gcd(2, p) = 1$ (for a more detailed proof of $\text{ord}(hk) = 2p$, see the lemma presented in Solution 1 of 2.25.2). This implies that the cyclic subgroup generated by $hk$ is $G$. Therefore, $G$ is a cyclic group, which is what we want.                                    ∎


**2.** Note first that $G$ has no elements of even order, in particular of order 2, because $G$ is of odd order. We shall show that the map $\psi : G \to G$ defined by $\psi(g) = g^2$ is an automorphism of $G$. To see that $\psi$ is one-to-one, suppose $\psi(a) = \psi(b)$ for some $a, b \in G$. It easily follows that $(ab^{-1})^2 = e$, where $e$ is the identity element of $G$. But $\text{ord}(ab^{-1}) \neq 2$. So $ab^{-1} = e$, yielding $a = b$. That is, $\psi$ is one-to-one. It now follows that $\psi$ is an automorphism of $G$ because $G$ is finite and abelian. Thus, for an arbitrary $g \in G$, there exists an $a \in G$ such that $g = a^2$. Set $x = a\phi(a)$ and $y = a\phi(a^{-1})$. Noting that $G$ is abelian, we can write

$$xy \;=\; \big(a\phi(a)\big)\big(a\phi(a^{-1})\big) = a^2\phi(aa^{-1}) = a^2 = g.$$

On the other hand,

$$\phi(x) \;=\; \phi\big(a\phi(a)\big) = \phi(a)\phi^2(a) = a\phi(a) = x,$$
$$\phi(y) \;=\; \phi\big(a\phi(a^{-1})\big) = \phi(a)\phi^2(a^{-1}) = a^{-1}\phi(a) = y^{-1},$$

as desired. To show that such $x, y$ are unique, suppose that for $g \in G$, there are $x, y, z, t \in G$ such that $g = xy = zt$, $\phi(x) = x$, $\phi(z) = z$, $\phi(y) = y^{-1}$, and $\phi(t) = t^{-1}$. We have

$$\phi(g) = \phi(xy) = \phi(zt) \implies \phi(x)\phi(y) = \phi(z)\phi(t)$$
$$\implies xy^{-1} = zt^{-1} \implies xyy^{-2} = zt^{-1}$$
$$\implies zty^{-2} = zt^{-1} \implies y^2 = t^2$$
$$\implies \psi(y) = \psi(t) \implies y = t,$$

which, in view of $xy = zt$, yields $x = z$. Therefore, any element $g \in G$ can, uniquely, be written as $g = xy$, where $\phi(x) = x$ and $\phi(y) = y^{-1}$, finishing the proof. ∎

**3.** As every Boolean ring is commutative, it suffices to show that $R$ is a Boolean ring. To this end, pick $x \in R$ such that $x \neq 0, 1$. As $\operatorname{ch}(R) = 2$, there exists $x_1 \in R \setminus \{0, 1\}$ such that $x = 1 + x_1$. It follows from the hypothesis that $xy = xy^2$ for all $y \in R$. In other words, $(1 + x_1)y = (1 + x_1)y^2$, from which, in view of $x_1 y = x_1 y^2$, we obtain $y = y^2$ for all $y \in R$. That is, $R$ is a Boolean ring, and hence commutative, which is what we want. ∎

**4.** First, we claim that every ascending chain of ideals of $R$ terminates. To this end, suppose

$$I_1 \subseteq I_2 \subseteq \cdots$$

is an ascending chain of ideals of $R$. Set $I = \bigcup_{n=1}^{+\infty} I_n$. It is easily seen that $I$ is an ideal of $R$, and hence there exists an $x_0 \in R$ such that $I = \langle x_0 \rangle = x_0 R$. As $x \in I$, it follows that there exists an $N \in \mathbb{N}$ such that $x_0 \in I_N$. We can write

$$I_N \subseteq I = \bigcup_{k=1}^{+\infty} I_k = x_0 R \subseteq I_N \implies I_N = \bigcup_{k=1}^{+\infty} I_k$$
$$\implies \forall n \in \mathbb{N} : I_n \subseteq I_N.$$

On the other hand, $I_N \subseteq I_n$ for all $n \geq N$, from which, in view of the above inclusion, we obtain $I_N = I_n$ for all $n \geq N$. This proves the claim. Now, suppose that $f : R \to R$ is a surjective homomorphism. To see that $f$ is one-to-one, we note that

$$\ker f \subseteq \ker f^2 \subseteq \ker f^3 \subseteq \cdots$$

is an ascending chain of ideals of $R$. It follows from the claim that there exists an $N \in \mathbb{N}$ such that $\ker f^n = \ker f^N$ for all $n \geq N$. In particular, $\ker f^{N+1} = \ker f^N$. Suppose $x \in \ker f$ is arbitrary so that $f(x) = 0$. As $f$ is onto, so is $f^n$ for all $n \in \mathbb{N}$, and hence there exists an $x_0 \in R$ such that $x = f^N(x_0)$. Consequently, $f^{N+1}(x_0) = 0$, yielding $x_0 \in \ker f^{N+1} = \ker f^N$. Therefore, $x = f^N(x_0) = 0$, implying $\ker f = \{0\}$. That is, $f$ is one-to-one, which is what we want. ∎

**5. First solution:** Let $\lambda_i$'s $(1 \leq i \leq 3)$ be the eigenvalues of the matrix $A$ in the algebraic closure of $F$. As $A$ is invertible, $\lambda_i^{-1}$'s $(1 \leq i \leq 3)$ are the eigenvalues of $A^{-1}$. If $f$ is the characteristic polynomial of $A$, we have

$$
\begin{aligned}
f &= (x - \lambda_1)(x - \lambda_2)(x - \lambda_3) \\
&= x^3 - (\lambda_1 + \lambda_2 + \lambda_3)x^2 + (\lambda_1\lambda_2 + \lambda_1\lambda_3 + \lambda_2\lambda_3)x - \lambda_1\lambda_2\lambda_3.
\end{aligned}
$$

By the hypothesis,

$$
\begin{aligned}
\operatorname{tr}(A) &= \lambda_1 + \lambda_2 + \lambda_3 = 0, \\
\operatorname{tr}(A^{-1}) &= \lambda_1^{-1} + \lambda_2^{-1} + \lambda_3^{-1} = 0, \\
\det(A) &= \lambda_1\lambda_2\lambda_3 = 1.
\end{aligned}
$$

On the other hand, $\lambda_1\lambda_2 + \lambda_1\lambda_3 + \lambda_2\lambda_3 = \lambda_1^{-1} + \lambda_2^{-1} + \lambda_3^{-1} = 0$. So, we must have $f(x) = x^3 - 1$. Now, by the Cayley-Hamilton Theorem, $A^3 = I$, which is what we want.

**Second solution:** Let $\lambda_i$'s $(1 \le i \le 3)$ be as in the first solution. We see that $\lambda_i$'s $(1 \le i \le 3)$ are distinct because $\lambda_1 + \lambda_2 + \lambda_3 = 0 = \lambda_1^{-1} + \lambda_2^{-1} + \lambda_3^{-1}$ and $\lambda_1\lambda_2\lambda_3 = 1$. With that in mind, suppose $f(x) = x^3 + ax^2 + bx + c$ is the characteristic polynomial of $A$. We have $a = \operatorname{tr}(A) = 0$ and $c = \det(A) = -1$. By the Cayley-Hamilton Theorem, $A^3 + bA - I = 0$, yielding $(A^{-1})^3 - b(A^{-1})^2 - I = 0$. As $\lambda_i$'s $(1 \le i \le 3)$ are distinct, the characteristic polynomial of $A^{-1}$, denoted by $g$, is equal to $g(x) = x^3 - bx^2 - 1$. But $b = \operatorname{tr}(A^{-1}) = 0$. Therefore, $(A^{-1})^3 - I = 0$, implying $A^3 = I$, which is what we want.  ∎

**6.** It is obvious that $\alpha$ and $\beta$ are roots of the polynomial

$$
f(x) = p! + p!x + \frac{p!}{2!}x^2 + \cdots + x^p,
$$

which is irreducible over $\mathbb{Q}$ by Eisenstein's Criterion. Thus, $f$ is the minimal polynomial of $\alpha$ and $\beta$ over $\mathbb{Q}$. We prove the assertion by way of contradiction.

First suppose that $\alpha - \beta = r \in \mathbb{Q}$. We have $f(\beta + r) = f(\alpha) = 0$. That is, $\beta$ is a root of the polynomial $f(x + r)$ whose coefficients are in $\mathbb{Q}$. But $f(x)$ is the minimal polynomial of $\beta$ over $\mathbb{Q}$. It follows that $f(x)|f(x + r)$. As $f(x)$ and $f(x + r)$ are both monic polynomials of the same degree, we obtain $f(x + r) = f(x)$. Consequently, if $\alpha_1, \dots, \alpha_p$ are all of the roots of $f(x) = 0$, then so are $\alpha_1 + r, \dots, \alpha_p + r$. In particular, we must have

$$
\begin{aligned}
(\alpha_1 + r) + \cdots + (\alpha_p + r) = \alpha_1 + \cdots + \alpha_p &\implies pr = 0 \\
&\implies r = 0 \\
&\implies \alpha = \beta,
\end{aligned}
$$

which is a contradiction. Thus, $\alpha - \beta = r \notin \mathbb{Q}$, as desired.

Now, assume that $\alpha + \beta = r \in \mathbb{Q}$. We have $-f(r - \beta) = f(\alpha) = 0$. That is, $\beta$ is a root of the polynomial $-f(r - x)$ whose coefficients are in $\mathbb{Q}$. But $f(x)$ is the minimal polynomial of $\beta$ over $\mathbb{Q}$. Thus, $f(x)| - f(r - x)$. As $f(x)$ and $-f(r - x)$ are both monic polynomials of the same degree, we obtain $-f(r - x) = f(x)$. Consequently, letting $x = \frac{r}{2}$, we obtain $-f(\frac{r}{2}) = f(\frac{r}{2})$, which yields $f(\frac{r}{2}) = 0$. Hence, $f(x)$ is divisible by $x - \frac{r}{2}$, which is in contradiction with the fact that $f(x)$ is irreducible over $\mathbb{Q}$. Therefore, $\alpha + \beta = r \notin \mathbb{Q}$, as desired.

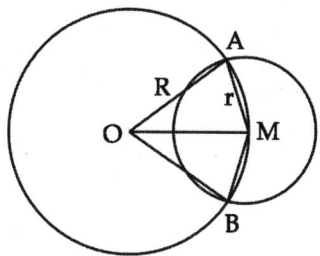

Figure 11

Next, suppose that $\alpha\beta = r \in \mathbb{Q}$. Obviously, $\alpha, \beta \neq 0$. We can write $f(\frac{r}{\beta}) = f(\alpha)$. It thus follows that $\beta$ is a root of the polynomial $\frac{x^p}{p!}f(\frac{r}{x})$ whose coefficients are in $\mathbb{Q}$. Again, we obtain $f(x)\big|\frac{x^p}{p!}f(\frac{r}{x})$, from which, as $f(x)$ and $\frac{x^p}{p!}f(\frac{r}{x})$ are both monic polynomials of the same degree, we see that $\frac{x^p}{p!}f(\frac{r}{x}) = f(x)$. There are two cases to consider: (i) $r > 0$ and (ii) $r < 0$. If $r > 0$, substituting $x = \sqrt{r}$ into $\frac{x^p}{p!}f(\frac{r}{x}) = f(x)$, we obtain $f(\sqrt{r}) = 0$. If $\sqrt{r} \in \mathbb{Q}$, then $f(x)$ is divisible by $x - \sqrt{r}$, which is in contradiction with the irreducibility of $f$ over $\mathbb{Q}$. And if $\sqrt{r} \notin \mathbb{Q}$, then $x^2 - r$ would be the minimal polynomial of $\sqrt{r}$ over $\mathbb{Q}$. From this, in view of $f(\sqrt{r}) = 0$, we see that $x^2 - r\big|f(x)$, implying that $f$ is not irreducible over $\mathbb{Q}$, a contradiction again. If $r < 0$, substituting $x = i\sqrt{-r}$ into $\frac{x^p}{p!}f(\frac{r}{x}) = f(x)$, we obtain $f(i\sqrt{-r}) = 0$. It follows that $x^2 + r$ is the minimal polynomial of $\sqrt{r}$ over $\mathbb{Q}$. From this, in view of $f(i\sqrt{-r}) = 0$, we see that $x^2 + r\big|f(x)$, implying that $f$ is not irreducible over $\mathbb{Q}$, which is again a contradiction. Therefore, $\alpha\beta = r \notin \mathbb{Q}$, which is what we want. It might be worth mentioning that one can show, in a similar fashion, that $\frac{\alpha}{\beta} \notin \mathbb{Q}$. ∎

**2.12.3. General. 1.** Let $R$ and $a$, respectively, be the radius of the circle $C$ centered at $O$ and the side length of the square whose center $M$ is on the circle. From the hypothesis, we see that $a^2 < \frac{\pi R^2}{2}$, whence $a\sqrt{2} \leq R\sqrt{\pi} < 2R$. That is, the diameter of the square is smaller than that of the circle.

It follows that the vertices of the square, with center $M$, are on the circle centered at $M$ with radius $r = \frac{a\sqrt{2}}{2} < R$. Let $A$ and $B$ be the points at which these two circle intersect one another. Consider two isosceles triangles $OAM$ and $OBM$. We have $AM = MB < OA = OB = OM$. It is plain that $\angle AMO > 60°$ and $\angle BMO > 60°$, and hence $\angle AMB > 120°$. Therefore, the arc $AB$ from the circle centered at $M$ is greater than $120°$, and hence some vertex of the square must be on this arc which itself lies inside the circle with radius $R$ centered at $O$. This proves the assertion. ∎

**2.** The answer is: yes, it does. To see this, we count the number of the desired answer sheets in a test in which $n$ questions are given. Suppose that there are $n$ questions in the test. Use 1 and 0 to denote a false and a true answer, respectively, and $A_n$ to denote the number of desired answer sheets. Then to any answer sheet, there corresponds an $n$-tuple of 0's and 1's. Let's call an answer sheet or $n$-tuple *admissible* if there are no consecutive 1's in it. First, for a given $k$ $(0 \le k \le n)$, we count the number of admissible $n$-tuples each of which contains $k$ ones and use $N_k$ to denote this number. To do so, note that in any such admissible $n$-tuple, there are $k$ ones and $n-k$ zeros and that we want to insert a zero between any two ones unless the zero occurs in the 1st position or the $n$th position. With that in mind, there are $n - k - 1 + 2 = n - k + 1$ positions and of those we need to choose $k$ positions to each of which assign a one. Thus, the number $N_k$ is equal to the number of ways of choosing $k$ positions out of $n - k + 1$ positions which is equal to $\binom{n-k+1}{k}$. Obviously, $k \le p := \left[\frac{n+1}{2}\right]$ and that we have

$$
\begin{aligned}
A_n &= \sum_{k=0}^{p} N_k = \sum_{k=0}^{p} \binom{n-k+1}{k} \\
&= \binom{n+1}{0} + \binom{n}{1} + \cdots + \binom{n-p+1}{p}.
\end{aligned}
$$

It thus follows that the number of desired answer sheets in the test in which 15 questions are given is equal to

$$
A_{15} = \binom{16}{0} + \binom{15}{1} + \cdots + \binom{8}{8} = 1597.
$$

As 1700 people have participated in the test, there are necessarily two equal answer sheets. ∎

**3.** We have

$$
(1+i)^n = \sum_{k=0}^{n} C_n^k i^{n-k},
$$

where $i \in \mathbb{C}$ with $i^2 = -1$. On the other hand, $1 + i = \sqrt{2}\left(\cos \frac{\pi}{4} + i \sin \frac{\pi}{4}\right)$, and hence

$$
(1+i)^n = 2^{\frac{n}{2}} \left(\cos \frac{n\pi}{4} + i \sin \frac{n\pi}{4}\right).
$$

So we have

$$
\sum_{k=0}^{n} C_n^k i^{n-k} = 2^{\frac{n}{2}} \left(\cos \frac{n\pi}{4} + i \sin \frac{n\pi}{4}\right),
$$

whence

$$
\left(C_n^0 - C_n^2 + C_n^4 - \cdots\right) + i\left(C_n^1 - C_n^3 + C_n^5 - \cdots\right) = 2^{\frac{n}{2}} \cos \frac{n\pi}{4} + i2^{\frac{n}{2}} \sin \frac{n\pi}{4},
$$

form which, we obtain

$$
C_n^0 - C_n^2 + C_n^4 - \cdots = 2^{\frac{n}{2}} \cos \frac{n\pi}{4},
$$

$$
C_n^1 - C_n^3 + C_n^5 - \cdots = 2^{\frac{n}{2}} \sin \frac{n\pi}{4},
$$

which is what we want. ∎

## 2.13. Thirteenth Competition

**2.13.1. Analysis. 1. First solution:** As $f : \mathbb{R} \to \mathbb{R}$ is integrable on any closed interval, for all $x, y \in \mathbb{R}$, we can write

$$\int_0^x f(t+y)dt = \int_0^x f(t)dt + \int_0^x f(y)dt,$$

which easily yields

$$
\begin{aligned}
xf(y) &= \int_0^x f(t+y)dt - \int_0^x f(t)dt \\
&= \int_y^{x+y} f(t)dt - \int_0^x f(t)dt \\
&= \int_0^{x+y} f(t)dt - \int_0^x f(t)dt - \int_0^y f(t)dt,
\end{aligned}
$$

and hence $xf(y) = yf(x)$ for all $x, y \in \mathbb{R}$. Therefore, $\frac{f(x)}{x} = \frac{f(1)}{1} = f(1)$ for all $x \in \mathbb{R}$ with $x \neq 0$. This implies $f(x) = f(1)x$ for all $x \in \mathbb{R}$, as desired.

**Second solution:** As $f : \mathbb{R} \to \mathbb{R}$ is integrable, it follows from the lemma presented in Solution 1 of 2.5.1 that the function $f$ is continuous at infinitely many points of any closed interval of $\mathbb{R}$. On the other hand, $f(x+y) = f(x) + f(y)$ for all $x, y \in \mathbb{R}$ and hence $f(r) = rf(1)$ for all $r \in \mathbb{Q}$. From this, it is easily seen that $f$ is continuous on $\mathbb{R}$ and that $f(x) = xf(1)$ for all $x \in \mathbb{R}$, which is what we want. ∎

**2. First solution:** We need the following well-known lemma which is known as Dedekind's Extension of Abel's Theorem.

**Lemma.** *Let $(a_n)_{n=1}^{+\infty}$ and $(b_n)_{n=1}^{+\infty}$ be sequences of real numbers. If the series $\sum_{n=1}^{+\infty} a_n$ and $\sum_{n=1}^{+\infty} |b_n - b_{n+1}|$ converge, then so does the series $\sum_{n=1}^{+\infty} a_n b_n$.*
**Proof.** It is easily verified that

$$
\begin{aligned}
a_{n+1}b_{n+1} + \cdots + a_{n+p}b_{n+p} = \ & (s_{n+1} - s_n)(b_{n+1} - b_{n+2}) + \cdots + \\
& (s_{n+p-1} - s_n)(b_{n+p-1} - b_{n+p}) + \\
& (s_{n+p} - s_n)b_{n+p},
\end{aligned}
$$

where $s_n = \sum_{k=1}^{n} a_k$ and $n, p \in \mathbb{N}$. Note that as $\sum_{n=1}^{+\infty} |b_n - b_{n+1}|$ converges, so does the telescopic series $\sum_{n=1}^{+\infty} (b_n - b_{n+1})$, from which, we see that $\lim_n b_n$ exists, and hence the sequence $(b_n)_{n=1}^{+\infty}$ is bounded. Now, the above equality, together with the facts that $(s_n)_{n=1}^{+\infty}$ and $\left( \sum_{k=1}^{n} |b_k - b_{k+1}| \right)_{n=1}^{+\infty}$ are Cauchy and that $(b_n)_{n=1}^{+\infty}$ is bounded, implies that $\left( \sum_{k=1}^{n} a_k b_k \right)_{n=1}^{+\infty}$ is Cauchy, and hence the series $\sum_{n=1}^{+\infty} a_n b_n$ is convergent, which is what we want. □

To prove the assertion, letting $c_n = \frac{a_n}{b_n}$, we can write

$$\frac{a_n}{a_n + b_n} = \frac{c_n - c_n^2}{1 - c_n^2} = (c_n - c_n^2)\left(1 + \frac{c_n^2}{1 - c_n^2}\right).$$

Let $A_n = c_n - c_n^2$ and $B_n = \frac{c_n^2}{1-c_n^2}$. We have

$$\frac{a_n}{a_n + b_n} = A_n + A_n B_n.$$

As $\sum_{n=1}^{+\infty} c_n^2$ is convergent, $\lim_n c_n^2 = 0$, and hence there exists an $N \in \mathbb{N}$ such that $c_n^2 < \frac{1}{2}$ for all $n \geq N$. It follows that for all $n \geq N$ we have

$$0 \leq B_n = \frac{c_n^2}{1-c_n^2} < 2c_n^2,$$

from which, in view of the Comparison Test, we see that $\sum_{n=1}^{+\infty} B_n$ is convergent. Note that

$$\left| B_n - B_{n+1} \right| \leq |B_n| + |B_{n+1}| = B_n + B_{n+1},$$

for all $n \geq N$. Again the Comparison Test together with the above inequality, in view of the convergence of $\sum_{n=1}^{+\infty} B_n$, implies that $\sum_{n=1}^{+\infty} |B_n - B_{n+1}|$ is convergent. On the other hand, it follows from the hypothesis that $\sum_{n=1}^{+\infty} A_n$ is convergent. Now, using Dedekind's Extension of Abel's Theorem, we see that $\sum_{n=1}^{+\infty} A_n B_n$ converges. This together with the convergence of $\sum_{n=1}^{+\infty} A_n$ implies that $\sum_{n=1}^{+\infty} (A_n + A_n B_n)$ converges. Thus, so does the series $\sum_{n=1}^{+\infty} \frac{a_n}{a_n+b_n}$, which is what we want.

**Second solution:** We can write

$$\frac{a_n}{a_n + b_n} = \frac{c_n + c_n^2 - c_n^2}{1 + c_n} = c_n - \frac{c_n^2}{1+c_n},$$

where $c_n = \frac{a_n}{b_n}$. As $\lim_n c_n = 0$, it follows that there exists an $N \in \mathbb{N}$ such that $\frac{c_n^2}{1+c_n} > 0$ for all $n \geq N$. This together with the convergence of $\sum_{n=1}^{+\infty} c_n^2$, in view of the Limit Comparison Test, implies that the series $\sum_{n=1}^{+\infty} \frac{c_n^2}{1+c_n}$ converges. From the above equality and the hypothesis that $\sum_{n=1}^{+\infty} c_n$ is convergent, we see that the series $\sum_{n=1}^{+\infty} \frac{a_n}{a_n+b_n}$ is convergent and that we have

$$\sum_{n=1}^{+\infty} \frac{a_n}{a_n + b_n} = \sum_{n=1}^{+\infty} c_n - \sum_{n=1}^{+\infty} \frac{c_n^2}{1 + c_n},$$

finishing the proof.                                                                                    ∎

**3. First solution:** To prove the assertion by contradiction, suppose that the sequence $(f_n)_{n=1}^{+\infty}$ does not converge uniformly to zero on $[0, 1]$. It follows that there are an $\epsilon > 0$, a sequence $(n_k)_{k=1}^{+\infty}$, with $n_k \geq k$, of natural numbers, and a sequence $(x_k)_{k=1}^{+\infty}$ in $[0, 1]$ such that $|f_{n_k}(x_k)| \geq \epsilon$ for all $k \in \mathbb{N}$. If necessary, by passing to a subsequence of $(x_k)_{k=1}^{+\infty}$ and replacing $f_{n_k}$ by $-f_{n_k}$, without loss of generality, we may assume that $f_{n_k}(x_k) > 0$ for all $k \in \mathbb{N}$ and that there is an $x_0 \in [0, 1]$ such that $\lim_k x_k = x_0$. We can write

$$f_{n_k}(x_k) - f_{n_k}(x_0) = \int_{x_0}^{x_k} f_{n_k}'(t)dt,$$

from which, in view of $\|f_n'\|_\infty \leq 1$ for all $n \in \mathbb{N}$, we obtain

$$\left| f_{n_k}(x_k) - f_{n_k}(x_0) \right| \leq |x_k - x_0|,$$

for all $k \in \mathbb{N}$. Let $K_1 \in \mathbb{N}$ be such that $|x_k - x_0| < \frac{\epsilon}{2}$ for all $k \geq K_1$. We must have

$$\left| f_{n_k}(x_0) \right| \geq \left| f_{n_k}(x_k) \right| - \left| f_{n_k}(x_k) - f_{n_k}(x_0) \right| \geq \epsilon - \frac{\epsilon}{2} = \frac{\epsilon}{2},$$

for all $k \geq K_1$. Now, from this and

$$f_{n_k}(x) - f_{n_k}(x_0) = \int_{x_0}^{x} f'_{n_k}(t)dt,$$

in a similar fashion, we see that

$$\left| f_{n_k}(x) \right| \geq \frac{\epsilon}{4},$$

for all $x \in [0,1]$ with $|x - x_0| < \frac{\epsilon}{4}$ and $k \geq K_2$, where $K_2$ is such that $|x_k - x_0| < \frac{\epsilon}{4}$ for all $k \geq K_2$. Since $f_{n_k}$'s ($k \geq K_2$) are continuous functions, we conclude that $f_{n_k}(x) > 0$, and hence,

$$f_{n_k}(x) \geq \frac{\epsilon}{4},$$

for all $x \in [0,1]$ with $|x - x_0| < \frac{\epsilon}{4}$ and $k \geq K_2$, for otherwise, by the Intermediate Value Theorem, there must exist $z_k \in [0,1]$ with $|z_k - x_0| < \frac{\epsilon}{4}$ and $k \geq K_2$ such that $f_{n_k}(z_k) = 0$, which is impossible. Let $[a,b] = [0,1] \cap [x_0 - \frac{\epsilon}{4}, x_0 + \frac{\epsilon}{4}]$. Now define $g : [0,1] \to \mathbb{R}$ by

$$g(x) = \begin{cases} 0 & 0 \leq x \leq a, \\ \frac{3}{b-a}(x-a) & a \leq x \leq a + \frac{b-a}{3}, \\ 1 & a + \frac{b-a}{3} \leq x \leq a + 2\frac{b-a}{3}, \\ \frac{-3}{b-a}(x-b) & a + 2\frac{b-a}{3} \leq x \leq b, \\ 0 & b \leq x \leq 1. \end{cases}$$

The function $g$ is continuous and nonnegative on $[0,1]$. For all $k \geq K_2$, we can write

$$\int_0^1 f_{n_k}g = \int_a^{a+\frac{b-a}{3}} f_{n_k}g + \int_{a+\frac{b-a}{3}}^{a+2\frac{b-a}{3}} f_{n_k}g + \int_{a+2\frac{b-a}{3}}^b f_{n_k}g$$

$$\geq \int_{a+\frac{b-a}{3}}^{a+2\frac{b-a}{3}} f_{n_k} \geq \frac{(b-a)\epsilon}{12},$$

contradicting the hypothesis that $\lim_n \int_0^1 f_{n_k}g = 0$. Therefore, the assertion follows by contradiction.

**Second solution:** It suffices to show that every subsequence $(g_k)_{k=1}^{+\infty}$ of $(f_n)_{n=1}^{+\infty}$, where $g_k = f_{n_k}$ ($k \in \mathbb{N}$), in turn, has a subsequence $(h_j)_{j=1}^{+\infty}$, where $h_j = g_{k_j}$ ($j \in \mathbb{N}$), which converges uniformly to zero on $[0,1]$. We first show that the sequence $(f_n)_{n=1}^{+\infty}$ is uniformly bounded on $[0,1]$. To this end, noting that $f_n : [0,1] \to \mathbb{R}$ and $||f'_n||_\infty \leq 1$ for all $n \in \mathbb{N}$, from the Mean Value Theorem, we see that

$$|f_n(x)| \leq 1 + |f_n(0)|,$$

for all $x \in [0,1]$ and $n \in \mathbb{N}$. We claim that the sequence $\left( f_n(0) \right)_{n=1}^{+\infty}$ is bounded. Suppose to the contrary that there exists a subsequence $\left( f_{n_k}(0) \right)_{n=1}^{+\infty}$ of $\left( f_n(0) \right)_{n=1}^{+\infty}$ such that $\lim_k f_{n_k}(0) = +\infty$ or $\lim_k f_{n_k}(0) = -\infty$. If necessary, replacing $f_n$ by $-f_n$, we may assume that $\lim_k f_{n_k}(0) = +\infty$. Thus, there exists a $K \in \mathbb{N}$ such that $f_{n_k}(0) \geq 2$ for all $k \geq K$. The sequence $(f_n)_{n=1}^{+\infty}$ is uniformly equicontinuous on $[0,1]$ because $\|f_n'\|_\infty \leq 1$ for all $n \in \mathbb{N}$. It follows that the sequence $(f_{n_k})_{n=1}^{+\infty}$ is uniformly equicontinuous on $[0,1]$ as well. So choosing $\varepsilon = 1$, we obtain $0 < 2\delta < 1$ such that

$$\left| f_{n_k}(x) - f_{n_k}(0) \right| \; < \; 1,$$

whenever $0 < x < 2\delta$ and $k \in \mathbb{N}$. This easily yields

$$f_{n_k}(x) \; \geq \; 2 - 1 = 1,$$

for all $0 < x < 2\delta$ and $k \in \mathbb{N}$ with $k \geq K$. Now, define $g : [0,1] \to \mathbb{R}$ by

$$g(x) \; = \; \begin{cases} 1 & 0 \leq x \leq \delta, \\ -\frac{x}{\delta} + 2 & \delta \leq x \leq 2\delta, \\ 0 & 2\delta < x \leq 1. \end{cases}$$

Obviously, $g$ is continuous and nonnegative on $[0,1]$. So it follows from the hypothesis that $\lim_n \int_0^1 f_n g = 0$, from which, we obtain

$$\lim_k \int_0^1 f_{n_k} g \; = \; 0.$$

Note that as $g \geq 0$ on $[\delta, 2\delta]$, for all $k \geq K$, we have

$$\int_0^1 f_{n_k} g \; = \; \int_0^{2\delta} f_{n_k} g$$
$$= \; \int_0^\delta f_{n_k} + \int_\delta^{2\delta} f_{n_k} g$$
$$\geq \; 1 \times (\delta - 0) = \delta.$$

That is, $\int_0^1 f_{n_k} g \geq \delta$ for all $k \geq K$, which is in contradiction with $\lim_k \int_0^1 f_{n_k} g = 0$ and $0 < \delta$. Therefore, $\left( f_n(0) \right)_{n=1}^{+\infty}$ is bounded and hence so is $(f_n)_{n=1}^{+\infty}$ because $|f_n(x)| \leq 1 + |f_n(0)|$ for all $x \in [0,1]$ and $n \in \mathbb{N}$.

Now suppose $(g_k)_{k=1}^{+\infty}$, with $g_k = f_{n_k}$ $(k \in \mathbb{N})$, is an arbitrary subsequence of $(f_n)_{n=1}^{+\infty}$. The sequence $(g_k)_{k=1}^{+\infty}$ is uniformly bounded and equicontinuous on $[0,1]$ because so is $(f_n)_{n=1}^{+\infty}$ on $[0,1]$. It thus follows from Arzela's Theorem that the sequence $(g_k)_{k=1}^{+\infty}$ has a subsequence $(h_j)_{j=1}^{+\infty}$, with $h_j = g_{k_j}$ $(j \in \mathbb{N})$, such that $(h_j)_{j=1}^{+\infty}$ converges uniformly to a function $h : [0,1] \to \mathbb{R}$ on $[0,1]$. The function $h$ is continuous because it is a uniform limit of continuous functions. This together with the hypothesis yields $\lim_j \int_0^1 h_j h = 0$. But it is plain that $(h_j h)_{j=1}^{+\infty}$ converges uniformly to $h^2$ on $[0,1]$. So, we can write

$$\int_0^1 h^2 \; = \; \int_0^1 \lim_j h_j h = \lim_j \int_0^1 h_j h = 0,$$

implying $\int_0^1 h^2 = 0$, which, in turn, implies $h = 0$ because $h$ is continuous on $[0,1]$. In other words, we have shown that every subsequence of $(f_n)_{n=1}^{+\infty}$, in turn, has a subsequence converging uniformly to zero on $[0,1]$. This completes the proof. ∎

**2.13.2. Algebra. 1.** Note first that every nonzero ideal $I$ of $R$ is un-countable. To see this, as $\frac{R}{I}$ is countable, we have

$$\frac{R}{I} = \{a_n + I : a_n \in R\}.$$

It follows that $R = \bigcup_{n=1}^{+\infty}(a_n + I)$, implying that $I$ is uncountable because otherwise $R$ would be countable, which is impossible. Now, to prove the assertion by contradiction, suppose that $0 \neq a \in R$ is a divisor of zero in $R$. Define

$$I := \{r \in R : ar = 0\}.$$

It is plain that $I$ is a nonzero ideal of $R$ and hence it is uncountable. Also note that $aR$ is a nonzero ideal of $R$. Define the map $f : aR \to \frac{R}{I}$ by $f(ax) = x + I$. The map $f$ is obviously onto. Suppose that $f(ax_1) = f(ax_2)$. We have $x_1 + I = x_2 + I$, implying that $x_1 - x_2 \in I$, whence $a(x_1 - x_2) = 0$. That is, $ax_1 = ax_2$ which means $f$ is one-to-one. Therefore, $f$ is a one-to-one correspondence between the uncountable set $aR$ and the countable set $\frac{R}{I}$, a contradiction. Thus, $R$ has no divisor of zero, and hence $R$ is an integral domain, which is what we want. ∎

**2.** Let $S = M_2(\mathbb{Q})$. We have $\mathbb{Q}I_2 \leq R \leq S$ and $\dim_{\mathbb{Q}} R \leq \dim_{\mathbb{Q}} S \leq 4$, where $I_2$ denotes the $2 \times 2$ identity matrix. Suppose that $I$ is a nonzero left ideal of $R$. As $R$ includes $\mathbb{Q}I_2$, the left ideal $I$ can be viewed as a vector space over $\mathbb{Q}$ and we have $\dim_{\mathbb{Q}} I \leq 4$. Thus, there are $x_i \in I$ $(1 \leq i \leq k \leq 4)$ such that $I = \mathbb{Q}x_1 + \cdots + \mathbb{Q}x_k$. This together with $\mathbb{Q}I_2 \leq R$ implies that $I = Rx_1 + \cdots + Rx_k$. In other words, $I$ is finitely generated. Likewise, one can see that every right ideal of $R$ is finitely generated as well. To show that the condition $(*)$ cannot be dropped, define

$$J = \begin{pmatrix} 0 & G \\ 0 & 0 \end{pmatrix},$$

where $G = \langle \frac{1}{2^i} \rangle_{i=1}^{+\infty} \leq \mathbb{Q}$ is the additive group generated by $\frac{1}{2^i}$'s in $\mathbb{Q}$. It is easily verified that $J$ is a left ideal of the ring $T$ which is not finitely generated, which is what we want. ∎

**3. First solution:** Let $G = GL_2(\mathbb{Q})$ be the multiplicative group of all $2 \times 2$ invertible matrices over the field $\mathbb{Q}$ and

$$a = \begin{pmatrix} 1 & 1 \\ 0 & -1 \end{pmatrix}, \quad b = \begin{pmatrix} -1 & 1 \\ 0 & 1 \end{pmatrix}.$$

It is easily seen that $a, b \in G$, $a^2 = b^2 = I_2$, and $ab = -I_2 + 2N$, where

$$I_2 = \begin{pmatrix} 1 & 0 \\ 0 & 1 \end{pmatrix}, \quad N = \begin{pmatrix} 0 & 1 \\ 0 & 0 \end{pmatrix}.$$

We have

$$(ab)^n = (-1)^n I_2 + (-1)^{n-1} 2nN \neq I_2,$$

for all $n \in \mathbb{N}$. That is, $ab$ has order infinity and yet both $a$ and $b$ have order two, which is what we want.

**Second solution:** Let $G$ be the group of isometries of the Euclidean plane under composition of isometries. Let $A, B \in \mathbb{R}^2$ be two distinct points in the plane and $a, b$ the half turns around the points $A$ and $B$, respectively. If $e$ denotes the identity isometry, we have $a^2 = b^2 = e$. On the other hand, it is easy to see that $ab$ is a translation isometry along the vector $2\overrightarrow{AB}$, and hence $ab$ has order infinity because it is a translation. This is what we want, finishing the proof. ∎

**4. (a)** It is well-known that

$$|GL_n(\mathbb{Z}_p)| = (p^n - 1)(p^n - p) \cdots (p^n - p^{n-1}).$$

To see this, note that a matrix $A \in GL_n(\mathbb{Z}_p)$ if and only if the columns of $A$ are linearly independent. In view of this, there are $p^n - 1$ choices for the first column of an arbitrary element, say $A$, of $GL_n(\mathbb{Z}_p)$ (the zero column vector being excluded). For an $1 \leq i < n$, assuming that the first $i$ columns of an arbitrary element, say $A$, of $GL_n(\mathbb{Z}_p)$ are chosen, there are exactly $p^n - p^i$ choices for the $(i+1)$st column of $A$, because the $(i+1)$st column of $A$ cannot be a linear combination of the first $i$ columns of $A$. Therefore, by the product rule of combinatorics, $GL_n(\mathbb{Z}_p)$ has exactly $(p^n - 1)(p^n - p) \cdots (p^n - p^{n-1})$ elements, as desired. Thus, $|G| = (3^2 - 1)(3^2 - 3) = 48$. As $K = Z(G)$ includes only scalar matrices, we see that $K = \{I_2, 2I_2\}$, where $I_2$ is the identity matrix, and hence $|K| = 2$. It is now obvious that $K \leq H \leq G$. To calculate $|H|$, we note that $a, b, c \in \mathbb{Z}_3$ can be chosen independent of one another and that $a, c$ can be 1 or 2 and $b$ can be 0, 1, 2. Thus, using the product rule of combinatorics, we have

$$|H| = 2 \times 2 \times 3 = 12.$$

**(b)** Let $N = \bigcap_{x \in G} x^{-1} H x$. First, we claim that $N$ is the largest subgroup of $H$ that is normal in $G$. We present two proofs for the claim.

**First proof.** Firstly, $N \subseteq e^{-1} H e = H$. Secondly, $N$ is a subgroup of $G$ because it is an intersection of subgroups of $G$, namely, $x^{-1} H x$'s where $x \in G$. To see that $N$ is a normal subgroup of $G$, suppose that $a \in N$ and $g \in G$ are arbitrary. As $a \in N = \bigcap_{x \in G} x^{-1} H x$ and $xg^{-1} \in G$, we see that $a \in gx^{-1} H x g^{-1}$, yielding $g^{-1} a g \in x^{-1} H x$. It follows that $g^{-1} a g \in \bigcap_{x \in G} x^{-1} H x = N$. That is, $N$ is a normal subgroup of $G$. Now, suppose that $N_1 \subseteq H$ is a normal subgroup of $G$. We can write

$$N_1 = x^{-1} N_1 x \subseteq x^{-1} H x,$$

for all $x \in G$, implying that $N_1 \subseteq \bigcap_{x \in G} x^{-1} H x = N$, as desired.

**Second proof.** We see from (a) that $[G : H] = 4$. Consequently, if $X$ is used to denote the set of left cosets of $H$ in $G$, then there exist $g_1 = e, g_2, g_3, g_4 \in H$ such that

$$X = \{g_1 H, \ldots, g_4 H\}.$$

Observe that $G$ acts on $X$ via left multiplication. That is, $g(g_i H) = g g_i H \in X$, where $g \in G$. Now, for a fixed $g \in G$, define the map $\ell_g : X \to X$ by $\ell_g(g_i H) = g g_i H$. It is easily verified that $\ell_g$ is a one-to-one map which is onto because $|X| = 4 < \infty$. In other words, $\ell_g$ defines a permutation on $X$, i.e., $\ell_g \in S(X) \cong S_4$, where $S_4$ denotes the symmetric group of degree four. Now, define the map $\phi : G \to S(X) \cong S_4$ by $\phi(g) = \ell_g$. As $G$ acts on $X$ via left multiplication, we see that $\ell_{g_1 g_2} = \ell_{g_1} \ell_{g_2}$ for all $g_1, g_2 \in G$. This means

$$\phi(g_1 g_2) = \phi(g_1)\phi(g_2),$$

for all $g_1, g_2 \in G$. Thus, $\phi$ is a homomorphism of groups. We claim that

$$\ker \phi = N = \bigcap_{x \in G} x^{-1} H x \subseteq e^{-1} H e = H,$$

from which, we see that $N$ is a normal subgroup of $G$ which is contained in $H$. To prove this last claim, suppose $g \in \ker \phi$. We have $\phi(g) = \ell_g = \ell_e$, where $e$ is the identity element of $G$. It follows that $\ell_g(x^{-1}H) = \ell_e(x^{-1}H)$ for all $x \in G$. That is, $gx^{-1}H = x^{-1}H$, yielding $xgx^{-1} \in H$, and hence $g \in x^{-1}Hx$ for all $x \in G$. This implies $g \in \bigcap_{x \in G} x^{-1} H x = N$. Conversely, if $g \in \bigcap_{x \in G} x^{-1} H x = N$, we conclude that $g \in \ker \phi$. Therefore, $\ker \phi = \bigcap_{x \in G} x^{-1} H x = N$, proving the claim.

We now show that $N = K$. To this end, let $n \in N$ be arbitrary. As $N \subseteq H$, we have $n = \begin{pmatrix} a_1 & b_1 \\ 0 & c_1 \end{pmatrix}$, where $a_1, b_1, c_1 \in \mathbb{Z}_3$ and $a_1 c_1 \neq 0$. Suppose $g = \begin{pmatrix} a & b \\ c & d \end{pmatrix} \in G$, where $a, b, c, d \in \mathbb{Z}_3$ with $ad - bc \neq 0$, is arbitrary. As $N$ is normal in $G$, we must have $h = g^{-1} n g \in N \subseteq H$. This implies that $h_{21} = \frac{-aca_1 - c^2 b_1 + acc_1}{ad - bc} = 0$, where $h_{21}$ denotes the 21 entry of the matrix $h$. Simplifying, we obtain $ac(c_1 - a_1) + c^2(-b_1) = 0$ for all $a, b, c, d \in \mathbb{Z}_3$ with $ad - bc \neq 0$. This easily yields $a_1 = c_1$ and $b_1 = 0$, proving that $n \in K$, which is what we want.

(c) We have shown that the map $\phi : G \to S(X) \cong S_4$ defined by $\phi(g) = \ell_g$ is a homomorphism of groups and that

$$\ker \phi = \bigcap_{x \in G} x^{-1} H x = N = K.$$

In view of the First Isomorphism Theorem for groups, we can write

$$\frac{G}{\ker \phi} \cong \mathrm{im}(\phi) \leq S_4.$$

That is, $\frac{G}{K} \cong \mathrm{im}(\phi) \leq S_4$. On the other hand, $|\mathrm{im}(\phi)| = [G : K] = 24 = |S_4|$. Thus, $\mathrm{im}(\phi) \cong S_4$, and hence $\frac{G}{K} \cong S_4$, which is what we want. ∎

**5.** Let $\big(A\nu(\xi)\big)_{i1}$ $(1 \leq i \leq n)$ denote the $i1$ entry of the column matrix $A\nu(\xi)$. We have

$$\big(A\nu(\xi)\big)_{i1} = \sum_{k=1}^{n} a_{ik}\big(\nu(\xi)\big)_{k1} = a_{i1}\big(\nu(\xi)\big)_{11} + \sum_{k=2}^{n} a_{ik}\big(\nu(\xi)\big)_{k1}$$

$$= \delta_{i,n} + \sum_{k=2}^{n} \delta_{i,k-1}\xi^{k-1} = \begin{cases} 1 & i = n, \\ \xi^i & i < n. \end{cases}$$

In other words,

$$
A\nu(\xi) = \begin{pmatrix} \xi \\ \xi^2 \\ \vdots \\ \xi^{n-1} \\ 1 \end{pmatrix} = \begin{pmatrix} \xi \\ \xi^2 \\ \vdots \\ \xi^{n-1} \\ \xi^n \end{pmatrix} = \xi \begin{pmatrix} 1 \\ \xi \\ \xi^2 \\ \vdots \\ \xi^{n-1} \end{pmatrix} = \xi\nu(\xi).
$$

That is, $\nu(\xi)$ is an eigenvector of $A$ whose corresponding eigenvalue is $\xi$. ∎

**2.13.3. General. 1.** Let $M$ be an arbitrary point on the unit circle in the complex plane centered at the origin and $A_1, \ldots, A_n$ denote the vertices of a regular $n$-gon which is inscribed in the circle. We have

$$
\begin{aligned}
M &= \exp i\theta = \cos\theta + i\sin\theta, \\
A_j &= \exp i\theta_j = \cos\theta_j + i\sin\theta_j,
\end{aligned}
$$

where $\theta, \theta_1 \in \mathbb{R}$, and $\theta_j = \theta_1 + (j-1)\frac{2\pi}{n}$ $(1 \le j \le n)$. We can write

$$
\begin{aligned}
\sum_{j=1}^{n} MA_j^2 &= \sum_{j=1}^{n} \left| \exp i\theta - \exp i\theta_j \right|^2 \\
&= \sum_{j=1}^{n} \left( 2 - \exp i(\theta - \theta_j) - \exp i(\theta_j - \theta) \right) \\
&= 2n - \exp i(\theta - \theta_1) \sum_{j=1}^{n} \left( \exp -i\frac{2\pi}{n} \right)^{j-1} - \\
&\quad \exp i(\theta_1 - \theta) \sum_{j=1}^{n} \left( \exp i\frac{2\pi}{n} \right)^{j-1} \\
&= 2n - \exp i(\theta - \theta_1) \frac{\left( \exp -i\frac{2\pi}{n} \right)^n - 1}{\exp -i\frac{2\pi}{n} - 1} - \\
&\quad \exp i(\theta_1 - \theta) \frac{\left( \exp i\frac{2\pi}{n} \right)^n - 1}{\exp i\frac{2\pi}{n} - 1} \\
&= 2n - \exp i(\theta - \theta_1) \times 0 - \exp i(\theta_1 - \theta) \times 0 = 2n.
\end{aligned}
$$

That is, $\sum_{j=1}^{n} MA_j^2 = 2n$, which is what we want. ∎

**2.** Let

$$
f(x) = a_n x^n + a_{n-1} x^{n-1} + \cdots + a_1 x + a_0,
$$

where $a_i \in \mathbb{Q}$ $(0 \le i \le n)$ and $a_n > 0$, be a polynomial with the property that $f(x)$ is irrational whenever $x$ is irrational. Let $p$ be a prime and

$$
\begin{aligned}
g(x) &= \mathrm{lcm}(a_1, \ldots, a_n) f(x) - \mathrm{lcm}(a_1, \ldots, a_n) a_0 - p \\
&= g_n x^n + g_{n-1} x^{n-1} + \cdots + g_1 x - p,
\end{aligned}
$$

where $g_i = \mathrm{lcm}(a_1, \ldots, a_n) a_i$ $(1 \le i \le n)$. As the leading coefficient of $g$ is positive and $g(0) = -p < 0$, it follows from the Intermediate Value Theorem that $g$ has a positive root which must be rational because, in view of the above

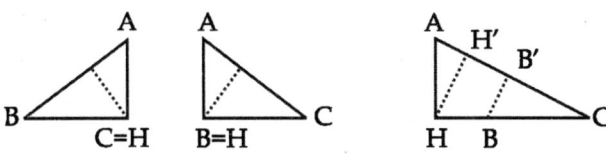

Figure 12

equality, the polynomial $g$ has the property that $g(x)$ is irrational whenever $x$ is irrational. Now by the Rational Root Theorem, the positive root is of the form $\frac{1}{d}$ or $\frac{p}{d}$ for some $d$ dividing $g_n$. If $n > 1$, it is obvious that by choosing $p$ large enough none of the numbers $\frac{1}{d}$ and $\frac{p}{d}$, where $d$ divides $g_n$, can be a root of $g(x) = 0$. Therefore, $n \leq 1$, yielding $f(x) = ax + b$ for some $a, b \in \mathbb{Q}$ with $a \neq 0$. $\blacksquare$

**3.** To prove the assertion, we need the following theorem which is known as Sylvester's problem.

**Theorem.** *Let $S$ be a finite set of the points of a plane such that the line passing through any two points of $S$ contains a third point of $S$. Then, all of the points of $S$ lie on a straight line.*

**Proof.** To prove the assertion by contradiction, suppose that all of the points of $S$ are not collinear. As $S$ is finite, there exist $A, B, C \in S$ such that the smallest height of the triangle $ABC$ is minimal among all triangles with vertices from $S$. Without loss of generality, assume that the altitude $AH$ drawn from the vertex $A$ perpendicular to the base $BC$ is the smallest height of the triangle $ABC$.

If $H = B$ or $H = C$, then the altitude drawn from $H$ to the base opposite to it is smaller than $AH$, which is impossible. If not, then $H$ is between $B$ and $C$ and there is a third point $D$ of $S$ on the line $BC$. Thus, if necessary by renaming the points $B$, $C$, and $D$, we may assume that $B$ and $C$ are both on the left or the right of $AH$ such that $BH < CH$. Assume, with no loss of generality, that $B$ and $C$ are on the right of $AH$ such that $BH < CH$. It is now easily seen that the altitude drawn from $B$ to $AC$ is smaller than $AH$, a contradiction. Thus, all of the points of $S$ are collinear, as desired. $\square$

**Remark.** Motivated by Sylvester's problem, we pose the following problem. *Let $S$ be a finite set of the points of a plane containing no three collinear points such that the circle passing through any three points of $S$ contains a fourth point of $S$. Then, all of the points of $S$ lie on a circle.* Hint. Perform an inversion with center at one of the given points and any radius, say, radius one.

First, we show that there exists a plane which contains only two lines of the given $n$ parallel lines. To this end, intersect the $n$ parallel lines with a plane to obtain $n$ points of intersection on the plane. It follows from the above theorem that in the plane there is a line which contains exactly two points of the $n$ points. Now, this line and two lines of the given $n$ parallel lines form a plane which passes through only two lines of the $n$ parallel lines. With this in mind, we prove the assertion by induction on $n$. If $n = 3$, the assertion is obvious. Assuming that the assertion holds for $n - 1$ parallel lines, to prove the assertion for $n$ parallel lines, as explained in the above find a plane that contains exactly two lines of the given $n$ lines. Now, one of these two lines and the $n - 2$ remaining lines are not coplanar. It thus follows from the induction hypothesis that there exist at least $n - 1$ distinct planes each of which includes at least two lines of these $n - 1$ lines. These $n - 1$ distinct planes together with the plane that contains only two lines of the given $n$ lines form $n$ planes satisfying the desired property, which is what we want. ∎

## 2.14. Fourteenth Competition

**2.14.1. Analysis. 1.** To prove the assertion by contradiction, suppose that $f'(x_0)$ exists. As $f'(x) < 0$ on $(-\infty, x_0)$, $f$ is strictly decreasing on $(-\infty, x_0]$, from which, we obtain $f(x) > f(x_0)$ for all $x < x_0$. Likewise, since $f'(x) > 0$ on $(x_0, +\infty)$, we see that $f(x) > f(x_0)$ for all $x > x_0$. Thus, $f$ assumes it absolute minimum, and hence local minimum, at $x_0$, yielding $f'(x_0) = 0$. Now, $f''(x) < 0$ on $(x_0, +\infty)$ implies $f'$ is strictly decreasing on $[x_0, +\infty)$, which, in turn, implies $f'(x) < f'(x_0) = 0$ for all $x > x_0$, which is in contradiction with the hypothesis. This shows that $f'(x_0)$ does not exist, which is what we want. ∎

**2.** We need the following lemma.

**Lemma.** Let $g : \mathbb{R} \to \mathbb{R}$ be a uniformly continuous and nonnegative function on $\mathbb{R}$ such that $\int_{-\infty}^{+\infty} g < \infty$. Then

$$\lim_{x \to -\infty} g(x) = 0, \quad \lim_{x \to +\infty} g(x) = 0.$$

**Proof.** We prove that $\lim_{x \to +\infty} g(x) = 0$. One can prove $\lim_{x \to -\infty} g(x) = 0$ in a similar fashion. Proceed by way of contradiction. Thus, there exists an $\varepsilon_0 > 0$ such that for all $n \in \mathbb{N}$, there is an $x_n > n$ for which

$$g(x_n) = |g(x_n)| \geq \varepsilon_0.$$

From the hypothesis that $g$ is uniformly continuous on $\mathbb{R}$, we see that for $\frac{\varepsilon_0}{2} > 0$, there exists a $0 < \delta_0 < 1$ such that

$$|g(x) - g(y)| < \frac{\varepsilon_0}{2},$$

for all $x, y \in \mathbb{R}$ with $|x - y| < \delta_0$. In particular, for all $i \in \mathbb{N}$ and $x \in \mathbb{R}$ with $|x - x_i| < \delta_0$, we have

$$|g(x) - g(x_i)| < \frac{\varepsilon_0}{2},$$

from which, in view of $|g(x_i)| - |g(x)| \leq |g(x) - g(x_i)|$, we obtain

$$g(x) = |g(x)| \geq |g(x_i)| - \frac{\varepsilon_0}{2} \geq \frac{|g(x_i)|}{2} = \frac{g(x_i)}{2} \geq \frac{\varepsilon_0}{2}.$$

Thus, $g(x) \geq \frac{\varepsilon_0}{2}$ on $(x_i - \delta_0, x_i + \delta_0)$ for all $i \in \mathbb{N}$, yielding

$$\int_{x_i - \delta_0}^{x_i + \delta_0} g \geq 2\delta_0 \frac{\varepsilon_0}{2} = \varepsilon_0 \delta_0.$$

On the other hand, $\int_{-\infty}^{x_i + \delta_0} g - \int_{-\infty}^{x_i - \delta_0} g = \int_{x_i - \delta_0}^{x_i + \delta_0} g$. Thus, letting $i \to +\infty$ in the above, in view of the hypothesis, we obtain $0 \geq \varepsilon_0 \delta_0$, which is a contradiction. Therefore,

$$\lim_{x \to +\infty} g(x) = 0,$$

which is what we want. $\qquad\qquad\qquad\qquad\qquad\qquad\qquad\qquad\qquad\qquad\square$

**First solution:** Suppose by way of contradiction that there exists an $\varepsilon_0 > 0$ such that for all $n \in \mathbb{N}$ there is an $x_n \in \mathbb{R}$ with $|x_n| > n$ for which $|f(x_n)| \geq \varepsilon_0$. As $f$ is uniformly continuous on $\mathbb{R}$, for $\frac{\varepsilon_0}{2} > 0$ there exists a $\delta_0 > 0$ such that

$$\left| f(x) - f(y) \right| < \frac{\varepsilon_0}{2},$$

whenever $x, y \in \mathbb{R}$ and $|x - y| < \delta_0$. In particular, for all $i \in \mathbb{N}$ and $|x - x_i| < \delta_0$, we have

$$\left| f(x) - f(x_i) \right| < \frac{\varepsilon_0}{2}.$$

But

$$\left| f(x_i) \right| - \left| f(x) \right| \leq \left| f(x) - f(x_i) \right|$$

from which, we see that

$$\left| f(x) \right| \geq \left| f(x_i) \right| - \frac{\varepsilon_0}{2} \geq \frac{|f(x_i)|}{2} \geq \frac{\varepsilon_0}{2}. \qquad (*)$$

On the other hand, $f$ is bounded on $\mathbb{R}$. So there exists an $M > 0$ such that $|f(x)| \leq M$ for all $x \in \mathbb{R}$. By the continuity of $g$ on $\mathbb{R}$, $g$ is continuous on the compact interval $[-M, M]$, and hence it is uniformly continuous on $[-M, M]$. Consequently, for given $\varepsilon > 0$, there exists a $\delta_1 > 0$ such that

$$\left| g(x) - g(y) \right| < \varepsilon, \qquad (*')$$

whenever $x, y \in [-M, M]$ and $|x - y| < \delta_1$. It follows from the uniform continuity of $f$ on $\mathbb{R}$ that for the above $\delta_1 > 0$, which was obtained from the given $\varepsilon > 0$, there exists a $\delta > 0$ such that

$$\left| f(x) - f(y) \right| < \delta_1,$$

whenever $x, y \in \mathbb{R}$ and $|x - y| < \delta$. Now, for all $x, y \in \mathbb{R}$ with $|x - y| < \delta$, we have $|f(x) - f(y)| < \delta_1$. This together with the fact that $f(x), f(y) \in [-M, M]$ for all $x, y \in \mathbb{R}$ with $|x - y| < \delta$, in view of $(*')$, implies

$$\left| g(f(x)) - g(f(y)) \right| < \varepsilon.$$

That is, $g \circ f$ is uniformly continuous on $\mathbb{R}$. On the other hand, $g \circ f$ is integrable on $\mathbb{R}$. It thus follows from the lemma that

$$\lim_{x \to -\infty} g \circ f(x) = 0, \quad \lim_{x \to +\infty} g \circ f(x) = 0.$$

Now, set

$$m = \inf \left\{ g(x) \,\middle|\, x \in \left[ -M, -\frac{\varepsilon_0}{2} \right] \cup \left[ \frac{\varepsilon_0}{2}, M \right] \right\}.$$

As $g(x) > 0$ whenever $x \neq 0$ and that $g$ is continuous on the compact set $[-M, -\frac{\varepsilon_0}{2}] \cup [\frac{\varepsilon_0}{2}, M]$, we see that there exists a $c \in [-M, -\frac{\varepsilon_0}{2}] \cup [\frac{\varepsilon_0}{2}, M]$ such that $m = g(c)$, and hence $m = g(c) > 0$. Now, from $\lim_{x \to \infty} g \circ f(x) = 0$, it follows that for $m > 0$, there exists an $N > 0$ such that

$$g\big(f(x)\big) = |g(f(x)) - 0| < m,$$

whenever $x \in \mathbb{R}$ and $|x| > N$. On the other hand, we have $\lim_i |x_i| = +\infty$, and hence for the above $N > 0$, there exists an $I > 0$ such that $|x_i| > N$ whenever $i \geq I$. In particular, $|x_I| > N$, implying $g\big(f(x_I)\big) < m$. But, by $(*)$, we have $\frac{\varepsilon_0}{2} < f(x_I) \leq M$, yielding $g\big(f(x_I)\big) \geq m$, a contradiction. Therefore, $\lim_{x \to \infty} f(x) = 0$, which is what we want.

**Second solution:** Again we proceed by contradiction and continue up to $(*)$ and set

$$m = \inf \left\{ g(x) | x \in [-M, -\frac{\varepsilon_0}{2}] \cup [\frac{\varepsilon_0}{2}, M] \right\}.$$

As we saw in the first solution, there exists a $c \in [-M, -\frac{\varepsilon_0}{2}] \cup [\frac{\varepsilon_0}{2}, M]$ such that $m = g(c)$, and hence $m = g(c) > 0$. It follows from $(*)$ that for all $i \in \mathbb{N}$ and $x \in \mathbb{R}$ such that $x \in (x_i - \delta_0, x_i + \delta_0)$, we have $M \geq |f(x)| \geq \frac{\varepsilon_0}{2}$, implying $f(x) \in [-M, -\frac{\varepsilon_0}{2}] \cup [\frac{\varepsilon_0}{2}, M]$, whence $g\big(f(x)\big) \geq m > 0$. Thus, we can write

$$\int_{x_i - \delta_0}^{x_i + \delta_0} g \circ f \geq m(x_i - \delta_0 - x_i + \delta_0) = 2\delta_0 m \qquad (*'')$$

On the other hand,

$$\int_{x_i - \delta_0}^{x_i + \delta_0} g \circ f = \int_{-\infty}^{x_i + \delta_0} g \circ f - \int_{-\infty}^{x_i - \delta_0} g \circ f.$$

Thus, letting $i \to +\infty$ in $(*'')$, in view of the hypothesis, we obtain $0 \geq 2\delta_0 m$, which is a contradiction. Therefore, $\lim_{x \to \infty} f(x) = 0$, as desired. ∎

**3.** Define the sets $A$ and $B$ as follows

$$A := \{x \in [c, d] : f(x) = a\} = g^{-1}(\{a\}),$$
$$B := \{x \in [c, d] : f(x) = b\} = g^{-1}(\{b\}),$$

where $g = f|_{[c,d]}$, which is continuous on $[c, d]$ because so is $f$ on $[a, b]$. It follows that $A = g^{-1}(\{a\})$ and $B = g^{-1}(\{b\})$ are closed, and hence compact, subsets of the compact interval $[c, d]$. Thus, there exist $r \in A$ and $s \in B$ such that

$$|s - r| = \inf \{|x - y| : x \in A, y \in B\}.$$

It is plain that $r, s \in [c, d]$ and $r \neq s$. Assume, without loss of generality, that $r < s$. By showing that $f([r, s]) = [a, b]$, we settle the proof. To this end, note first that $f(r) = a$ and $f(s) = b$. Next, suppose that $\lambda \in [a, b]$ is arbitrary. We have $f(r) = a \leq \lambda \leq b = f(s)$. From this and the Intermediate Value Theorem, we see that there exists a $c \in [r, s]$ such that $f(c) = \lambda$. That is, $\lambda \in f([r, s])$, implying $[a, b] \subseteq f([r, s])$. Now, suppose that $\lambda = f(c) \in f([r, s])$, where $c \in [r, s]$, is arbitrary. We need to show that $\lambda \in [a, b]$. Suppose to the contrary that $\lambda \notin [a, b]$. This implies $\lambda < a$ or $\lambda > b$. If $\lambda < a$, we can write

$$f(c) = \lambda < a < b = f(s).$$

From this, in view of the Intermediate Value Theorem, we see that there exists $r_1 \in (c,s)$ such that $f(r_1) = a$. On the other hand, $r \leq c < r_1 < s$, $f(r) = f(r_1) = a$, $f(s) = b$, and $|s - r_1| < |s - r|$, which is in contradiction with our choice of $r$ and $s$. This implies $a \leq \lambda$. Likewise, one can see that $\lambda \leq b$. Thus, $a \leq \lambda \leq b$, yielding $f([r,s]) \subseteq [a,b]$ because $\lambda \in f([r,s])$ was arbitrary. Therefore, $f([r,s]) = [a,b]$, which is what we want. ∎

**2.14.2. Algebra. 1. First solution:** As $G$ is a finite $p$-group, we have $|G| = p^n$ for some $n \in \mathbb{N}$. It now follows from Sylow's First Theorem that $G$ has a normal subgroup $H$ such that $|H| = p^{n-1}$. We have $\frac{|G|}{|H|} = p$. Thus, $\frac{G}{H}$ is a cyclic, and hence an ableian, group because $p$ is a prime number. Since $\frac{G}{H}$ is abelian, we obtain $G' \subseteq H \subsetneq G$, implying that $G' \neq G$, as desired.

**Second solution:** Recall that a group is solvable if and only if for some $n \in \mathbb{N} \cup \{0\}$, we have $G^{(n)} = \{e\}$, where $G^{(0)} = G$ and $G^{(n+1)} = (G^{(n)})'$ ($n \in \mathbb{N} \cup \{0\}$). It is well-known that every finite $p$-group is solvable. With all that in mind, proceed by contradiction. Suppose $G' = G$. It follows that $G^{(n)} = G$ for all $n \in \mathbb{N} \cup \{0\}$. Consequently, $G^{(n)} \neq \{e\}$ for all $n \in \mathbb{N} \cup \{0\}$, and hence $G$ cannot be solvable, which is in contradiction with the fact that every finite $p$-group is solvable. Thus, $G' \neq G$, which is what we want. ∎

**2.** Let $p$ be the smallest prime dividing $|G|$ and $K$ a normal subgroup of $G$ with $p$ elements. As $p$ is prime, it follows that $K$ is cyclic so that $K = \langle a \rangle$ for some $a \in G$ with $\mathrm{ord}(a) = p$. Obviously, it suffices to prove that $a \in Z(G)$. To this end, suppose that $g \in G$ is arbitrary. It follows that $g^{-1}ag \in K$ because $K$ is a normal subgroup of $G$. Hence, there is an integer $j$ with $0 \leq j \leq p-1$ such that $g^{-1}ag = a^j$. Firstly, $j \neq 0$ because otherwise $a = e$, which is impossible, for $\mathrm{ord}(a) = p$. So $0 < j \leq p-1$ and we can write

$$a^j = g^{-1}ag \implies (a^j)^j = g^{-1}a^jg,$$

whence $a^{j^2} = g^{-2}ag^2$. By an easy induction on $n$, we see that $a^{j^n} = g^{-n}ag^n$ for all $n \in \mathbb{N}$. In particular, if $n = p-1$, we obtain $a^{j^{p-1}} = g^{-(p-1)}ag^{p-1}$. Now, since $0 < j \leq p-1$ and $p$ is prime, it follows that $\gcd(j, p-1) = 1$. This together with Fermat's Little Theorem yields $j^{p-1} \stackrel{p}{\equiv} 1$. That is, there exists an integer $k$ such that $j^{p-1} = 1 + kp$. As $\mathrm{ord}(a) = p$, we have

$$(g^{p-1})^{-1}ag^{p-1} = a^{j^{p-1}} = a^{1+kp} = a(a^p)^k = ae^k = a,$$

which yields

$$(g^{p-1})^{-1}ag^{p-1} = a.$$

In other words, $ag^{p-1} = g^{p-1}a$ for all $g \in G$. To finish the proof, we need to show that the map $\varphi : G \to G$ defined by $\varphi(g) = g^{p-1}$ is onto. As $|G|$ is finite, it suffices to show that the map $\varphi$ is one-to-one. To this end, suppose $\varphi(g_1) = \varphi(g_2)$ for some $g_1, g_2 \in G$. We have $g_1^{p-1} = g_2^{p-1}$ and that $\gcd(p-1, |G|) = 1$, for $p$ is the smallest prime that divides $|G|$. It follows that there exist $r, s \in \mathbb{Z}$

such that $r(p-1) + s|G| = 1$. Now, noting that $g^{|G|} = e$ for all $g \in G$, we can write

$$g_1 = g_1^{r(p-1)+s|G|} = \left(g_1^{(p-1)}\right)^r \left(g_1^{|G|}\right)^s = \left(g_2^{(p-1)}\right)^r \left(g_2^{|G|}\right)^s = g_2^{r(p-1)+s|G|} = g_2.$$

That is, $g_1 = g_2$. In other words, $\varphi : G \to G$ is one-to-one, and hence onto, finishing the proof. ∎

**3.** Recall that if $R$ is a commutative ring and $a, b \in R$ are nonzero elements of it, then the greatest common divisor of $a$ and $b$ in $R$, denoted by $\gcd(a, b)$, is a nonzero element $d \in R$ satisfying the following. (i) $d|a$ and $d|b$, that is, there exist $k_1, k_2 \in R$ such that $a = k_1 d$ and $b = k_2 d$. (ii) If $d'|a$ and $d'|b$ for some $0 \neq d' \in R$, then $d'|d$. Also, the least common multiple of $a$ and $b$, denoted by $\mathrm{lcm}(a, b)$, is a nonzero element $c \in R$ satisfying the following. (i) $a|c$ and $b|c$. (ii) If $a|c'$ and $b|c'$ for some $0 \neq c' \in R$, then $c|c'$.

We claim that in $\mathbb{Z}_6$, the ring of integers mod 6, the greatest common divisor of 2 and 3 is 1, whereas 2 and 3 have no common multiple. First, it is obvious that $1|2$ and $1|3$. Next, suppose $d'|2$ and $d'|3$. It follows that $d'|3-2 = 1$. Hence, 1 is a greatest common divisor of 2 and 3. Use contradiction to prove that 2 and 3 has no common multiple. Suppose to the contrary that for $0 \neq c \in \mathbb{Z}_6$ we have $2|c$ and $3|c$. It follows that there exist $k_1, k_2 \in \mathbb{Z}_6$ such that $c = 2k_1$ and $c = 3k_2$. We can write

$$3c = 6k_1 = 0, \quad 2c = 6k_2 = 0,$$

whence $c = 3c - 2c = 0 - 0 = 0$. That is, $c = 0$, which is impossible. This proves the claim by contradiction. ∎

**4.** First, if $n \in \mathbb{N}$ and $n > 1$, we show that $x^n y + x^{n-1} + 1$ is irreducible in $F[x, y]$. Suppose to the contrary that there exist $f, g \in F[x, y]$ such that $x^n y + x^{n-1} + 1 = fg$. As $\deg_y(x^n y + x^{n-1} + 1) = 1$, in view of the above equality, without loss of generality, we may assume that $\deg_y(f) = 0$ and $\deg_y(g) = 1$. Consequently, there exist a non-invertible $f_0 \in F[x]$ and $g_0, g_1 \in F[x]$ such that $f = f_0$ and $g = g_0 + yg_1$. We can write

$$x^n y + x^{n-1} + 1 = f_0 g_0 + f_0 g_1 y,$$

yielding $f_0 g_1 = x^n$ and $f_0 g_0 = x^{n-1} + 1$. As $f_0$ is not invertible, there exists $1 < j_0 < n$ such that $f_0 = x^{j_0}$ and $g_1 = x^{n-j_0}$. From this, we obtain $g_0 x^{j_0} = x^{n-1} + 1$, which is impossible because $j_0 > 1$ and $n > 1$. Thus, the polynomial $x^n y + x^{n-1} + 1$ is irreducible in $F[x, y]$ whenever $n > 1$.

Next, if $n = 1$, we show that $xy + 1 + 1 = xy + 2$ is irreducible in $F[x, y]$ if and only if $\mathrm{ch}(F) \neq 2$. To this end, note first that if $\mathrm{ch}(F) = 2$, then $xy + 2 = xy$ is reducible in $F[x, y]$. Now, suppose that $xy + 2$ is reducible in $F[x, y]$. We show that $\mathrm{ch}(F) = 2$. Suppose to the contrary that $\mathrm{ch}(F) \neq 2$. It follows that 2 has an inverse in $F$. Since $xy + 2$ is reducible in $F[x, y]$, just as we saw in the above, there exist a non-invertible $f_0 \in F[x]$ and $g_0, g_1 \in F[x]$ such that $f = f_0$, $g = g_0 + yg_1$, and $xy + 2 = fg$. We must have

$$xy + 2 = f_0 g_0 + f_0 g_1 y,$$

from which, we obtain $f_0 g_0 = 2$ and $f_0 g_1 = x$. Now, since 2 is invertible in $F$, it follows that so is $f_0$ in $F[x]$, a contradiction. Thus, $\text{ch}(F) = 2$, which is what we want. ■

**5.** We may assume that $V_1 \not\subseteq V_2$ and $V_2 \not\subseteq V_1$ because otherwise, in view of $\dim V_1 = \dim V_2$, we see that $V_1 = V_2$, in which case the assertion is easy to prove. More precisely, pick a basis $\{\alpha_i\}_{i=1}^m$ for $V_1 = V_2$, where $m = \dim V_1$, and enlarge it to a basis $\{\alpha_i\}_{i=1}^m \cup \{\beta_i\}_{i=1}^n$ for $V$, where $m + n = \dim V$. It is plain that

$$U \oplus V_1 = U \oplus V_2 = V,$$

where $U = \langle \beta_1, \ldots, \beta_m \rangle$ is the vector subspace spanned by $\beta_i$'s $(1 \leq i \leq n)$. We prove the assertion by induction on $\dim V - \dim V_1 + 1$. If $\dim V - \dim V_1 + 1 = 1$, then $\dim V_1 = \dim V_2 = \dim V$. Hence, for $U = \{0\}$, we obviously have $U \oplus V_1 = U \oplus V_2 = V$. Assuming that the assertion holds whenever $\dim V - \dim V_1 + 1 = n$, we prove it whenever $\dim V - \dim V_1 + 1 = n + 1$. As $V_1 \not\subseteq V_2$ and $V_2 \not\subseteq V_1$, we see that $V_1 \cup V_2 \subsetneq V$, and hence there exists a vector $\beta \in V \setminus (V_1 \cup V_2)$. Set

$$V_1' = V_1 \oplus \langle \beta \rangle, \ V_2' = V_2 \oplus \langle \beta \rangle.$$

We have $\dim V_1' = \dim V_2' = \dim V_1 + 1$, yielding $\dim V - \dim V_1' + 1 = \dim V - \dim V_1 = n$. It thus follows from the induction hypothesis that there exists a subspace $U_1$ such that

$$U_1 \oplus V_1' = U_1 \oplus V_2' = V.$$

Letting $U = U_1 \oplus \langle \beta \rangle$, we obviously obtain

$$U \oplus V_1 = U \oplus V_2 = V,$$

proving the induction assertion, which is what we want. ■

**2.14.3. General. 1. First solution:** Let $A = \{A_1, \ldots, A_n\}$ be a family containing $n$ distinct subsets of the set $S$ with $n$ elements. Define the edge-labeled graph $G$ with vertices in $A$ as follows. For $1 \leq i, j \leq n$ with $i \neq j$, connect the distinct vertices $A_i$ and $A_j$ with an edge labeled $s \in S$ whenever $A_i = A_j \cup \{s\}$ or $A_j = A_i \cup \{s\}$. Now, define the relation $\sim$ on $A$ as follows. For $1 \leq i, j \leq n$, we write $A_i \sim A_j$ whenever $i = j$ or there exists a path in the graph $G$ joining the vertices $A_i$ and $A_j$ of the graph. It is easily verified that $\sim$ is an equivalence relation on $A$ and that the equivalence classes of the relation $\sim$ are connected components of the graph $G$. Let $\{G_i\}_{i=1}^k$ $(1 \leq k \leq n)$ denote the connected components of $G$. Also, let $T_i$ $(1 \leq i \leq k)$ be a maximal tree of $G_i$ (recall that a tree is connected graph not having any cycle). That is, $T_i$ is a subgraph of $G_i$ which is a tree and that it is maximal as a tree in $G_i$ (from this, we immediately see that $T_i$ is a spanning tree of $G_i$, i.e., $T_i$ a subgraph of $G_i$ that contains all the vertices of $G_i$ and is a tree). For a subgraph $H$ of $G$, use $L(H)$ to denote the set of labels of the edges of $H$. We claim that $L(T_i) = L(G_i)$ for all $1 \leq i \leq k$. Obviously, it suffices to show that $L(G_i) \subseteq L(T_i)$. To this end, let $s \in L(G_i)$ be arbitrary. It follows that $s \in S$ is the label of an edge $A_i A_j$ of the subgraph $G_i$. If the edge $A_i A_j$ is already in the tree $T_i$, there is nothing to prove. If not, as $T_i$ is a maximal tree of $G_i$, we see that the graph obtain by adding the edge $A_i A_j$ to the tree $T_i$ is not a tree. It thus follows that the edge $A_i A_j$ participates in a cycle whose all edges but the edge $A_i A_j$ come from the tree $T_i$. Now, $s \in S$ being the label of $A_i A_j$, we see that $A_i = A_j \cup \{s\}$ or $A_j = A_i \cup \{s\}$. By symmetry, we may assume without loss of generality, that $A_i = A_j \cup \{s\}$. Note that $s \notin A_j$, for $i \neq j$. Now, going from $A_j$ to $A_i$ along the other edges of the cycle, in which the edge $A_i A_j$ participates, and noting that $s \notin A_j$ but $s \in A_i$, we see that there must be an edge $A_k A_l \in T_i$, where $1 \leq k, l \leq n$, having $s$ as its label, because otherwise $s \notin A_i$, which is impossible. Thus, $s \in L(T_i)$, whence $L(G_i) \subseteq L(T_i)$, implying $L(G_i) = L(T_i)$, as desired. It is plain that $L(T_i) \leq \varepsilon(T_i)$, where $\varepsilon(T_i)$ denotes the number of the edges of $T_i$. On the other hand, since $T_i$ is a tree, by a standard theorem from the theory of graphs, we have

$$\varepsilon(T_i) \; = \; \nu(T_i) - 1,$$

where $\nu(T_i)$ denotes the number of vertices of $T_i$. The number $k$ being the number of connected components of $G$, we can write

$$L(G) \; = \; \sum_{i=1}^{k} L(G_i) \leq \sum_{i=1}^{k} \varepsilon(T_i) = \sum_{i=1}^{k} \nu(T_i) - k,$$

implying

$$L(G) \; \leq \; \sum_{i=1}^{k} \nu(T_i) - k = \nu(G) - k,$$

where $\nu(G)$ denotes the number of vertices of $G$. This implies $L(G) \leq \nu(G) - k < \nu(G)$. Thus, there exists an $s_0 \in S$ such that $s_0 \notin L(G)$. It is now plain that the sets

$$A_1 \cup \{s_0\}, \; \ldots, \; A_n \cup \{s_0\}$$

are distinct, for otherwise we must have $A_i = A_j \cup \{s_0\}$ or $A_j = A_i \cup \{s_0\}$ for some $1 \leq i, j \leq n$ with $i \neq j$, yielding $s_0 \in L(G)$, which is impossible. This completes the proof of the assertion.

**Second solution:** This solution is taken from Linear Algebra Methods in Combinatorics by L. Babai and P. Frankl. Without loss of generality, assume that $S = \{1, \ldots, n\}$. It suffices to show that there exists an element $x \in S$ such that the sets

$$A_1 \setminus \{x\}, \ldots, A_n \setminus \{x\}$$

are all distinct because so will then be the sets

$$A_1 \cup \{x\}, \ldots, A_n \cup \{x\}.$$

To see this, just note that for some $x \in S$ and $1 \leq i, j \leq n$, $A_i \cup \{x\} = A_j \cup \{x\}$ if and only if $A_i \setminus \{x\} = A_j \setminus \{x\}$. Define the $n \times n$ matrix $M = (m_{ij})$ with $m_{ij} \in \{0, 1\}$ $(1 \leq i, j \leq n)$ as follows

$$m_{ij} \; = \; \begin{cases} 1 & j \in A_i, \\ 0 & j \notin A_i. \end{cases}$$

By the hypothesis, the rows of $M$ are all distinct. We need to show that this remains to be the case after omitting an appropriate column from $M$. There are two cases to consider. (i) $\det M = 0$; and (ii) $\det M \neq 0$. If $\det M = 0$, there is a column of $M$, say, column $j$ for some $1 \le j \le n$, which is linearly dependent on the other columns of $M$. We see that after deleting column $j$ of $M$, the remaining rows are all distinct. Suppose on the contrary that this is not the case. That is, the matrix obtained from deleting column $j$ of $M$ has two equal rows, say rows $i_1$ and $i_2$ for some $1 \le i_1 < i_2 \le n$. It follows that rows $i_1$ and $i_2$ of $M$ are equal as well because column $j$ of $M$ depends linearly on the other columns of $M$. This is a contradiction, proving the assertion in this case. If $\det M \ne 0$, if necessary by interchanging two rows of $M$, we may assume that the first row of $M$ is a row with the minimum number of ones. Note that there might be several rows with the minimum number of ones. Expanding $\det M$ by the first row, we see that $m_{1j} \det M_{1j} \ne 0$, where $M_{1j}$ is the matrix obtained by eliminating row 1 and column $j$ from the matrix $M$. Consequently, $m_{1j} = 1$ and no two rows of $M_{1j}$ are the same. It thus follows that deleting column $j$ leaves no two equal rows. Suppose on the contrary that two rows of the matrix obtained from deleting column $j$ of $M$ are equal. Since no two rows of $M_{1j}$ are the same, we see that for some $1 < i \le n$, row 1 and row $i$ of the matrix obtained from deleting column $j$ of $M$ must be equal. This easily implies that $m_{ij} = 0$, for otherwise rows 1 and $i$ of $M$ would be equal, which is impossible. Thus, the number of ones of row $i$ of $M$ is less than that of row 1, which is a contradiction. Therefore, no two rows of the matrix obtained from deleting column $j$ of $M$ are equal, which is what we want. ∎

**2.** We prove that the answer is 14 days. It is obvious that the maximum number of the exhibition days is attained provided that there is exactly one common book for any two days of the exhibition. With that in mind, first, we obtain the number of books needed to exhibit the books for $k$ days in such a way that 100 books are to be exhibited in each day and that the exhibited books of any two days have exactly one book in common. Then, we find the maximum number of the days subject to the condition that the number of the books is less than or equal to 1369. In doing so, suppose that we would like to run the exhibition for $k$ days. First, choose $k-1$ books $(b_i^1)_{i=2}^k$ for the first day of the exhibition and set aside the book $b_j^1$ to be exhibited for a second time in the $j$th day. Then, choose $k-2$ books $(b_i^2)_{i=3}^k$, from the books not already chosen, for the second day and set aside the book $b_j^2$ $(3 \le j \le k)$ to be exhibited for a second time in the $j$th day. Then, choose $k-3$ books $(b_i^3)_{i=4}^k$, from the books not already chosen, for the third day and set aside the book $b_j^3$ $(4 \le j \le k)$ to be exhibited for a second time in the $j$th day. Continue this way to finally choose one book $b_k^{k-1}$, from the books not already chosen, for the $(k-1)$st day and set aside the book $b_k^{k-1}$ to be exhibited for a second time in the $k$th day. So far, for each day we have $k-1$ books to be exhibited and that the exhibited books of any two days, say day $i$ and day $j$ where $i < j$, have exactly one book in common, namely, the book $b_j^i$. In order to exhibit 100 books per day, for each day we need another $100 - (k-1) = 101 - k$ books. So the number of the books needed is

$$(k-1) + (k-2) + \cdots + 2 + 1 + k(101-k) \quad = \quad \frac{k(k-1)}{2} + k(101-k)$$

$$= \frac{k(201 - k)}{2}.$$

Thus, for $k = 14$, the number of the books needed is $\frac{14(201-14)}{2} = 1309$ and for $k = 15$ the number of the books needed is $\frac{15(201-15)}{2} = 1395$. Since we have only 1369 books, it follows that the maximum number of the days of the exhibition is 14, finishing the proof. ∎

**3.** The assertion is a quick consequence of the following lemma.

**Lemma.** *Let $F$ be a field and with $\mathrm{ch}(F) = 0$ or $> n$, $K$ an extension of $F$, and $\{\lambda_1, \ldots, \lambda_n\} \subseteq K$. Also, let $m \in \mathbb{N} \cup \{0\}$ and $c \in F$. If*

$$\lambda_1^k + \cdots + \lambda_n^k = c^{k-m}(\lambda_1^m + \cdots + \lambda_n^m),$$

*for each $k = m, m + 1, \ldots, m + n$, then $\lambda_i = 0$ or $c$ for all $1 \leq i \leq n$.*

**Proof.** We prove the assertion by induction on $n$. If $n = 1$, the proof is obvious. Suppose that the assertion holds for all $k < n$. Let $\{\lambda_1, \ldots, \lambda_n\} \subseteq K$ be a given subset as described in the lemma. There are two cases to consider.

(i) First, suppose $c = 0$ or $\lambda_1^m + \cdots + \lambda_n^m = 0$. Then

$$\lambda_1^k + \cdots + \lambda_n^k = 0,$$

for each $k = m+1, \ldots, m+n$. We claim that $\lambda_i = 0$ for all $1 \leq i \leq n$. Suppose, by contradiction, that this is not the case. Let $\mu_i \in K$ $(1 \leq i \leq l)$ be nonzero and distinct such that $\{\mu_1, \ldots, \mu_l\} \cup \{0\} = \{\lambda_1, \ldots, \lambda_n\} \cup \{0\}$. It follows that there exist $n_1, \ldots, n_l \in \mathbb{N}$ with $n_1 + \cdots + n_l \leq n$ such that

$$n_1\mu_1^k + \cdots + n_l\mu_l^k = \lambda_1^k + \cdots + \lambda_n^k = 0,$$

for each $k = m + 1, \ldots, m + n$. In other words,

$$\begin{pmatrix} \mu_1^{m+1} & \cdots & \mu_l^{m+1} \\ \vdots & \ddots & \vdots \\ \mu_1^{m+n} & \cdots & \mu_l^{m+n} \end{pmatrix} \begin{pmatrix} n_1 \\ \vdots \\ n_l \end{pmatrix} = \begin{pmatrix} 0 \\ \vdots \\ 0 \end{pmatrix},$$

implying that

$$\mu_1^{m+1} \cdots \mu_l^{m+1} \det \begin{pmatrix} 1 & \cdots & 1 \\ \vdots & \ddots & \vdots \\ \mu_1^{n-1} & \cdots & \mu_l^{n-1} \end{pmatrix} = 0.$$

This in turn, in view of Vandermonde's determinant formula, implies $\mu_i = 0$ for some $1 \leq i \leq l$ or $\mu_i = \mu_j$ for some $1 \leq i < j \leq l$, a contradiction in any event. Thus, $\lambda_i = 0$ for all $1 \leq i \leq n$, which is what we want.

(ii) Next, suppose $c \neq 0$, and hence $\lambda_1^m + \cdots + \lambda_n^m \neq 0$. For each $k = 1, \ldots, n$, let $S_k$ denote the elementary symmetric polynomial in $\lambda_1, \ldots, \lambda_n$ of degree $k$, i.e., $S_1 = \lambda_1 + \cdots + \lambda_n, \ldots, S_n = \lambda_1 \cdots \lambda_n$. Obviously, we can write

$$x^n - S_1 x^{n-1} + \cdots + (-1)^n S_n = (x - \lambda_1) \cdots (x - \lambda_n), \qquad (*)$$

from which, we obtain

$$\lambda_i^{n+m} - S_1\lambda_i^{n+m-1} + \cdots + (-1)^n S_n \lambda_i^m = 0,$$

for all $1 \leq i \leq n$. Adding up these equations, we get

$$\sum_{i=1}^{n} \lambda_i^{n+m} - S_1 \sum_{i=1}^{n} \lambda_i^{n+m-1} + \cdots + (-1)^n S_n \sum_{i=1}^{n} \lambda_i^m = 0.$$

It now follows from the hypothesis that

$$c^n \sum_{i=1}^{n} \lambda_i^m - S_1 c^{n-1} \sum_{i=1}^{n} \lambda_i^m + \cdots + (-1)^n S_n \sum_{i=1}^{n} \lambda_i^m$$

$$= \sum_{i=1}^{n} \lambda_i^{n+m} - S_1 \sum_{i=1}^{n} \lambda_i^{n+m-1} + \cdots + (-1)^n S_n \sum_{i=1}^{n} \lambda_i^m = 0.$$

Thus,

$$\Big(\sum_{i=1}^{n} \lambda_i^m\Big)\big(c^n - S_1 c^{n-1} + \cdots + (-1)^n S_n\big) = 0,$$

yielding

$$c^n - S_1 c^{n-1} + \cdots + (-1)^n S_n = 0.$$

This together with (∗) implies that $c = \lambda_i$ for some $1 \leq i \leq n$. It is now plain that the induction hypothesis can be applied to the set $\{\lambda_1, \ldots, \lambda_n\} \setminus \{c\}$, completing the proof. □

Now, to prove the assertion, just let $F = K = \mathbb{R}$ and $c = m = 1$ in the lemma above. ∎

## 2.15. Fifteenth Competition

**2.15.1. Analysis. 1.** (a) Letting $0 \leq x < y$, we can write

$$2\varphi\left(\frac{x+y}{2}\right) = 2\int_0^{\frac{x+y}{2}} f = \int_0^x f + \int_0^{\frac{x+y}{2}} f + \int_x^{\frac{x+y}{2}} f.$$

As $f$ is increasing on $[0, +\infty)$ and $\dfrac{y-x}{2}$ is positive, we obtain

$$\int_x^{\frac{x+y}{2}} f \leq \int_x^{\frac{x+y}{2}} f\big(t + \frac{y-x}{2}\big) dt = \int_{\frac{x+y}{2}}^y f.$$

So, we have

$$2\varphi\left(\frac{x+y}{2}\right) = \int_0^x f + \int_0^{\frac{x+y}{2}} f + \int_x^{\frac{x+y}{2}} f$$

$$\leq \int_0^x f + \int_0^{\frac{x+y}{2}} f + \int_{\frac{x+y}{2}}^y f$$

$$= \int_0^x f + \int_0^y f$$

$$= \varphi(x) + \varphi(y).$$

This proves (a).

(b) Note first that $f$ is integrable on any closed and bounded subinterval of $[0, +\infty)$ because $f$ is increasing. Consequently, $\varphi$ is continuous on $[0, +\infty)$. On the other hand, it easily follows from (a) that

$$\varphi(\lambda x + (1-\lambda)y) \leq \lambda\varphi(x) + (1-\lambda)\varphi(y), \qquad (*)$$

where $\lambda = \frac{k}{2^k}$, $1 \leq k \leq 2^n$, and $k, n \in \mathbb{N}$. Now, as the set $\{\frac{k}{2^k} : k, n \in \mathbb{N}, 1 \leq k \leq 2^n\}$ is dense in $[0, 1]$, in view of the continuity of $\varphi$, we see that (*) holds for all $\lambda \in [0, 1]$. In other words, $\varphi$ is convex, which is what we want.  ∎

**2.** It is worth mentioning that the hypothesis $g(0) = 0$ is redundant. The function $g$ is continuous on the closed interval $[0, 1]$, so it is bounded on $[0, 1]$. Hence, there exists an $M > 0$ such that $|g(x)| \leq M$ for all $x \in [0, 1]$. Noting that $0 \leq \sin x \leq 1$ for all $x \in [0, 1]$, we can write

$$|f_n(x)| = \left| \frac{g(x)(\sin x)^n}{1 + nx} \right| \leq \frac{M \sin x}{1 + nx} \leq \frac{M}{n}.$$

From this, we conclude that the sequence $(f_n)_{n=1}^{+\infty}$ uniformly converges to the zero function on $[0, 1]$. So the assertion follows.  ∎

**3.** (a) For all $x, z \in (0, +\infty)$, we have

$$2xz \leq xf(x) + zf^{-1}(z).$$

Letting $z = f(y)$, we obtain

$$2xf(y) \leq xf(x) + yf(y),$$

for all $x, y \in (0, +\infty)$. We can write

$$xf(y) - xf(x) \leq yf(y) - xf(y) \implies f(y) - f(x) \leq \frac{y-x}{x}f(y),$$

for all $x, y \in (0, +\infty)$. Interchanging $x, y$ in the above, we obtain

$$f(x) - f(y) \leq \frac{x-y}{y}f(x) \implies f(y) - f(x) \geq \frac{y-x}{y}f(x)$$

for all $x, y \in (0, +\infty)$. This together with the preceding inequality proves (a).

(b) First, we prove that $f$ is increasing on $(0, +\infty)$. To this end, we have

$$2xf(y) \leq xf(x) + yf(y),$$
$$2yf(x) \leq yf(y) + xf(x),$$

yielding

$$2(xf(y) + yf(x)) \leq 2(xf(x) + yf(y)),$$

for all $x, y \in (0, +\infty)$. Consequently,

$$(y-x)(f(y) - f(x)) \geq 0,$$

for all $x, y \in (0, +\infty)$. This shows that $f$ is increasing on $(0, +\infty)$. It thus follows that $f$ has limit from the left and limit from the right at any point and that the two limits at any point coincide because $f$ is increasing and onto. Thus, $f$ is continuous on $(0, +\infty)$. From (a), we have

$$\frac{y-x}{y}f(x) \leq f(y) - f(x) \leq \frac{y-x}{x}f(y),$$

for all $x, y \in (0, +\infty)$. Dividing by $y - x$ where $x, y \in (0, +\infty)$ with $x < y$, we can write

$$\frac{f(x)}{y} \leq \frac{f(y) - f(x)}{y - x} \leq \frac{f(y)}{x}.$$

Letting $y \to x^+$ in the above, we obtain

$$f'(x^+) = \frac{f(x)}{x},$$

where $f'(x^+)$ denotes the right derivative of $f$ at $x$. Likewise, dividing by $y - x$ where $x, y \in (0, +\infty)$ with $x > y$, we can write

$$\frac{f(x)}{y} \geq \frac{f(y) - f(x)}{y - x} \geq \frac{f(y)}{x},$$

from which by letting $y \to x^-$, we obtain

$$f'(x^-) = \frac{f(x)}{x},$$

where $f'(x^-)$ denotes the left derivative of $f$ at $x$. That is, $f'(x)$ exists for all $x \in (0, +\infty)$ and moreover $f'(x) = \frac{f(x)}{x}$. This implies $xf'(x) - f(x) = 0$ and hence $\frac{d}{dx}\left(\frac{f(x)}{x}\right) = \frac{xf'(x) - f(x)}{x^2} = 0$ for all $x \in (0, +\infty)$. Therefore, there exists a $c \in \mathbb{R}$ such that $\frac{f(x)}{x} = c$, yielding $f(x) = cx$ for all $x \in (0, +\infty)$, which is what we want.                    ■

**2.15.2. Algebra. 1.** Let $\Omega = \{P_1, \ldots, P_{p+1}\}$ denote the set of all (distinct) Sylow $p$-subgroups of $G$. The group $G$ acts on $\Omega$ by conjugation. The kernel of the action is

$$H := \{g \in G| \ \forall i = 1, \ldots, p+1: \ g^{-1}P_ig = P_i\} = \bigcap_{i=1}^{p+1} N_G(P_i),$$

where $N_G(P_i)$ denotes the normalizer of $P_i$ in $G$. Recall that for a subset $S$ of $G$, the normalizer of $S$ in $G$, denoted by $N_G(S)$, is defined by

$$N_G(S) = \{g \in G : g^{-1}Sg = S\}.$$

We note that by Sylow's Second Theorem $H$ is a normal subgroup of $G$. From the First Isomorphism Theorem for groups, we see that the group $\frac{G}{H}$ is isomorphic to a subgroup of the symmetric group $S_{p+1}$, and since $p|(p+1)!$ but $p^2 \nmid (p+1)!$, it follows that $p^\alpha| \ |H|$ or $p^{\alpha-1}| \ |H|$. Note that the subgroup $H \cap P_i$ is a Sylow $p$-subgroup of $H$ for each $i = 1, \ldots, p+1$, for $\frac{|HP_i|}{|P_i|} = \frac{|H|}{|H\cap P_i|}$.

Now, as $H \cap P_i$ and $H \cap P_j$ are Sylow $p$-subgroups of $H$, by Sylow's Second Theorem, they are conjugate in $H$, and hence there exists an $h \in H$ such that $h^{-1}(H \cap P_i)h = H \cap P_j$. On the other hand, $h \in H$, implies $h \in N_G(P_i)$ and hence

$$H \cap P_i = (h^{-1}Hh) \cap (h^{-1}P_ih) = h^{-1}(H \cap P_i)h = H \cap P_j.$$

That is,

$$H \cap P_i = H \cap P_j,$$

for all $1 \leq i, j \leq p+1$. From this, we obtain $H \cap P_i \subseteq P_j$ for all $1 \leq i, j \leq p+1$, yielding

$$H \cap P_i \subseteq \bigcap_{j=1}^{p+1} P_j,$$

for all $1 \leq i \leq p+1$. On the other hand,

$$\bigcap_{i=1}^{p+1} P_i \subseteq H \cap P_i,$$

for all $1 \leq i \leq p+1$. Therefore,

$$\bigcap_{j=1}^{p+1} P_j = H \cap P_i,$$

for all $1 \leq i \leq p+1$. This implies that $\bigcap_{j=1}^{p+1} P_j$ is a Sylow $p$-subgroup of $H$. Now, if $p^\alpha \mid |H|$, then $\left| \bigcap_{j=1}^{p+1} P_j \right| = p^\alpha = |P_i|$, and hence $\bigcap_{j=1}^{p+1} P_j = P_i$, implying that $P_i = P_j$, a contradiction. Thus, $p^{\alpha-1} \mid |H|$. Consequently, a Sylow $p$-subgroup of $H$ must be of order $p^{\alpha-1}$, and hence $\left| \bigcap_{i=1}^{p+1} P_i \right| = p^{\alpha-1}$, which is what we want. ∎

**2.** First, we claim that if $P$ is a prime ideal of $R$ and $ab \in P$ for some $a, b \in R$, then $a \in P$ or $b \in P$. To see this, from $ab \in P$, we get $Rab \subseteq P$. Now, as the left ideal $Ra$ is also a right ideal of $R$, we see that $RaR \subseteq Ra$, implying $RaRb \subseteq Rab$. This yields $RaRb \subseteq P$, from which, we obtain $Ra \subseteq P$ or $Rb \subseteq P$, for $P$ is a prime ideal of $R$. This, in turn, implies $a \in P$ or $b \in P$ because $R$ is unital.

Now to prove the assertion, first, suppose that $x \in R$ is nilpotent. So there exists an $n \in \mathbb{N}$ such that $x^n = 0$. We have $x^n = 0 \in P$ for all prime ideals of $R$. In view of the above claim, we see that $x \in P$ for all prime ideals of $R$. In other words, $x \in \bigcap \{P : P \triangleleft R, P \text{ is prime}\}$. Conversely, suppose that $x \in R$ and that $x^n \neq 0$ for all $n \in \mathbb{N}$. In other words, for the multiplicative set $S := \{x^n : n \in \mathbb{N}\}$, we have $S \cap \{0\} = \emptyset$. A standard argument using Zorn's Lemma shows that there exists a prime ideal $P$ such that $P \cap S = \emptyset$. This yields $x \notin P$. In other words, we have proved that if $x \in \bigcap \{P : P \triangleleft R, P \text{ is prime}\}$, then $x$ must be nilpotent. Therefore, the intersection of all prime ideals of $R$ is equal to the set of the nilpotent elements of $R$, which is what we want. ∎

**3.** If we do not require the matrix $AB$ to be a nonzero idempotent, then the assertion is trivial. Just let $B = 0$. Then, $AB = 0$ is an idempotent, settling the proof. However, we state and prove the following nontrivial problem.

*Let $D$ be a division ring, $n \in \mathbb{N}$, and $A \in M_n(D)$. Then, there exists a matrix $B \in M_n(D)$ such that $AB$ is an idempotent whose rank is equal to that of $A$.*

Let $D^n$ denote the right vector space of all $n \times 1$ column vectors with entries in $D$; that is, the addition $x + y$ is defined componentwise and the multiplication of the scalar $\lambda \in D$ into the vector $x = (x_i)_{i=1}^n \in D^n$ is defined by $x\lambda := (x_i \lambda)_{i=1}^n$. The members of $M_n(D)$ can be viewed as linear transformations acting on the left of $D^n$ via the usual matrix multiplication; that is, we can write

$M_n(D) = \mathcal{L}(D^n)$, where $\mathcal{L}(D^n)$ is the ring of all right linear transformations acting on the left of $D^n$. For $x \in D^n$ and $f \in (D^n)'$, where $(D^n)' = D_n$ denotes the dual of $D^n$ which is the left vector space of all $1 \times n$ row vectors with entries in $D$, define the rank-one linear transformation $x \otimes f \in \mathcal{L}(D^n)$ by $(x \otimes f)(y) := xf(y)$. Choose $y_i \in D^n$ $(1 \leq i \leq r)$ such that $\{Ay_i\}_{1 \leq i \leq r}$ is a basis for the range of $A$. Set $x_i := A_i y_i$ and enlarge $\{x_i\}_{1 \leq i \leq r}$ to a basis $\mathcal{B} \cup \{x_i\}_{1 \leq i \leq r}$ for $D^n$, where the set $\mathcal{B}$ is linearly independent. Now, let $\{f_i\}_{1 \leq i \leq r}$ be a dual subset with respect to $\mathcal{B} \cup \{x_i\}_{1 \leq i \leq r}$ so that $\langle \mathcal{B} \rangle \subseteq \ker f_i$ and $f_i(x_j) = \delta_{ij}$ for each $i, j = 1, \ldots, r$. Let $B = y_1 \otimes f_1 + \cdots + y_r \otimes f_r$. Since $\langle \mathcal{B} \rangle \subseteq \ker f_i$ and $f_i(x_j) = \delta_{ij}$ for each $i, j = 1, \ldots, r$, we easily see that $AB = A(y_1 \otimes f_1 + \cdots + y_r \otimes f_r) = x_1 \otimes f_1 + \cdots + x_r \otimes f_r$ is an idempotent whose rank is $r = \text{rank}(A)$, proving the assertion. ∎

**2.15.3. General. 1.** We prove that a necessary and sufficient condition for the product of two integers $a, b$ to be divisible by their sum is that there exist integers $x, y, z$ such that $x, y$ are relatively prime and that

$$a = x(x + y)z, \quad b = y(x + y)z.$$

Sufficiency is easy. To prove necessity, from $a + b \mid ab$, we see that

$$ab = m(a + b),$$

for some $m \in \mathbb{Z}$. If one of $a$ or $b$ is zero, say, $a = 0$, then letting $x = 0$, $y = 1$, $z = b$ proves the assertion. So without loss of generality, assume that $a$ and $b$ are nonzero. Set $d = \gcd(a, b)$ and

$$x = \frac{a}{d}, \quad y = \frac{b}{d}.$$

It is obvious that $x$ and $y$ are relatively prime. From $ab = m(a + b)$, we obtain $dxy = m(x + y)$, implying $x + y \mid dxy$. But $xy$ and $x + y$ are relatively prime because so are $x$ and $y$. Thus, $x + y \mid d$, and hence there exists an integer $z$ such that $d = (x + y)z$. Form this, we obtain

$$a = x(x + y)z, \quad b = y(x + y)z,$$

which is what we want. ∎

**2.** Set $\beta = \int_0^1 xf(x)dx$. We prove that

$$\int_0^1 (x - \beta)^2 f(x)dx \leq \int_0^1 (x - \alpha)^2 f(x)dx, \tag{*}$$

for all $\alpha \in \mathbb{R}$. To this end, using $\int_0^1 f(x)dx = 1$ and $\int_0^1 xf(x)dx = \beta$, we can write

$$
\begin{aligned}
\int_0^1 (x - \alpha)^2 f(x)dx &= \int_0^1 \left((x - \beta) + (\beta - \alpha)\right)^2 f(x)dx \\
&= \int_0^1 (x - \beta)^2 f(x)dx + (\beta - \alpha)^2 + 2(\beta - \alpha)\beta - 2(\beta - \alpha)\beta \\
&= \int_0^1 (x - \beta)^2 f(x)dx + (\beta - \alpha)^2 \\
&\geq \int_0^1 (x - \beta)^2 f(x)dx.
\end{aligned}
$$

As $(x - \frac{1}{2})^2 \leq \frac{1}{4}$ on $[0, 1]$, we can write

$$
\int_0^1 \left(x - \frac{1}{2}\right)^2 f(x)dx \leq \int_0^1 \frac{1}{4} f(x)dx = \frac{1}{4}.
$$

Now, in view of $(*)$, we have

$$
\int_0^1 (x - \beta)^2 f(x)dx \leq \int_0^1 \left(x - \frac{1}{2}\right)^2 f(x)dx \leq \frac{1}{4},
$$

which is what we want.  ∎

**3.** (a) The desired probability is the ratio of the number of "favorable cases" to the number of "total cases". The number of "total cases" is obviously $n^p$. To calculate the number of "favorable cases", let $A_i$ be the event that no one gets on the $i$th wagon. It follows that the number of "disfavorable cases" is equal to $|A_1 \cup \cdots \cup A_n|$. Using the inclusion-exclusion principle from combinatorics, we can write

$$
\begin{aligned}
|A_1 \cup \cdots \cup A_n| &= \sum_i |A_i| - \sum_{i,j} |A_i \cap A_j| + \cdots + (-1)^{n-1}|A_1 \cap \cdots \cap A_n| \\
&= \sum_i (n - 1)^p - \sum_{i,j} (n - 2)^p + \cdots + (-1)^{n-1} 0^p \\
&= \binom{n}{1}(n - 1)^p - \binom{n}{2}(n - 2)^p + \cdots + (-1)^{n-2}\binom{n}{n-1}1^p.
\end{aligned}
$$

Thus, the number of "favorable cases" is equal to

$$
n^p - \binom{n}{1}(n - 1)^p + \binom{n}{2}(n - 2)^p + \cdots + (-1)^{n-1}\binom{n}{n-1},
$$

and hence the desired probability is

$$
\frac{1}{n^p}\left(n^p - \binom{n}{1}(n - 1)^p + \binom{n}{2}(n - 2)^p + \cdots + (-1)^{n-1}\binom{n}{n-1}\right),
$$

which is what we want.

(b) First, if $p < n$, the probability calculated in (a) is obviously zero, implying that the number of "favorable cases" is zero. This, in view of $\binom{n}{r} =$

$\binom{n}{n-r}$, yields

$$\binom{n}{1}1^p - \binom{n}{2}2^p + \binom{n}{3}3^p - \cdots + (-1)^{n-1}\binom{n}{n}n^p = 0.$$

Next, if $p = n$, then the number of wagons is equal to that of the passengers. In this case, in order not to have any empty wagon, we must have one passenger in each wagon. Thus, the number of "favorable cases" is equal $n!$. On the other hand, if we let $p = n$ in the above formula, which we obtained for the number of "favorable cases", and equate these two values, we obtain

$$\binom{p}{1}1^p - \binom{p}{2}2^p + \binom{p}{3}3^p - \cdots + (-1)^{p-1}\binom{p}{p}p^p = (-1)^{p-1}p!,$$

finishing the proof. ∎

## 2.16. Sixteenth Competition

**2.16.1. Analysis. 1.** As $g$ is continuous on the compact interval $[0,1]$, it follows that there exists an $M > 0$ such that $|g(x)| \le M$ for all $x \in [0,1]$. From the continuity of $g$ at 1 and $g(1) = 0$, we see that for given $\varepsilon > 0$, there exists a $0 < \delta < 1$ such that

$$|g(x)| < \varepsilon,$$

whenever $x \in [0,1]$ and $0 < 1 - x < \delta$. On the other hand, for given $\varepsilon > 0$, there exists an $N \in \mathbb{N}$ such that $(1-\delta)^n M < \varepsilon$ for all $n \ge N$. It thus follows that for a given $\varepsilon > 0$, for all $x \in [0,1]$ and $n \ge N$, we have

$$|f_n(x) - 0| = |x^n||g(x)| \le \max\left((1-\delta)^n M, \sup_{1-\delta \le x \le 1}|g(x)|\right) < \varepsilon,$$

proving the assertion. ∎

**2.** (a) Let $x = [x] + 0.a_1 a_2 a_3 \ldots$. We have

$$x + 0.1 = [x] + a_0.a_1' a_2 a_3 \ldots,$$

where $a_0 = 0$ or 1 and $a_1' \in \{0,1,\ldots,9\}$ and hence

$$f(x) = f(x + 0.1) = a_2.$$

Hence, 0.1 is a period for $f$. To prove that 0.1 is the period of $f$, suppose to the contrary that $0 < \alpha < 0.1$ is a period of $f$. If we let $x_1 = 0.1 - (0.1)\alpha$, we have $0.09 < x_1 < 0.1$. Consequently, $x_1 = 0.09\alpha_1\alpha_2\ldots$, where all of $\alpha_i$'s are not equal to 9. It follows that $f(x_1) = 9$. On the other hand, if we let $x_2 = \alpha + x_1 = 0.1 + (0.9)\alpha$, we have

$$0.0999\ldots = 0.1 < x_2 < 0.19 = 0.18999\ldots.$$

Thus, $x_2 = 0.1\beta_1\beta_2\ldots$, where $\beta_1 \le 8$. Hence, $f(x_2) = \beta_2 \le 8$, whence $f(x_1) \ne f(x_2)$, which is a contradiction. This shows that $\alpha$ cannot be the period of $f$.

(b) Using integration by parts, we can write

$$\int_0^c x df(x) \;=\; \int_0^{0.1} x df(x) = 0.1 f(0.1) - \int_0^{0.1} f(x)\,dx$$

$$= \;(0.1).9 - \sum_{k=0}^{9} k\frac{1}{100} = 0.45,$$

which is what we want.                                              ∎

**3.** We prove the assertion in the normed linear space setting. *Let $X$ be a real or complex normed linear space and $f : X \to X$ a uniformly continuous function. Then, there exists $a, b \in \mathbb{R}^+$ such that*

$$||f(x)|| \;\leq\; a||x|| + b,$$

*for all $x \in X$, where $||.||$ denotes the norm of $X$.* As $f$ is uniformly continuous on $X$, for $\varepsilon = 1$, there exists $\delta_0 > 0$ such that $||f(x) - f(y)|| < 1$ whenever $x, y \in \mathbb{R}^n$ and $||x - y|| \leq \delta_0$. We show that for $a = \frac{1}{\delta_0}$ and $b = |f(0)| + 1$, we have $||f(x)|| \leq a||x|| + b$ for all $x \in X$. To this end, for $x \in X$, set $N = \left[\frac{||x||}{\delta_0}\right]$, where bracket stands for the integer part function. We can write

$$
\begin{aligned}
||f(x)|| \;\leq\;& ||f(0)|| + ||f(x) - f(0)|| \\
\leq\;& ||f(0)|| \\
&+ \sum_{k=1}^{N}\left\|f\left(k\delta_0\frac{x}{||x||}\right) - f\left((k-1)\delta_0\frac{x}{||x||}\right)\right\| + \left\|f(x) - f\left(N\delta_0\frac{x}{||x||}\right)\right\|.
\end{aligned}
$$

From $N = \left[\frac{||x||}{\delta_0}\right]$, we obtain

$$N \leq \frac{||x||}{\delta_0} < N + 1,$$

yielding

$$\left\|x - N\delta_0\frac{x}{||x||}\right\| \;<\; \frac{||x||}{N+1} < \delta_0,$$

$$\left\|k\delta_0\frac{x}{||x||} - (k-1)\delta_0\frac{x}{||x||}\right\| \;=\; \left\|\frac{\delta_0}{||x||}x\right\| = \delta_0 \leq \delta_0,$$

for all $1 \leq k \leq N$. So, for all $x \in X$, we can write

$$||f(x)|| \leq ||f(0)|| + N + 1 \leq \frac{1}{\delta_0}||x|| + (||f(0)|| + 1),$$

whence

$$||f(x)|| \;\leq\; a||x|| + b,$$

for all $x \in X$, where $a, b$ are as in the above, which is what we want.    ∎

**2.16.2. Algebra. 1.** First, recall that $\text{Inn}(G) \cong \frac{G}{Z(G)}$, where $Z(G)$ denotes the center of the group $G$. To see this, it is easily checked that the map $f : G \longrightarrow \text{Aut}(G)$ defined by $f_g(x) = gxg^{-1}$ is a homomorphism of groups and that $\ker f = Z(G)$. It thus follows from the First Isomorphism Theorem for groups that $\text{Inn}(G) \cong \frac{G}{Z(G)}$. Next, we need to recall that *if $p$ is a prime, then every group $G$ of order $p^2$ is abelian*. To prove this by contradiction, suppose that $G$ is not abelian. Note first that $|Z(G)| > 1$ because $G$ is a $p$-group (see Solution 3 of 2.11.2). Consequently, $|Z(G)| = p$. But then $\left|\frac{G}{Z(G)}\right| = p$, implying that $\frac{G}{Z(G)}$ is cyclic and hence $G$ is abelian, a contradiction. Thus, $G$ is abelian, as desired.

We now prove the assertion. To this end, from the hypothesis that $[G : A] = [G : B] = p$, it follows that $|A| = |B|$. And since $p$ is the smallest prime that divides $|G|$, from Problem 1 of 2.9.2, we see that $A$ and $B$ are both normal in $G$. It thus follows that $AB \leq G$. By the Second Isomorphism Theorem for groups, we have $\frac{AB}{A} = \frac{B}{A \cap B}$. So we can write

$$[G : AB] = \frac{|G|}{|AB|} = \frac{[G : A]}{[B : A \cap B]} = \frac{p}{[B : A \cap B]}.$$

If $[B : A \cap B] = 1$, we obtain $A = B$, which is a contradiction. Thus, $[B : A \cap B] = p$, and hence $G = AB$. We now prove that $A \cap B = Z(G)$. To this end, for $x \in A \cap B$, as is usual, use $C_G(x)$ to denote the centralizer of the element $x$ of $G$. As $A$ and $B$ are abelian, we have $A, B \leq C_G(x)$, and hence

$$p = [G : A] = [G : C_G(x)][C_G(x) : A],$$

$$p = [G : B] = [G : C_G(x)][C_G(x) : B],$$

from which, we see that $C_G(x) = G$. That is, $x \in Z(G)$, and hence $A \cap B \leq Z(G)$. Now, as

$$[G : A \cap B] = [G : B][B : A \cap B] = [G : B][AB : A] = [G : B][G : A] = p^2,$$

we conclude that $[G : Z(G)] = 1$ or $p$ or $p^2$ because $[G : Z(G)]\,|\,[G : A \cap B]$. As $G$ is nonabelian, the cases $[G : Z(G)] = 1$ or $p$ are impossible. Thus, $[G : Z(G)] = p^2$, from which, in view of $[G : A \cap B] = p^2$, we see that $A \cap B = Z(G)$. Consequently, $\left|\text{Inn}(G)\right| = \left|\frac{G}{Z(G)}\right| = p^2$, and hence $\text{Inn}(G)$ is abelian. This together with the Fundamental Theorem of finite abelian groups implies $\text{Inn}(G) \cong \frac{G}{Z(G)} \cong \mathbb{Z}_{p^2}$ or $\mathbb{Z}_p \oplus \mathbb{Z}_p$. From $\text{Inn}(G) \cong \frac{G}{Z(G)} \cong \mathbb{Z}_{p^2}$, it follows that $G$ is abelian, which is impossible. Therefore, $\text{Inn}(G) \cong \mathbb{Z}_p \oplus \mathbb{Z}_p$, which is what we want. ∎

**2.** There exists an $n \in \mathbb{N}$ such that $(r^2 - r)^n = 0$, for $r^2 - r$ is nilpotent. It is plain that there exists a $g \in \mathbb{Z}[x]$ such that

$$0 = (r^2 - r)^n = r^n - r^{n+1}g(r),$$

whence $r^n = r^{n+1}g(r)$. Setting $f(x) = x^n g(x)^n$, we have

$$f(r)^2 = r^{2n}g(r)^{2n},$$

and hence we can write

$$
\begin{aligned}
r^n &= r^{n+1}g(r) = rg(r)r^n = rg(r)r^{n+1}g(r) = r^{n+2}g(r)^2 = r^2 g(r)^2 r^n \\
&= r^2 g(r)^2 r^{n+1}g(r) = r^{n+3}g(r)^3 = \cdots = r^{n+n}g(r)^n = r^{2n}g(r)^n.
\end{aligned}
$$

That is, $r^n = r^{2n}g(r)^n$, yielding

$$
f(r)^2 = r^{2n}g(r)^{2n} = r^{2n}g(r)^n g(r)^n = r^n g(r)^n = f(r).
$$

If $f(r) = 0$, then

$$
0 = r^n f(r) = r^n r^n g(r)^n = r^{2n}g(r)^n = r^n,
$$

implying $r^n = 0$, which contradicts the hypothesis that $r$ is not nilpotent. Therefore, $f(r)$ is a nonzero idempotent element of $R$, finishing the proof. ∎

**3. Remark.** Adjusting the proof below one can prove that

$$
\det\big(\mathrm{lcm}(i,j)\big) = n! f(1) \cdots f(n),
$$

where $f(n) = \sum_{d|n} d\mu(d)$ and $\mu$ denotes the Möbius function. Hint. Use $\mathrm{lcm}(i,j) = \frac{ij}{\gcd(i,j)}$.

We show that the matrix $A$ is invertible. To this end, define the matrices $B = (b_{ij})$ and $C = (c_{ij})$ in $M_n(\mathbb{Q})$ as follows

$$
b_{ij} = \begin{cases} \varphi(i) & i = j \\ 0 & i \neq j \end{cases}, \quad c_{ij} = \begin{cases} 1 & j | i \\ 0 & j \nmid i \end{cases},
$$

where $\varphi$ is the Euler's totient function. Let $C^t$ denote the transpose of $C$. We have

$$
\begin{aligned}
(CBC^t)_{ij} &= \sum_{k=1}^n c_{ik}(BC^t)_{kj} = \sum_{k=1}^n c_{ik} \sum_{k'=1}^n b_{kk'}c_{jk'} \\
&= \sum_{k=1}^n c_{ik}\varphi(k)c_{jk} = \sum_{k|i,\ k|j} \varphi(k)c_{ik}c_{jk} = \sum_{k|\gcd(i,j)} \varphi(k) \\
&= \gcd(i,j) = a_{ij} = (A)_{ij},
\end{aligned}
$$

for all $1 \leq i, j \leq n$ (for a proof of $\sum_{d|n} \phi(d) = n$ for all $n \in \mathbb{N}$, see Solution 1 of 2.2.2). Consequently, $A = CBC^t$. So we can write

$$
\begin{aligned}
\det A &= \det CBC^t = \det C \det B \det C^t \\
&= 1 . \det B . 1 = \det B = \varphi(1)\varphi(2)\cdots\varphi(n),
\end{aligned}
$$

implying that $\det A \neq 0$. Thus, $A$ is invertible, which is what we want. ∎

## 2.17. Seventeenth Competition

**2.17.1. Analysis. 1.** To prove the assertion by contradiction, suppose that $|f'(x)| \leq \left|\frac{f(b)-f(a)}{b-a}\right|$ for all $x \in (a,b)$. There are two cases to consider.

(i) $\frac{f(b)-f(a)}{b-a} \geq 0$.

Define the function $g : [a,b] \to \mathbb{R}$ by $g(x) = f(x) - \frac{f(b)-f(a)}{b-a}(x-a)$. Now, from the contradiction hypothesis, we see that

$$g'(x) = f'(x) - \frac{f(b)-f(a)}{b-a} \leq 0,$$

for all $x \in (a,b)$. Thus, $g$ is non-increasing on $[a,b]$. On the other hand, we have $g(a) = f(a) = g(b)$, and hence $f(a) = g(b) \leq g(x) \leq g(a) = f(a)$ for all $x \in [a,b]$. Therefore, $g(x) = f(a)$ for all $x \in [a,b]$. In other words, $f(x) = f(a) + \frac{f(b)-f(a)}{b-a}(x-a)$ for all $x \in [a,b]$. That is, the graph of $f$ is a line segment, which is a contradiction. This proves the assertion in this case.

(ii) $\frac{f(b)-f(a)}{b-a} \leq 0$.

Adjusting the above proof or replacing $f$ by $-f$ and repeating the above argument, one can prove the assertion in this case as well. ∎

**2.** Let

$$p(x) = \int_{-1}^{x} p_1 p_3 \int_{-1}^{x} p_2 p_4 - \int_{-1}^{x} p_1 p_4 \int_{-1}^{x} p_2 p_3.$$

It is plain that $p \in \mathbb{R}[x]$ and $p(-1) = 0$. Form this, it follows that $p$ is divisible by $x + 1$. We can write

$$p'(x) = p_1(x)p_3(x) \int_{-1}^{x} p_2 p_4 + p_2(x)p_4(x) \int_{-1}^{x} p_1 p_3$$

$$-p_1(x)p_4(x) \int_{-1}^{x} p_2 p_3 - p_2(x)p_3(x) \int_{-1}^{x} p_1 p_4,$$

from which, we obtain $p'(-1) = 0$. Thus, $p(x)$ is divisible by $(x+1)^2$. Taking derivative of the both sides of the above equality yields

$$p''(x) = (p_1 p_3)'(x) \int_{-1}^{x} p_2 p_4 + (p_1 p_3 p_2 p_4)(x) + (p_2 p_4)'(x) \int_{-1}^{x} p_1 p_3$$

$$+(p_2 p_4 p_1 p_3)(x) - (p_1 p_4)'(x) \int_{-1}^{x} p_2 p_3 - (p_1 p_4 p_2 p_3)(x)$$

$$+(p_2 p_3)'(x) \int_{-1}^{x} p_1 p_4 - (p_2 p_3 p_1 p_4)(x),$$

implying $p''(-1) = 0$, and hence $p(x)$ is divisible by $(x+1)^3$. Finally, we can write

$$p'''(x) = (p_1 p_3)''(x) \int_{-1}^{x} p_2 p_4 + ((p_1 p_3)' p_2 p_4)(x) + (p_2 p_4)''(x) \int_{-1}^{x} p_1 p_3$$

$$+((p_2 p_4)' p_1 p_3)(x) - (p_1 p_4)''(x) \int_{-1}^{x} p_2 p_3 - ((p_1 p_4)' p_2 p_3)(x)$$

$$-(p_2 p_3)''(x) \int_{-1}^{x} p_1 p_4 - ((p_2 p_3)' p_1 p_4)(x),$$

which obtains

$$p'''(-1) = ((p_1p_3)'p_2p_4 + (p_2p_4)'p_1p_3)(-1) - ((p_1p_4)'p_2p_3 + (p_2p_3)'p_1p_4)(-1)$$
$$= (p_1p_2p_3p_4)'(-1) - (p_1p_2p_3p_4)'(-1) = 0.$$

Therefore, $p(x)$ is divisible by $(x+1)^4$, which is what we want. ∎

**3.** " $\Longrightarrow$ " Suppose that $Z(f) := \{x \in X : f(x) = 0\} = f^{-1}(\{0\})$ is an open subset of $X$. It follows that $Z(f)$ is both open and close, and hence so is $Z(f)^c = X \setminus Z(f)$. Now, since $X = Z(f) \cup Z(f)^c$ and $Z(f)$ and $Z(f)^c$ are both open sets, the function $g : X \to \mathbb{R}$ defined by

$$g(x) = \begin{cases} \frac{1}{f(x)} & x \in Z(f)^c \\ 0 & x \in Z(f) \end{cases}$$

is continuous on $X$ and furthermore satisfies $f = gf^2$ on $X$. This is what we want.

" $\Longleftarrow$ " Suppose that there exists a continuous function $g : X \to \mathbb{R}$ such that $f = gf^2$. To show that $Z(f) := \{x \in X : f(x) = 0\} = f^{-1}(\{0\})$ is an open subset of $X$, we prove that $Z(f)^c := X \setminus Z(f)$ is closed. It suffices to show that for any sequence $(x_n)_{n=1}^{+\infty}$, with $x_n \in Z(f)^c$, we have $x_\infty = \lim_n x_n \in Z(f)^c$. To this end, first note that $f(x_n) = g(x_n)f(x_n)^2$ for all $n \in \mathbb{N}$. As $f(x_n) \neq 0$ for all $n \in \mathbb{N}$, we see that $f(x_n)g(x_n) = 1$ for all $n \in \mathbb{N}$. Now, in view of the continuity of $f$ and $g$, letting $n \to +\infty$, we obtain $f(x_\infty)g(x_\infty) = 1$, yielding $f(x_\infty) \neq 0$. In other words, $x_\infty \in Z(f)^c$, which is what we want. ∎

**4.** Define the function $g : [0, \frac{1}{2}] \to \mathbb{R}$ by $g(x) = f(x + \frac{1}{2}) - f(x)$. We have

$$g(0) = f(\frac{1}{2}) - f(0) = -\left(f(1) - f(\frac{1}{2})\right) = -g(\frac{1}{2}),$$

yielding $g(0)g(\frac{1}{2}) = -g(0)^2 \leq 0$. It thus follows from the Intermediate Value Theorem that there exists a $c \in [0, \frac{1}{2}]$ such that $g(c) = 0$. That is, $f(c + \frac{1}{2}) = f(c)$. Letting $a = c$ and $b = c + \frac{1}{2}$, we have $b - a = \frac{1}{2}$ and $f(a) = f(b)$, which is what we want. ∎

**2.17.2. Algebra. 1.** For $a, b \in G$, define $a \sim b$ if and only if $h_1 a h_2 = b$ for some $h_1, h_2 \in H$. It is readily verify that $\sim$ is indeed an equivalence relation on $G$. Also, for any $x \in G$, we have $[x] = HxH$, where $[x]$ denotes the equivalence class of $x$. With all that in mind, let $e$ denote the identity element of $G$ and $\{x_1 = e, x_1, \ldots, x_n\}$ be a maximal set of nonequivalent elements of $G$. It follows from the hypothesis that $n > 1$. We have

$$G = \bigcup_{i=1}^{n} Hx_iH.$$

This yields

$$|G| = \sum_{i=1}^{n} \frac{|H||x_i^{-1}Hx_i|}{|H \cap (x_i^{-1}Hx_i)|} = \sum_{i=1}^{n} \frac{|H|^2}{|H \cap (x_i^{-1}Hx_i)|},$$

from which, together with the hypothesis, we obtain

$$|G| = |H| + \sum_{i=2}^{n} |H|^2 = |H| + (n-1)|H|^2.$$

This implies $[G : H] = \frac{|G|}{|H|} = 1 + (n-1)|H|$. In other words, $[G : H] - (n-1)|H| = 1$, which easily yields $\gcd([G : H], |H|) = 1$, which is what we want. ∎

**2.** The assertion is a special case of the following. *Let $R$ be a unital ring with the property that $ab = 1$ implies $ba = 1$ whenever $a, b \in R$. Then, the ring $R[x]$ has the same property, i.e., if $a, b \in R[x]$ and $ab = 1$, then $ba = 1$.* If $D$ is a division ring and $n \in \mathbb{N}$, it follows from the Rank-Nullity Theorem that for $A, B \in M_n(D)$, we have $AB = I_n$ if and only if $BA = I_n$, where $I_n$ denotes the identity matrix. Thus, the ring $R = M_n(F)$ satisfies the above property. As a matter of fact, it can be shown that any left or right Noetherian ring satisfies the above property. To prove the above more general assertion, suppose that for $f, g \in R[x]$ we have $fg = 1$. We prove that $gf = 1$. Suppose to the contrary that $gf \neq 1$. Note first that $(gf)^2 = g(fg)f = gf$. Assuming that $f = a_0 + a_1 x + \cdots + a_n x^n$ and $g = b_0 + b_1 x + \cdots + b_m x^m$, we obtain $a_0 b_0 = 1$ because $fg = 1$. Thus, $b_0 a_0 = 1$. So we can write $gf = 1 + cx^k + \cdots$, where $k$ is the least exponent of $x$ such that $c \neq 0$. Now, from $(gf)^2 = gf$, we see that

$$(1 + cx^k + \cdots)^2 = 1 + cx^k + \cdots,$$

from which, we obtain $2c = c$, yielding $c = 0$, which is a contradiction. Thus $gf = 1$, which is what we want. ∎

**3.** We prove the counterpart of the problem for left (resp. right) finite-dimensional vector spaces over a division ring $D$. So assume that $V$ is a left (resp. right) vector space over a division ring $D$ and $T : V \longrightarrow V$ a left (resp. right) linear transformation on $V$. As $T^2 V \subseteq TV$, the linear transformation $S : TV \to TV$ defined by $S(Tx) = T^2 x$ defines a left (resp. right) linear transformation on the left (resp. right) vector space $TV$. By the Rank-Nullity Theorem, we can write

$$\dim TV = \dim S(TV) + \dim \ker S.$$

As $S(TV) = T^2 V$ and $\ker S = \ker T \cap TV$, we see that

$$\dim(\ker T \cap TV) = \dim TV - \dim T^2 V.$$

Therefore,

$$\dim(\ker T \cap TV) = \text{rank}(T) - \text{rank}(T^2),$$

as desired. ∎

## 2.18. Eighteenth Competition

**2.18.1. Analysis. 1.** We have $a_n = A_n - A_{n-1}$. This together with the hypothesis easily yields $\lim_n \frac{A_{n-1}}{A_n} = 1$, and hence $\lim_n \frac{A_n}{A_{n-1}} = 1$ because $A_n > 0$ for all $n \in \mathbb{Z}$ with $n \geq 0$. From this, it follows that the radius of convergence of the power series $\sum_{n=0}^{+\infty} A_n x^n$ is 1. In particular, the series absolutely converges for all $x \in (-1, 1)$. Now, as $a_0 = A_0$ and $a_n < A_n$ for all $n \in \mathbb{N}$, it follows that

$$0 \leq a_n |x|^n \leq A_n |x|^n,$$

for all $n \in \mathbb{N} \cup \{0\}$ and $x \in (-1, 1)$. Thus, by the Comparison Test, $\sum_{n=0}^{+\infty} a_n |x|^n$ converges for all $x \in (-1, 1)$ because so does $\sum_{n=0}^{+\infty} A_n |x|^n$ for all $x \in (-1, 1)$. So, if $R$ is the radius of convergence of the series $\sum_{n=0}^{+\infty} a_n x^n$, then $R \geq 1$. But $R > 1$ is impossible, for otherwise the series $\sum_{n=0}^{+\infty} a_n$ must be convergent, which contradicts the hypothesis that $\lim_n A_n = +\infty$. Therefore, $R = 1$, which is what we want. ∎

**2.** We prove the assertion under the weaker hypothesis that the functions $f$ and $g$ are Riemann integrable on $[0, 1]$. In the following integrals, perform the substitutions $nx = t$ and $s = t - k + 1$, respectively, to obtain

$$\int_0^1 f(x)g(nx)dx = \int_0^n \frac{1}{n} f\left(\frac{t}{n}\right) g(t) dt$$

$$= \sum_{k=1}^n \int_{k-1}^k \frac{1}{n} f\left(\frac{t}{n}\right) g(t) dt$$

$$= \sum_{k=1}^n \int_{k-1}^k \frac{1}{n} f\left(\frac{s}{n} + \frac{k-1}{n}\right) g(s + (k-1)) ds.$$

But the period of $g$ is one. So we can write

$$\int_0^1 f(x)g(nx)dx = \int_0^1 f_n(s)g(s)ds,$$

where $f_n(s) = \frac{1}{n} \sum_{k=1}^n f\left(\frac{s}{n} + \frac{k-1}{n}\right)$. Suppose we have proved that the sequence $(f_n)_{n=1}^{+\infty}$ uniformly converges to $\int_0^1 f$ on $[0, 1]$. Since $g$ is bounded on $[0, 1]$, we see that the sequence $(f_n g)_{n=1}^{+\infty}$ uniformly converges to $g \int_0^1 f$ on $[0, 1]$. Hence, we can write

$$\lim_n \int_0^1 f(x)g(nx)dx = \lim_n \int_0^1 f_n(s)g(s)ds$$

$$= \int_0^1 \lim_n f_n(s)g(s)ds = \int_0^1 \left(\int_0^1 f\right) g(s) ds$$

$$= \left(\int_0^1 f\right)\left(\int_0^1 g\right),$$

proving the assertion. It remains to show that the sequence $(f_n)_{n=1}^{+\infty}$ uniformly converges to $\int_0^1 f$ on $[0, 1]$. To this end, as $f$ is Riemann integrable, it follows from Riemann's criterion for integrability that for given $\varepsilon > 0$, there exists a

$\delta > 0$ such that for any partition $P \in \mathcal{P}[0, 1]$, where $\mathcal{P}[0, 1]$ denotes the set of all partitions of the interval $[0, 1]$, with $||P|| < \delta$, we have

$$U(P, f) - L(P, f) = \sum_{i=1}^{n}(M_i - m_i)\Delta x_i < \varepsilon,$$

where

$$P : x_0 = 0 < x_1 < \cdots < x_n = 1, \; ||P|| = \max_{1 \leq i \leq n} \Delta x_i$$

$$U(P, f) = \sum_{i=1}^{n} M_i \Delta x_i, \; L(P, f) = \sum_{i=1}^{n} m_i \Delta x_i,$$

$$m_i = \inf_{x \in [x_{i-1}, x_i]} f(x), \; M_i = \sup_{x \in [x_{i-1}, x_i]} f(x).$$

We also know that if $U(P, f) - L(P, f) < \varepsilon$ for some $P \in \mathcal{P}[0, 1]$, then for any $x_{i-1} \leq \xi_i \leq x_i$, we have

$$\left| \sum_{i=1}^{n} f(\xi_i)\Delta x_i - \int_0^1 f \right| < \varepsilon.$$

Note that $f_n(s)$ corresponds to the Riemann sum with respect to the uniform partition of $[0, 1]$ with $n$ subintervals of the same length, i.e., $\Delta x_i = \frac{1}{n}$ for all $1 \leq i \leq n$, and the mid points $\xi_i = \frac{s}{n} + \frac{i-1}{n}$. In view of this, for given $\varepsilon > 0$, find $\delta > 0$ from the above and let $N \in \mathbb{N}$ such that $\frac{1}{N} < \delta$. Now, for each $n \geq N$, let $P_n$ be the uniform partition of $[0, 1]$ with $n$ subintervals of the same length. We have $||P_n|| = \frac{1}{n} < \frac{1}{N} < \delta$. It thus follows that

$$\left| f_n(s) - \int_0^1 f \right| = \left| \frac{1}{n}\sum_{i=1}^{n} f\left(\frac{s}{n} + \frac{i-1}{n}\right) - \int_0^1 f \right|$$

$$= \left| \sum_{i=1}^{n} f(\xi_i)\Delta x_i - \int_0^1 f \right|$$

$$\leq U(P_n, f) - L(P_n, f) < \varepsilon.$$

That is,

$$\left| f_n(s) - \int_0^1 f \right| < \varepsilon,$$

for all $s \in [0, 1]$ and $n \geq N$. In other words, the sequence $(f_n)_{n=1}^{+\infty}$ uniformly converges to $\int_0^1 f$ on $[0, 1]$, which is what we want. $\blacksquare$

**3.** We need the following lemma.

**Lemma.** *If $\alpha \in [0, 1]$, then there exists a sequence $(a_i)_{i=1}^{+\infty}$ with $a_i \in \{0, 1\}$ ($i \in \mathbb{N}$) such that $\lim\limits_{n \to +\infty} \dfrac{\sum_{i=1}^{n} a_i}{n} = \alpha$.*

**Proof.** If $\alpha = 0$, let $a_i = 0$ for all $i \in \mathbb{N}$. If not, let

$$a_i = \begin{cases} 1 & \text{if } i = \lfloor \frac{m}{\alpha} \rfloor \text{ for some } m \in \mathbb{N}, \\ 0 & \text{otherwise}, \end{cases}$$

where $\lfloor . \rfloor$ stands for the integer part function. We have

$$\frac{\sum_{i=1}^{n} a_i}{n} = \frac{1}{n}\text{card}\left\{m \in \mathbb{N} : \left\lfloor \frac{m}{\alpha} \right\rfloor \leq n \right\}$$

$$= \frac{1}{n}\text{card}\left\{m \in \mathbb{N} : m < \alpha(n+1) \right\} = \frac{\lfloor \alpha(n+1)\rfloor}{n}.$$

This obtains

$$\lim \frac{\sum_{i=1}^{n} a_i}{n} = \lim \frac{\lfloor \alpha(n+1)\rfloor}{n} = \alpha,$$

as desired.                                                                                       □

To prove the assertion, note first that if a sequence $(a_i)_{i=1}^{+\infty}$ has the property that $\lim\limits_{n \to +\infty} \frac{\sum_{i=1}^{n} a_i}{n} = \alpha$, then the property remains intact under changing a finite number of the terms of $(a_i)_{i=1}^{+\infty}$. Now, let $I$ be a nonempty open interval of $\mathbb{R}$ so that $I = (x - r, x + r)$ for some $x \in I$ and $r > 0$. Clearly, $g(I) \subseteq [0, 1]$. Let $\alpha \in [0, 1]$ be arbitrary. Choose $n \in \mathbb{N}$ such that $\frac{1}{2^n} < \varepsilon$. And choose $x' \in \mathbb{R}$ such that the first $n$ digits of its binary expansion are the same of those of $x$ and from its $(n+1)$st digit onward its digits are equal to $a_i$'s. That is,

$$x' = 0.x_1 \ldots x_n a_1 a_2 a_3 \ldots,$$

where $x = 0.x_1 \ldots x_n x_{n+1} \ldots$. It is obvious that $x' \in I$ and that $g(x') = \alpha$. Consequently, $[0, 1] \subseteq g(I)$, and hence $g(I) = [0, 1]$, as desired.

For the rest, note first that any open subset $G$ of $\mathbb{R}$ includes an open interval $I$, implying that $g(G) \supseteq g(I) = [0, 1]$. This implies $g(G) = [0, 1]$ because $g(G) \subseteq [0, 1]$. That is, for any open subset $G$ of $\mathbb{R}$, we have $g(G) = [0, 1]$. Now, define the function $h : \mathbb{R} \to \mathbb{R}$ by $h = f \circ g$, where the function $f : [0, 1] \to \mathbb{R}$ is defined by

$$f(x) = \begin{cases} x & 0 < x < 1, \\ \frac{1}{2} & x \in \{0, 1\}. \end{cases}$$

It is obvious that for any open subset $G$ of $\mathbb{R}$, we have

$$h(G) = f\big(g(G)\big) = f\big([0, 1]\big) = (0, 1).$$

In other words, $h$ is an open map. By proving that $h$ has no limit at any point of $\mathbb{R}$, we show that $h$ is not continuous on $\mathbb{R}$. Suppose to the contrary that there exist $x_0, \ell \in \mathbb{R}$ such that

$$\lim_{x \to x_0} h(x) = \ell.$$

It follows that for every $\varepsilon > 0$, there exists a $\delta > 0$ such that

$$\ell - \varepsilon < h(x) < \ell + \varepsilon,$$

whenever $x \in (x_0 - \delta, x_0 + \delta)$. We have $0 \leq l \leq 1$ because $0 < h < 1$. Consequently, we can choose $\varepsilon > 0$ such that $(\ell - \varepsilon, \ell + \varepsilon) \cap (0, 1) \subsetneq (0, 1)$. For this $\varepsilon > 0$, find $\delta > 0$ from the above. Thus, for all $x \in (x_0 - \delta, x_0 + \delta)$, we have

$$h(x) \in (\ell - \varepsilon, \ell + \varepsilon). \tag{$*$}$$

Letting $G = (x_0 - \delta, x_0 + \delta)$, we obtain

$$h(G) = f\big(g(G)\big) = (0, 1),$$

from which, in view of $(*)$, we see that

$$(0,1) \subseteq (\ell - \varepsilon, \ell + \varepsilon),$$

implying $(\ell - \varepsilon, \ell + \varepsilon) \cap (0,1) = (0,1)$, contradicting our choice of $\varepsilon > 0$. Therefore, the function $h$ has no limit at any point of $\mathbb{R}$, which is what we want. ∎

**2.18.2. Algebra. 1.** The subgroup $H \cap K$ is normal in $K$ because $K$ is normal in $G$. Since $K$ is simple, we have $H \cap K = K$ or $H \cap K = \{e\}$, where $e$ is the identity element of $G$. If $H \cap K = K$, then $K \subseteq H$, implying that $K = H$ because $|G| < \infty$ and $H \cong K$. To finish the proof, we show that $H \cap K \neq \{e\}$. Suppose to the contrary that $H \cap K = \{e\}$. As $K$ is normal in $G$, we have $HK = KH$, whence $HK \leq G$, and hence $|HK| \,\big|\, |G|$. But

$$|HK| = \frac{|H||K|}{|H \cap K|} = |K|^2.$$

In other words, $|K|^2 \,\big|\, |G|$, contradicting the hypothesis that the square of the order of $K$ does not divide that of $G$. Thus, $H \cap K \neq \{e\}$, which is what we want. ∎

**2.** It suffices to show that any right ideal of $R$ is finitely generated. To this end, let $I$ be a right ideal in $R$. If $I = 0$, the assertion is trivial. Suppose $I \neq 0$. As is usual, use $E_{ij}$ to denote the $2 \times 2$ matrix whose $ij$ entry is 1 and zero elsewhere. There are two cases to consider.

(i) There exists $A = a_0 E_{11} + b_0 E_{12} + c_0 E_{22} \in I$ such that $a_0 \neq 0$.

Using the well-ordering principle of natural numbers, let $a_m$ be the least positive integer $a$ for which there are $b, c \in \mathbb{Q}$ such that $a E_{11} + b E_{12} + c E_{22} \in I$. It is easily verified that if $a E_{11} + b E_{12} + c E_{22} \in I$ for some $a \in \mathbb{Z}$ and $b, c \in \mathbb{Q}$, then $a_m$ divides $a$. We claim that $I$ is generated by $\{a_m E_{11}, E_{12}, E_{22}\}$. Suppose $a_m E_{11} + b E_{12} + c E_{21} \in I$ for some $b, c \in \mathbb{Q}$. It follows that

$$(a_m E_{11} + b E_{12} + c E_{22}) E_{12} = a_m E_{12} \in I,$$

implying

$$(a_m E_{12})\left(\frac{1}{a_m} E_{22}\right) = E_{12} \in I.$$

That is, $E_{12} \in I$. We can write

$$(a_m E_{11} + b E_{12} + c E_{22}) E_{11} = a_m E_{11} \in I,$$

yielding $a_m E_{11} \in I$. Now, as $a_m E_{11}, E_{12} \in I$, we see that if $A = k a_m E_{11} + b E_{12} + c E_{22} \in I$ for some $k \in \mathbb{Z}$ and $b, c \in \mathbb{Q}$, then $c E_{22} \in I$. If $c = 0$ for all $A = k a_m E_{11} + b E_{12} + c E_{22} \in I$, then $I$ is generated by $\{a_m E_{11}, E_{12}\}$. Otherwise, just as we saw in the above, $c E_{22} \in I$ for some nonzero $c \in \mathbb{Q}$, yielding

$$(c E_{22})\left(\frac{1}{c} E_{22}\right) = E_{22} \in I.$$

That is, $E_{22} \in I$, in which case the right ideal $I$ is generated by $\{a_m E_{11}, E_{12}, E_{22}\}$. This proves the assertion in this case.

(ii) Every $A \in I$ is of the form $A = b E_{12} + c E_{22}$, where $b, c \in \mathbb{Q}$.

Suppose $A_0 = b_0 E_{12} + c_0 E_{22} \in I$, where $b_0, c_0 \in \mathbb{Q}$ are such that $(b_0, c_0) \neq (0,0)$. Set $J = \mathbb{Q} A_0$. It is easily verified that $J$ is a right ideal of $R$. If $I = J$, then $I$ is generated by $A_0$. If not, then there exist $b_1, c_1 \in \mathbb{Q}$ such that $b_1 c_0 - b_0 c_1 \neq 0$ and $b_1 E_{12} + c_1 E_{22} \in I$. It follows that $(b_0 c_1 - b_1 c_0) E_{12} = (b_0 E_{12} + c_0 E_{22})(b_1 E_{12} + c_1 E_{22}) - (b_1 E_{12} + c_1 E_{22})(b_0 E_{12} + c_0 E_{22}) \in I$, and hence

$$(b_0 c_1 - b_1 c_0) E_{12} \left( \frac{1}{b_0 c_1 - b_1 c_0} E_{22} \right) = E_{12} \in I.$$

That is, $E_{12} \in I$. So we have

$$E_{12}(b_0 E_{22}) = b_0 E_{12} \in I.$$

From this, we obtain $c_0 E_{22} \in I$ because $b_0 E_{12} + c_0 E_{22} \in I$. If $c_0 = 0$ whenever $b_0 E_{12} + c_0 E_{22} \in I$, then $I$ is generated by $E_{12}$. If not, then $c_0 E_{22} \in I$ for some nonzero $c_0 \in \mathbb{Q}$. This implies $E_{22} = (c_0 E_{22})(\frac{1}{c_0} E_{22}) = E_{22} \in I$. Thus, $I$ is generated by $\{E_{12}, E_{22}\}$ in this case. So in any event, $I$ is finitely generated.

We now prove that every ascending chain of right ideals of $R$ necessarily terminates. To this end, let $(I_n)_{n=1}^{+\infty}$ be an ascending sequence of right ideals of $R$. Set $I = \cup_{n=1}^{+\infty} I_n$. As $(I_n)_{n=1}^{+\infty}$ is an ascending sequence of right ideals of $R$, it is easily verified that $I$ is a right ideal of $R$. It follows from the above that there are at most three, not necessarily distinct, elements $A_1, A_2, A_3 \in I$ such that $I = \langle A_1, A_2, A_3 \rangle$. Since $I = \cup_{n=1}^{+\infty} I_n$, there exist $n_1, n_2, n_3 \in \mathbb{N}$ such that $A_i \in I_{n_i}$ for each $i = 1, 2, 3$. Letting $N = \max(n_1, n_2, n_3)$, we have $A_i \in I_n$ for all $n \geq N$ and $i = 1, 2, 3$. This implies

$$I_n \subseteq I = \langle A_1, A_2, A_3 \rangle \subseteq I_n,$$

for all $n \geq N$. Therefore, $I_n = I$ for all $n \geq N$. In other words, $I_n = I_N$ for all $n \geq N$, which is what we want. ∎

**3.** We need the following lemma.

**Lemma.** *Let $V$ be a finite-dimensional vector space over a field $F$ and $\{V_i\}_{i \in I}$ a family of proper subspaces of $V$ such that $|I| < |F|$. Then,*

$$\bigcup_{i \in I} V_i \subsetneqq V.$$

**Proof.** Without loss of generality, we may assume that $\{V_i\}_{i \in I}$ is a family of distinct proper subspaces of $V$. With that in mind, we prove the assertion by induction on $\dim V$. If $\dim V = 1$, we must have $\dim V_i = 0$, implying $V_i = \{0\}$ for all $i \in I$, in which case the assertion is trivial. Assuming that the assertion holds for any vector space $V$ with $\dim V < k$, we prove the assertion for any vector space $V$ with $\dim V = k$. Let $V_i$ $(i \in I)$ be as in the lemma. Set

$$S = \{ \dim V_i | i \in I \}.$$

Plainly, $S \subseteq \mathbb{N}$ and $S$ is nonempty and bounded from above by $k = \dim V$. It follows that $S$ has a terminal element. That is, there exists $i_0 \in I$ such that

$$\dim V_i \leq \dim V_{i_0},$$

for all $i \in I$. Set $W = V_{i_0}$, $J = I \setminus \{i_0\}$, and finally $W_j = V_j \cap V_{i_0}$. Firstly, for all $j \in J$ we have $W_j \subsetneqq V_{i_0}$. Suppose to the contrary that $W_j = V_j \cap V_{i_0} = V_{i_0}$

for some $j \in J$, then $V_{i_0} \subseteq V_j$, implying $\dim V_{i_0} \leq \dim V_j$. On the other hand, $\dim V_j \leq \dim V_{i_0}$, yielding $V_{i_0} = V_j$ which is impossible because $j \neq i_0$. Thus, $W_j \subsetneqq W = V_{i_0}$ for all $j \in J$. We have $\dim W = \dim V_{i_0} < \dim V = k$ and $|J| \leq |I| < |F|$. So it follows from the induction hypothesis that $\bigcup_{j \in J} W_j \subsetneqq W = V_{i_0}$. Consequently, there exists a vector $v_0 \in V_{i_0}$ such that $v_0 \notin W_j = V_j \cap V_{i_0}$ for all $j \in J$. That is, $v_0 \notin V_j$ for all $j \in I$ with $j \neq i_0$. On the other hand, since $V_{i_0} \subsetneqq V$, there exists a vector $v_1 \in V \setminus V_{i_0}$. We claim that there exists an $f_0 \in F$ such that $v_1 + f_0 v_0 \notin V_i$ for all $i \in I$. To prove this by contradiction, suppose that for each $f \in F$ there exists an $i_f \in I$ such that $v_1 + f v_0 \in V_{i_f}$. Note that if $f, f' \in F$ and $f \neq f'$, then $v_1 + f v_0 \neq v_1 + f' v_0$. Also note that the hypothesis $|I| < |F|$ implies that there exist $f, f' \in F$ with $f \neq f'$ such that $i_f = i_{f'}$. Firstly, we observe that $i_f \neq i_0$, for otherwise $v_1 + f v_0 \in V_{i_0}$, yielding $v_1 \in V_{i_0}$, because $v_0 \in V_{i_0}$, which is impossible. Secondly, from $v_1 + f v_0, v_1 + f' v_0 \in V_{i_f} = V_{i_{f'}}$, we obtain $(f - f') v_0 \in V_{i_f}$, yielding $v_0 \in V_{i_f}$, which, in turn, implies $i_f = i_0$, contradicting $i_f \neq i_0$ as observed in the above. That is, we obtain a contradiction in any event. Thus, there exists an $f_0 \in F$ such that $v_1 + f_0 v_0 \notin V_i$ for all $i \in I$. In other words, $v_1 + f_0 v_0 \notin \bigcup_{i \in I} V_i$, and hence $\bigcup_{i \in I} V_i \subsetneqq V$, which is what we want. $\quad\square$

We now use the lemma to prove the assertion. From this point on, the proof is almost identical to that of Problem 5 of 1.14.2.

Fix $i_0 \in I$. We prove the assertion by induction on $\dim V - \dim V_{i_0}$. If $\dim V - \dim V_{i_0} = 1$, then for $U = \{u\}$, where $u \in V \setminus (\bigcup_{i \in I} V_i)$, we obviously have $V_i \oplus U = V$ for all $i \in I$. Assuming that the assertion holds whenever $\dim V - \dim V_{i_0} = n$, we prove it whenever $\dim V - \dim V_{i_0} = n+1$. As, in view of the lemma, $\bigcup_{i \in I} V_i \subsetneqq V$, there exists a vector $\beta \in V \setminus (\bigcup_{i \in I} V_i)$. Set

$$V_i' = V_i \oplus \langle \beta \rangle.$$

We have $\dim V_i' = \dim V_j' = \dim V_{i_0} + 1$ for all $i, j \in I$, yielding $\dim V - \dim V_{i_0}' = \dim V - \dim V_i' = n$ for all $i \in I$. It thus follows from the induction hypothesis that there exists a subspace $U'$ such that

$$V_i' \oplus U' = V,$$

for all $i \in I$. Letting $U = \langle \beta \rangle \oplus U'$, we obviously obtain

$$V_i \oplus U = V,$$

for all $i \in I$, proving the induction assertion, which is what we want. $\quad\blacksquare$

## 2.19. Nineteenth Competition

### 2.19.1. Analysis. 1. Define the function $g : [0, 1] \to \mathbb{R}$ by

$$g(x) = 2x - 1 - \int_0^x f.$$

It suffices to show that $g$ has only one zero on the interval $[0, 1]$. As $f$ is continuous on $[0, 1]$, $g$ is differentiable on $(0, 1)$. Using the First Fundamental Theorem of Calculus, we can write

$$g'(x) = 2 - f(x),$$

for all $x \in (0,1)$. But $0 \le f(x) \le 1$ for all $x \in (0,1)$. In particular, we must have $g'(x) = 2 - f(x) \ge 1 > 0$ for all $x \in (0,1)$, implying that $g$ is strictly increasing on $[0,1]$. Therefore, $g$ has at most one zero on $[0,1]$. With that in mind, note that $g(0) = -1 < 0 \le 1 - \int_0^1 f = g(1)$. To see the second inequality, just note that from $f \le 1$ on $[0,1]$, it follows that $\int_0^1 f \le 1$, yielding $g(1) = 1 - \int_0^1 f \ge 0$. It thus follows from the Intermediate Value Theorem that there exists a $c \in (0,1]$ such that $g(c) = 0$. This completes the proof. ∎

**2. First Solution:** Note that $\lim_{x \to +\infty} e^{\frac{x}{h}} = +\infty$ whenever $h > 0$. With that in mind, using L'Hopital's rule, we can write

$$\lim_{x \to +\infty} f(x) = \lim_{x \to +\infty} \frac{e^{\frac{x}{h}} f(x)}{e^{\frac{x}{h}}} = \lim_{x \to +\infty} \frac{\left(e^{\frac{x}{h}} f(x)\right)'}{\left(e^{\frac{x}{h}}\right)'}$$

$$= \lim_{x \to +\infty} \frac{\frac{1}{h} e^{\frac{x}{h}} f(x) + e^{\frac{x}{h}} f'(x)}{\frac{1}{h} e^{\frac{x}{h}}} = \lim_{x \to +\infty} \left(f(x) + h f'(x)\right) = 0,$$

implying $\lim_{x \to +\infty} f(x) = 0$. This together with

$$\lim_{x \to +\infty} \left(f(x) + h f'(x)\right) = 0,$$

where $h \in \mathbb{R}^+$, yields $\lim_{x \to +\infty} f'(x) = 0$, which is what we want. (We note that the hypothesis that $f$ is continuously differentiable is redundant. The differentiability of $f$ is enough for the assertion to be true.)

**Second solution:** It follows from the hypothesis that for given $\varepsilon > 0$, there exists an $M = M(\varepsilon) > 0$ such that

$$\left|f(x) + h f'(x)\right| < \frac{\varepsilon}{2},$$

whenever $x \ge M$. This easily implies $\left|\frac{d}{dx}\left(e^{\frac{x}{h}} f(x)\right)\right| < \frac{\varepsilon}{2h} e^{\frac{x}{h}}$ whenever $x \ge M$. We can write

$$\left|e^{\frac{x}{h}} f(x) - e^{\frac{M}{h}} f(M)\right| = \left|\int_M^x \frac{d}{dt}\left(e^{\frac{t}{h}} f(t)\right) dt\right|$$

$$\le \frac{\varepsilon}{2h} \int_M^x e^{\frac{t}{h}} dt = \frac{\varepsilon}{2}\left(e^{\frac{x}{h}} - e^{\frac{M}{h}}\right),$$

for all $x \ge M$. Consequently, for all $x \ge M$,

$$\left|f(x) - e^{\frac{M-x}{h}} f(M)\right| < \frac{\varepsilon}{2}\left(1 - e^{\frac{M-x}{h}}\right),$$

which obtains

$$\left|f(x)\right| \le \left|f(x) - e^{\frac{M-x}{h}} f(M)\right| + \left|e^{\frac{M-x}{h}} f(M)\right|$$

$$\le \frac{\varepsilon}{2}\left(1 - e^{\frac{M-x}{h}}\right) + e^{\frac{M-x}{h}}\left|f(M)\right|$$

$$< \frac{\varepsilon}{2} + e^{\frac{M-x}{h}}\left|f(M)\right|,$$

for all $x \ge M$. Consequently,

$$\limsup_{x \to +\infty} f(x) \le \limsup_{x \to +\infty}\left(\frac{\varepsilon}{2} + e^{\frac{M-x}{h}}\left|f(M)\right|\right) = \frac{\varepsilon}{2},$$

for all $\varepsilon > 0$. Thus, $\limsup_{x \to +\infty} f(x) \leq 0$, and hence $\lim_{x \to +\infty} f(x) = 0$. As we pointed out in the first solution, this together with the hypothesis that $\lim_{x \to +\infty} \big( f(x) + h f'(x) \big) = 0$, where $h \in \mathbb{R}^+$, yields $\lim_{x \to +\infty} f'(x) = 0$, finishing the proof.  ∎

**3.** Recall that $A$ equipped with the uniform metric, induced by the uniform norm of $A$, denoted by $\|.\|_\infty$, which is defined by

$$d(f, g) = \|f - g\|_\infty = \sup_{x \in [0,1]} \big| f(x) - g(x) \big|,$$

is a complete normed space or a real Banach space. Also recall that by Problem 2 of 1.6.1 the following lemma holds.

**Lemma.** *Let $X$ and $Y$ be metric spaces and $f, f_n : X \to Y$ continuous functions such that the sequence $(f_n)_{n \in \mathbb{N}}$ uniformly converges to $f$ on $X$. Also let $x_0, x_n \in X$ be such that the sequence $(x_n)_{n \in \mathbb{N}}$ converges to $x_0$. Then, $\lim_n f_n(x_n) = f(x_0)$.*

With all that in mind, we now prove the assertion.

(a) Let $N \in \mathbb{N}$ be arbitrary. To show that $E_N$ is closed, suppose that $f : [0,1] \to \mathbb{R}$ is a limit point of $E_N$ with respect to the topology induced by the uniform norm of $A$. We show that $f \in E_N$. To this end, as $f$ is a limit point of $E_N$, we see that there exists a sequence $(f_n)_{n \in \mathbb{N}}$ in $E_N$ such that $f_n \to f$ in $A$ as $n \to +\infty$. In other words, $f_n$'s uniformly converge to $f$ on $[0,1]$. Now, from $f_n \in E_N$ for each $n \in \mathbb{N}$, it follows that there exists an $a_n \in [0,1]$ such that

$$\big| f_n(x) - f_n(a_n) \big| \leq N \big| x - a_n \big|,$$

for all $x \in [0,1]$. It is plain that the sequence $(a_n)_{n \in \mathbb{N}}$ in the compact interval $[0,1]$ has a subsequence $(a_{n_k})_{k \in \mathbb{N}}$ converging to some $a_0 \in [0,1]$. Let $b_k = a_{n_k}$ and $g_k = f_{n_k}$ for all $k \in \mathbb{N}$. We can write

$$\big| g_k(x) - g_k(b_k) \big| \leq N \big| x - b_k \big|,$$

for all $x \in [0,1]$ and $k \in \mathbb{N}$. Now, letting $k \to +\infty$ and using the above lemma, we see that

$$\big| f(x) - f(a) \big| \leq N \big| x - a \big|,$$

for all $x \in [0,1]$. That is, $f \in E_N$, and hence $E_N$ is closed for all $N \in \mathbb{N}$. To show that the interior of $E_N$ is empty for all $N \in \mathbb{N}$, we prove that the set $A \setminus E_N$ is dense in $A$. Suppose that $f \in A$ is arbitrary and $\varepsilon > 0$ is given. From the continuity of $f$ on the compact interval $[0,1]$, we see that $f$ is uniformly continuous on $[0,1]$. Thus, there exists a $\delta > 0$ such that

$$\big| f(x) - f(y) \big| < \frac{\varepsilon}{2},$$

whenever $x, y \in [0,1]$ and $|x - y| < \delta$. Choose $n \in \mathbb{N}$ such that $\frac{1}{n} < \delta$ and set $x_i = \frac{i}{n}$, where $1 \leq i \leq n$. By constructing a function $g$ satisfying $\|f - g\|_\infty < \varepsilon$ and $g \notin E_N$, we show that $A \setminus E_N$ is dense in $A$, finishing the proof. To construct $g$, it suffices to do so on any subinterval $[x_{i-1}, x_i]$ ($1 \leq i \leq n$) in such a way that $g$ is (piecewise) continuous, $\|f - g\|_\infty < \varepsilon$, and $g \notin E_N$. Here is a description of the graph of $g$ on the interval $[x_{i-1}, x_i]$. Let $g$ be any function on $[x_{i-1}, x_i]$ ($1 \leq i \leq n$) whose graph is a polygonal line in the rectangle

$[x_{i-1}, x_i] \times [f(x_{i-1}) - \frac{\varepsilon}{2}, f(x_i) + \frac{\varepsilon}{2}]$ joining the two points $(x_{i-1}, f(x_{i-1}))$ and $(x_i, f(x_i))$ satisfying the following properties. (i) The consecutive vertices of any segment of the polygonal line lie on the lines $y = f(x_{i-1}) - \frac{\varepsilon}{2}$ and $y = f(x_i) + \frac{\varepsilon}{2}$, respectively. (ii) The absolute value of the slope of any segment formed by any two consecutive vertices of the polygonal line is greater than $N$. It is plain that for any function $g : [0, 1] \to \mathbb{R}$ as described in the above, we have $\|f - g\|_\infty < \varepsilon$ and $g \notin E_N$, which is what we want.

(b) As $A$ is a complete metric space with respect to the metric induced by the uniform norm of $A$, it follows from the Baire Category Theorem that $A$ cannot be written as a countable union of nowhere dense subsets of $A$. Recall that is subset of a topological space $X$ is called nowhere dense if the complement of its closure is dense in $X$. It thus follows from (a) that $E_N$ is nowhere dense for all $N \in \mathbb{N}$, and hence, by the Baire Category Theorem, $\bigcup_{N \in \mathbb{N}} E_N \subsetneq A$. Define the set $E$ as follows

$$E := \{f \in A | \exists a \in [0, 1] \ni f \in D_a\} = \bigcup_{a \in [0,1]} D_a,$$

where $D_a$, with $a \in (0, 1)$, is the set of all functions that are differentiable at $a \in (0, 1)$; the set $D_0$ (resp. $D_1$) is defined to be the set of functions that are right (resp. left) differentiable at $0$ (resp. $1$). We claim that $E \subseteq \bigcup_{N \in \mathbb{N}} E_N$. To prove this, suppose $f \in E$ is given. It follows that there exists $a \in [0, 1]$ such that $f \in D_a$. Define the function $g : [0, 1] \to \mathbb{R}$ by

$$g(x) = \begin{cases} \frac{f(x) - f(a)}{x - a} & x \in [0, 1] \setminus \{a\}, \\ f'(a) & x = a. \end{cases}$$

We have $\lim_{x \to a} g(x) = \lim_{x \to a} \frac{f(x) - f(a)}{x - a} = f'(a) = g(a)$ for all $a \in (0, 1)$. Likewise, $\lim_{x \to 0+} g(x) = g(0)$ and $\lim_{x \to 1-} g(x) = g(1)$. Therefore, the function $g$ is continuous on the compact interval $[0, 1]$, and hence it is bounded on $[0, 1]$. So there exists an $M > 0$ such that

$$|g(x)| \leq M,$$

for all $x \in [0, 1]$. In other words,

$$|f(x) - f(a)| \leq M|x - a|,$$

for all $x \in [0, 1]$. Consequently, if we let $N \in \mathbb{N}$ be such that $N \geq M$, then

$$|f(x) - f(a)| \leq N|x - a|,$$

for all $x \in [0, 1]$. That is, $f \in E_N$, yielding $f \in \bigcup_{N \in \mathbb{N}} E_N$, proving the claim. Now, to prove the assertion, recall that $E \subseteq \bigcup_{N \in \mathbb{N}} E_N \subsetneq A$. This implies that there exists a function $f \in A \setminus E$. In other words, the function $f : [0, 1] \to \mathbb{R}$ is continuous and yet it is nowhere differentiable because $f \notin E$. This proves (b). ∎

**2.19.2.  Algebra. 1.** To prove the assertion by contradiction, suppose that $N_0 \neq \{e\}$ is a maximal element of $\mathcal{A}$. We have $\frac{G}{N_0} \cong G$. It follows that there exists an isomorphism $\phi : G \to \frac{G}{N_0}$. Since $N_0 \neq \{e\}$ is a normal subgroup of $G$ and $\phi$ is an isomorphism, we see that $\phi(N_0) \neq \{e\}$ is a normal subgroup of $\frac{G}{N_0}$. From this, we obtain a normal subgroup $M_0 \supsetneq N_0$ of $G$ such that $\phi(N_0) = \frac{M_0}{N_0}$. Thus, $\phi|_{N_0} : N_0 \to \frac{M_0}{N_0}$ is an isomorphism of groups, yielding $N_0 \cong \frac{M_0}{N_0}$. Now, the Third Isomorphism Theorem for groups  together with $\frac{G}{N_0} \cong G$ and $N_0 \cong \frac{M_0}{N_0}$ implies that

$$\frac{G}{M_0} \cong \frac{\frac{G}{N_0}}{\frac{M_0}{N_0}} \cong \frac{G}{N_0} \cong G.$$

In other words, $\frac{G}{M_0} \cong G$. But $N_0 \subsetneq M_0$, contradicting the hypothesis that $N_0$ is a maximal element of $\mathcal{A}$. Thus, the assertion follows by contradiction, which is what we want. $\blacksquare$

**2.** (a) Note first that for any nonzero $a \in R$ and any minimal left ideal $J$ of $R$, the left ideal $Ja$ is also a minimal left ideal of $R$. To prove this by contradiction, suppose that $Ja$ is not a minimal left ideal of $R$ for some nonzero element $a \in R$ and minimal left ideal $J$ of $R$. It follows that there exists a nonzero left ideal $I$ of $R$ such that $I \subsetneq Ja$. Define

$$J_1 := \{j \in J | ja \in I\}.$$

It is plain that $J_1$ is a left ideal of $R$, $J_1 \subsetneq J$, because $I \subsetneq Ja$, and that $J_1 \neq 0$, because $I \neq 0$. This is in contradiction with $J$ being a minimal left ideal of $R$. Thus, $Ja$ is a minimal left ideal of $R$ whenever $J$ is a minimal left ideal of $R$ and $a \in R$ is nonzero. To prove the assertion, if $R$ has no minimal left ideal, the assertion is trivial because, by definition, $S = 0$. If not, as $S = \sum\{J : J$ is a minimal left ideal of $R\}$, we see $S$ is a left ideal in $R$. To see that $S$ is a right ideal of $R$ as well, let $r \in R$ and $s \in S$ be arbitrary. We need to show that $sr \in S$. Since $s \in S$, it follows that there are minimal left ideals $J_1, \dots, J_n$, where $n \in \mathbb{N}$, and $j_k \in J_k$ $(1 \leq k \leq n)$ such that

$$s = j_1 + \cdots + j_n,$$

yielding

$$sr = j_1 r + \cdots + j_n r.$$

But $j_k r \in J_k r$ and $J_k r$ is a minimal left ideal of $R$ for all $1 \leq k \leq n$. It thus follows that $sr \in \sum\{J : J$ is a minimal left ideal of $R\} = S$. That is, $S$ is a right ideal of $R$ as well, and hence an ideal of $R$, as desired.

(b) It follows from (a) that

$$\bigcap_{I \lhd R} I \subseteq S,$$

for the two sided ideal $S$ appears in the intersection. To prove the inclusion in the opposite direction, it suffices to show that for any minimal left ideal $J$ of

$R$, we have

$$J \subseteq \bigcap_{I \triangleleft R} I.$$

To this end, assuming that $J$ and $I$ are, respectively, an arbitrary minimal left ideal and an arbitrary two-sided ideal of $R$, it is enough to show that $J \subseteq I$. To see this, note first that

$$IJ \subseteq J,$$

for $J$ is a left ideal of $R$. Secondly, $IJ \neq 0$ because otherwise $I(JR) = (IJ)R = 0$, where $I \neq 0$ and $JR \neq 0$ are two-sided ideals of $R$, contradicting the hypothesis. Thus, $IJ \neq 0$, and hence

$$0 \neq IJ \subseteq J.$$

That is, the nonzero left ideal $IJ$ is contained in the minimal left ideal $J$. This implies $IJ = J$. Now, since $I$ is a two-sided ideal of $R$, we see that $IJ \subseteq I \cap J$, whence $J \subseteq I \cap J$, yielding $J \subseteq I$, which is what we want.  ∎

**3.** Recall that the rank of a matrix is equal to its column rank as well as its row rank. With that in mind, let $A_j$ denote the $j$th column of the matrix $A$. We have

$$A_j = \begin{pmatrix} x_1 + y_j \\ x_2 + y_j \\ \vdots \\ x_n + y_j \end{pmatrix} = \begin{pmatrix} x_1 \\ x_2 \\ \vdots \\ x_n \end{pmatrix} + y_j \begin{pmatrix} 1 \\ 1 \\ \vdots \\ 1 \end{pmatrix} = \alpha + y_j \beta,$$

where $\alpha = \begin{pmatrix} x_1 \\ x_2 \\ \vdots \\ x_n \end{pmatrix}$ and $\beta = \begin{pmatrix} 1 \\ 1 \\ \vdots \\ 1 \end{pmatrix}$. That is, $A_j \in \langle \alpha, \beta \rangle$, yielding $\langle A_1, \ldots, A_n \rangle$
$\subseteq \langle \alpha, \beta \rangle$. Thus, $\text{rank}(A) = \dim \langle A_1, \ldots, A_n \rangle \leq \dim \langle \alpha, \beta \rangle \leq 2$. Therefore, $\text{rank}(A) \leq 2$, as desired.  ∎

**4. First solution:** Let $A \in M_n(F)$ be a rank one $n \times n$ matrix over $F$. Since $\text{rank}(A) = 1$, it follows from the Rank-Nullity Theorem that the rank and nullity of the linear transformation $T : M_{n \times 1}(F) \to M_{n \times 1}(F)$ defined by $TX = AX$, are, respectively, equal to $1$ and $n - 1$. Let $\{\alpha_i\}_{i=1}^{n-1}$ be a basis for $\ker T$. Enlarge $\{\alpha_i\}_{i=1}^{n-1}$ to a basis $\mathcal{B}' = \{\alpha_i\}_{i=1}^{n}$ for $M_{n \times 1}(F)$, where $\alpha_n$ is any vector in $M_{n \times 1}(F) \setminus \langle \alpha_i \rangle_{i=1}^{n-1}$, where $\langle \alpha_i \rangle_{i=1}^{n-1}$ denotes the vector subspace spanned by $\alpha_1, \ldots, \alpha_{n-1}$. It is plain that the matrix of $T$ with respect to the basis $\mathcal{B}'$, denoted by $B$, has the form

$$B = (b_{ij})_{n \times n} = \begin{pmatrix} 0 & \cdots & 0 & b_1 \\ 0 & \cdots & 0 & b_2 \\ \vdots & \cdots & \vdots & \vdots \\ 0 & \cdots & 0 & b_n \end{pmatrix}.$$

On the other hand, the matrix of $T$ with respect to the standard basis of $M_{n\times 1}(F)$ is equal to $A$. It thus follows that there exists an invertible matrix $P \in M_n(F)$ such that $B = P^{-1}AP$, whence $A = PBP^{-1}$. We can write

$$
\begin{aligned}
\det(I + A) &= \det(P(I + B)P^{-1}) = \det(I + B) \\
&= \det \begin{pmatrix} 1 & \cdots & 0 & b_1 \\ 0 & \cdots & 0 & b_2 \\ \vdots & \cdots & \vdots & \vdots \\ 0 & \cdots & 0 & 1+b_n \end{pmatrix} \\
&= 1 + b_n = 1 + \operatorname{tr}(B) = 1 + \operatorname{tr}(A).
\end{aligned}
$$

That is, $\det(I + A) = 1 + \operatorname{tr}(A)$, which is what we want.

**Second solution:** A proof identical to that of the second solution of Problem 7 of 1.9.3 shows that $\det(I + A) = 1 + \operatorname{tr}(A)$, which is what we want. ■

## 2.20. Twentieth Competition

**1.** If $\{(x,y),(z,t)\}$ or $\{(x',y'),(z',t')\}$ is linearly dependent, then there exist $\alpha, \beta \in \mathbb{C}$, which are not simultaneously zero, such that

$$\theta = \alpha(x,y) + \beta(z,t) = (0,0) \text{ or } \theta' = \alpha(x',y') + \beta(z',t') = (0,0).$$

In this case, the assertion is trivial because $\theta = 0$ or $\theta' = 0$. So suppose that $\{(x,y),(z,t)\}$ and $\{(x',y'),(z',t')\}$ are linearly independent. Set

$$A = \begin{pmatrix} x & y \\ z & t \end{pmatrix}, \quad A' = \begin{pmatrix} x' & y' \\ z' & t' \end{pmatrix}.$$

The matrices $A$ and $A'$ are invertible because their rows are linearly independent. It follows that the matrix $A^{-1}A' \in M_2(\mathbb{C})$ is also invertible. Now, suppose that $\lambda \in \mathbb{C}$ is a nonzero eigenvalue of the invertible matrix $A^{-1}A'$. We see that the matrix $\lambda I_2 - A^{-1}A'$ is not invertible. Hence, neither is $\lambda A - A' = A(\lambda I_2 - A^{-1}A')$. This implies that the rows of the noninvertible matrix

$$\lambda A - A' = \lambda \begin{pmatrix} x & y \\ z & t \end{pmatrix} - \begin{pmatrix} x' & y' \\ z' & t' \end{pmatrix}$$

are linearly dependent. Thus, there exist $\alpha, \beta \in \mathbb{C}$, which are not simultaneously zero, such that

$$\alpha(\lambda x - x', \lambda y - y') + \beta(\lambda z - z', \lambda t - t') = (0,0).$$

A straightforward calculation reveals that

$$\lambda\theta - \theta' = \alpha(\lambda x - x', \lambda y - y') + \beta(\lambda z - z', \lambda t - t') = (0,0),$$

where $\theta = \alpha(x,y) + \beta(z,t)$ and $\theta' = \alpha(x',y') + \beta(z',t')$. In other words, the vectors $\theta$ and $\theta'$ are linearly dependent, proving the assertion. ■

**2.** Set $x_n = \ln(1 + a_n)$. It follows from the hypothesis that

$$x_{m+n} \leq x_m + x_n,$$

for all $m, n \in \mathbb{N}$. We prove that $\lim_n \frac{x_n}{n}$ exists. To see this, letting $\alpha = \inf_{n \in \mathbb{N}} \frac{x_n}{n} \in \mathbb{R}$, we show that $\lim_n \frac{x_n}{n} = \alpha$. To this end, let $\varepsilon > 0$ be given. Since $\alpha = \inf_{n \in \mathbb{N}} \frac{x_n}{n}$, there exists a positive integer $N_1 = N_1(\varepsilon)$ such that

$$\alpha \leq \frac{x_{N_1}}{N_1} \leq \alpha + \frac{\varepsilon}{2}.$$

Now, for any $n \geq N_1$, we can write $n = N_1 q + r$, where $q \in \mathbb{N}$ and $r \in \mathbb{N} \cup \{0\}$ with $0 \leq r < N_1$. In view of the hypothesis, we can write

$$\alpha - \varepsilon < \alpha \leq \frac{x_n}{n} \leq \frac{q x_{N_1} + r x_1}{N_1 q + r} \leq \frac{q x_{N_1}}{N_1 q} + \frac{r x_1}{n}$$

$$< \frac{x_{N_1}}{N_1} + \frac{N_1 x_1}{n} \leq \alpha + \frac{\varepsilon}{2} + \frac{N_1}{n} x_1.$$

Now, pick $N_2 \in \mathbb{N}$ such that $N_2 > \frac{2}{\varepsilon} N_1 x_1$. Letting $N = \max(N_1, N_2)$, for all $n \geq N$, we have

$$\alpha - \varepsilon < \frac{x_n}{n} < \alpha + \frac{\varepsilon}{2} + \frac{N_1}{n} x_1 < \alpha + \varepsilon.$$

That is,

$$\left| \frac{x_n}{n} - \alpha \right| < \varepsilon,$$

for all $n \geq N$. In other words, $\lim_n \frac{x_n}{n} = \alpha$, implying that $\lim_n \sqrt[n]{1 + a_n} = e^\alpha$, which is what we want.  ∎

**3.** It follows from the hypothesis that $\sigma^2 = id$ for all $\sigma \in \text{aut}(G)$. Consequently, $\sigma^{-1} = \sigma$ for all $\sigma \in \text{aut}(G)$, from which we obtain

$$\sigma_1 \sigma_2 = (\sigma_1 \sigma_2)^{-1} = \sigma_2^{-1} \sigma_1^{-1} = \sigma_2 \sigma_1,$$

for all $\sigma_1, \sigma_2 \in \text{aut}(G)$. In other words, $\text{aut}(G)$ is abelian. On the other hand, just as we saw in Solution 1 of 2.16.2 using the First Isomorphism Theorem for groups, we have $\frac{G}{Z(G)} \cong \text{In}(G)$, where $\text{In}(G)$ denotes the set of all inner automorphisms of $G$, which is a normal subgroup of $\text{aut}(G)$. It follows that $\text{In}(G)$ is commutative and hence so is $\frac{G}{Z(G)}$. This yields $G' \subseteq Z(G)$, where $G'$ denotes the derived subgroup (a.k.a. the commutator subgroup) of $G$. From this, we see that $G'' = \{e\}$, implying that $G$ is solvable, as desired.  ∎

**4.** Note first that $f(x) > 0$ for all $x \in (\frac{1}{4}, 1)$ because $f(x) = x^{f(x)}$ for all such $x$'s. Next, since $f(x) > 0$, we see that $0 < (\frac{1}{4})^{f(x)} < x^{f(x)} < 1^{f(x)} = 1$, implying that $0 < f(x) < 1$ for all $x \in (\frac{1}{4}, 1)$. Now, noting that $x^{f(x)} = f(x)$ and $f(x) > 0$ for all $x \in (\frac{1}{4}, 1)$, and taking $\ln$, we obtain $f(x) \ln x = \ln f(x)$ for all $x \in (\frac{1}{4}, 1)$. This implies

$$x = e^{\frac{\ln f(x)}{f(x)}},$$

for all $x \in (\frac{1}{4}, 1)$. Define $g : (0, 1) \to (0, 1)$ by $g(x) = e^{\frac{\ln x}{x}}$. The function $g$ is differentiable and we have $g'(x) = \frac{1 - \ln x}{x^2} e^{\frac{\ln x}{x}}$. Thus, $g'(x) > 0$ for all $0 < x < 1$, and hence $g$ is strictly increasing on $(0, 1)$. The function $g : (0, 1) \to (0, 1)$ is surjective because $\lim_{x \to 0+} g(x) = 0$ and $\lim_{x \to 1-} g(x) = 1$. From this, it follows that $g$ has an inverse, say, $g^{-1} : (0, 1) \to (0, 1)$, and that $g^{-1}$ is differentiable and strictly increasing on $(0, 1)$. In particular, $g^{-1}$ is continuous on $(0,1)$. Also, it is easily verified that $g(\frac{1}{2}) = \frac{1}{4}$. With all that in mind, define the function $\widetilde{g} : [0,1] \to [0,1]$ by

$$\widetilde{g}(x) = \begin{cases} 0 & x = 0, \\ g(x) & 0 < x < 1, \\ 1 & x = 1. \end{cases}$$

It is now plain that the function $\widetilde{g} : [0,1] \to [0,1]$ is surjective, one-to-one and continuous and that its inverse $\widetilde{g}^{-1} : [0,1] \to [0,1]$ is also continuous, and hence uniformly continuous, because $[0,1]$ is a compact interval. Noting $0 < f(x) < 1$ for all $x \in (\frac{1}{4}, 1)$, we can write

$$\widetilde{g}\big(f(x)\big) = g\big(f(x)\big) = x,$$

from which, we obtain $f(x) = \widetilde{g}^{-1}(x)$ for all $x \in (\frac{1}{4}, 1)$. In other words, $f = \widetilde{g}^{-1}|_{(\frac{1}{4},1)} : (\frac{1}{4}, 1) \to (\frac{1}{2}, 1)$. From this, it follows that $f : (\frac{1}{4}, 1) \to (\frac{1}{2}, 1)$ is uniformly continuous because so is $\widetilde{g}^{-1} : [0,1] \to [0,1]$. This completes the proof. ∎

**5.** Let $I \neq 0$ and $Z$, respectively, denote the intersection of all nonzero ideals of $R$ and the set of all zero divisors of $R$ including zero. By showing that $Z$ is a maximal ideal of $R$ and that $Z = I$, we settle the proof, for, from this, it follows that $Z = I$ is the only nontrivial ideal of $R$. First, we show that $I \subseteq Z$. Suppose to the contrary there exists a nonzero $x_0 \in I$ such that $x_0 \notin Z$. We see that $x_0^2 \neq 0$ because $x_0$ is not a zero divisor of $R$. As $R$ is unital and $I$ is the intersection of all nonzero ideals of $R$, we obtain $I \subseteq x_0 R \neq 0$ and $I \subseteq x_0^2 R \neq 0$. On the other hand, $x_0 R \subseteq I$ and $x_0^2 R \subseteq I$, for $x_0 \in I$. It thus follows that $I = x_0 R = x_0^2 R$. In particular, we must have $x_0 \in x_0^2 R$, which yields $x_0 = x_0^2 r$ for some $r \in R$. This obtains $x_0(1 - x_0 r) = 0$, which, in turn, implies $1 = x_0 r \in I$, for $x_0 \notin Z$, a contradiction. Thus, $\{0\} \neq I \subseteq Z$, and hence $Z \neq \{0\}$.

Next, we show that $Z$ is a maximal ideal of $R$. To this end, let $x \in Z \setminus \{0\}$ and $\mathrm{Ann}(x) = \{r \in R : rx = 0\}$. Note that $Z \subseteq \mathrm{Ann}(x)$ because $Z^2 = 0$. On the other hand, $\mathrm{Ann}(x) \subseteq Z$ whenever $x \neq 0$. It thus follows that $\mathrm{Ann}(x) = Z$ whenever $x \in Z \setminus \{0\}$. Now, let $0 \neq x_0 \in I \subseteq Z = \mathrm{Ann}(x)$ be arbitrary. As we saw in the above $Rx_0 = I$. It is plain that the map $\phi : R \to Rx_0$ defined by $\phi(r) = rx_0$ is an epimorphism of the left modules $R$ and $Rx_0$. From the First Isomorphism Theorem for modules, we see that $\frac{R}{\ker \phi} \cong Rx_0$, implying that $Rx_0 \cong \frac{R}{\mathrm{Ann}(x)} \cong \frac{R}{Z}$ because $\ker \phi = \mathrm{Ann}(x) = Z$. We now show that $Rx_0$ is a simple left module, i.e., it has no nontrivial submodules. To this end, suppose that $M \neq 0$ is a left submodule of $Rx_0 = I$. Choose a nonzero $y_0 \in M$. As $M \leq Rx_0 = I$, there exists an $r_0 \in R$ such that $y_0 = r_0 x_0$. We have $y_0 R \subseteq I$ because $x_0 \in I$. On the other hand, $I \subseteq y_0 R$, for $R$ is unital and $I$ is the intersection of all nonzero ideals of $R$. Therefore, $I = y_0 R$. Consequently, we have $M \subseteq I$ and $I = y_0 R \subseteq M$ because $y_0 \in M$, yielding $M = I = y_0 R$. This proves that $Rx_0 = I$ is simple as a left module. This together with $Rx_0 \cong \frac{R}{Z}$ implies that $Z$ is a maximal left ideal, and hence a maximal ideal of $R$ because $R$ is commutative.

Finally, we show that $Z \subseteq I$. To this end, suppose that $0 \neq x_0 \in Z$ is arbitrary. We have $x_0 R \neq \{0\}$, from which, we obtain $\{0\} \neq I \subseteq x_0 R$. Now, choose an arbitrary nonzero element $y_0 \in I$. It follows that there exist $r_0 \in R$ such that $r_0 x_0 = y_0 \in I \subseteq Z$. Note that $r_0 \notin Z$, for otherwise $y_0 = r_0 x_0 = 0$ which is impossible. If $r_0 = 1$, then $x_0 = y_0 \in I$ yielding $Z \subseteq I$, as desired. If $r_0 \neq 1$, we obtain $r_0 R + Z = R$ because $Z$ is a maximal ideal and $r_0 \notin Z$. Consequently, there exist $r_1 \in R$ and $x_1 \in Z$ such that $r_0 r_1 + x_1 = 1$, whence $1 - r_1 r_0 = x_1 \in Z$. On the other hand, $x_0 \in Z$. Since $Z^2 = \{0\}$, we see that $x_0(1 - r_1 r_0) = 0$. This yields $x_0 = x_0 r_0 r_1 = y_0 r_1$, which, in turn, implies $x_0 \in I$ because $y_0 \in I$. That is, $x_0 \in I$, and hence $Z \subseteq I$, for $x_0$ was arbitrary. Therefore, $Z = I$ and $Z$ is a maximal ideal of $R$, proving the assertion. ∎

## 6. First solution: We need the following lemmas.

**Lemma 1 (Cauchy's criterion for the convergence of improper integrals).** *Let $f, \alpha : [a, +\infty) \to \mathbb{R}$ be two functions such that $\alpha$ is increasing on $[a, +\infty)$ and $f$ is integrable with respect to $\alpha$ in the sense of Riemann-Stieltjes, i.e., $f \in \mathcal{R}(\alpha)$, on any closed interval $[a, x]$, where $x \in \mathbb{R}$ with $x > a$. Then, a necessary and sufficient condition for the convergence of $\int_a^{+\infty} f d\alpha$ is that for every $\varepsilon > 0$, there exists an $x_0 > a$ such that*

$$\left| \int_{x_1}^{x_2} f d\alpha \right| < \varepsilon,$$

*whenever $x_2 > x_1 > x_0$.* □

As the lemma is standard, we omit its proof. The interested reader may consult standard books on analysis to see a proof of the lemma. The following lemma is also standard. We however include a proof for the reader's convenience.

**Lemma 2 (Second Mean Value Theorem for integrals).** *Let $g : [a, b] \to \mathbb{R}$ be a Riemann integrable function on $[a, b]$.*

*(a) If $f : [a, b] \to \mathbb{R}$ is a nonnegative and decreasing function on $[a, b]$, then there exists a number $\xi \in [a, b]$ such that*

$$\int_a^b fg = f(a) \int_a^{\xi} g.$$

*(b) If $f : [a, b] \to \mathbb{R}$ is a nonnegative and increasing function on $[a, b]$, then there exists a number $\eta \in [a, b]$ such that*

$$\int_a^b fg = f(b) \int_{\eta}^b g.$$

**Proof.** (a) Set $M = \sup_{x \in [a,b]} |g(x)|$. Obviously, $g(x) + M \geq 0$ for all $x \in [a, b]$. With that in mind, suppose that $P \in \mathcal{P}[a, b]$ is a partition of the closed interval $[a, b]$ which is given by

$$P : x_0 = a < x_1 < \cdots < x_n = b.$$

We can write

$$\int_a^b f(g+M) = \sum_{i=1}^n \int_{x_{i-1}}^{x_i} f(g+M) \le \sum_{i=1}^n f(x_{i-1}) \int_{x_{i-1}}^{x_i} (g+M)$$

$$= \sum_{i=1}^n f(x_{i-1}) \int_{x_{i-1}}^{x_i} g + M \sum_{i=1}^n f(x_{i-1})(x_i - x_{i-1}).$$

Now, let $G : [a,b] \to \mathbb{R}$ be defined by $G(x) = \int_a^x g$. As $G(a) = 0$ and $f$ is nonnegative and decreasing on $[a,b]$, we have

$$\sum_{i=1}^n f(x_{i-1}) \int_{x_{i-1}}^{x_i} g = \sum_{i=1}^n f(x_{i-1})\big(G(x_i) - G(x_{i-1})\big)$$

$$= \sum_{i=1}^n f(x_{i-1})G(x_i) - \sum_{i=1}^{n-1} f(x_i)G(x_i)$$

$$= \sum_{i=1}^n \big(f(x_{i-1}) - f(x_i)\big)G(x_i) + f(b)G(b)$$

$$\le \sum_{i=1}^n \big(f(x_{i-1}) - f(x_i)\big) \sup_{x \in [a,b]} G(x) + f(b) \sup_{x \in [a,b]} G(x)$$

$$= \big(f(a) - f(b)\big) \sup_{x \in [a,b]} G(x) + f(b) \sup_{x \in [a,b]} G(x)$$

$$= f(a) \sup_{x \in [a,b]} G(x).$$

So we can write

$$\int_a^b f(g+M) \le f(a) \sup_{x \in [a,b]} G(x) + M \sum_{i=1}^n f(x_{i-1})(x_i - x_{i-1}).$$

On the other hand, by a result due to Darboux, we have

$$\lim_{||P|| \to 0} \sum_{i=1}^n f(x_{i-1})(x_i - x_{i-1}) = \int_a^b f,$$

where $||P|| = \max_{1 \le i \le n}(x_i - x_{i-1})$. So letting $||P|| \to 0$ in the above inequality, we obtain

$$\int_a^b f(g+M) \le f(a) \sup_{x \in [a,b]} G(x) + M \int_a^b f,$$

implying

$$\int_a^b fg \le f(a) \sup_{x \in [a,b]} G(x).$$

Replacing $g$ by $-g$ in the above inequality, we obtain

$$\int_a^b f(-g) \le f(a) \sup_{x \in [a,b]} -G(x),$$

yielding

$$\int_a^b fg \ge f(a) \inf_{x \in [a,b]} G(x).$$

These inequalities together with the facts that $f(a) \geq 0$ and that $G$ is continuous, in view of the Intermediate Value Theorem, imply that there exists a number $\xi \in [a, b]$ such that

$$\int_a^b fg = f(a)G(\xi) = f(a) \int_a^\xi g,$$

which is what we want.

(b) A proof similar to that of (a) settles (b). However, here is a slick proof of (b) using (a). Define functions $f_1, g_1 : [0, b-a] \to \mathbb{R}$ as follows

$$f_1(x) = f(b-x), \quad g_1(x) = g(b-x).$$

It is plain that $f_1$ and $g_1$ satisfy the hypotheses of (a). It thus follows from (a) that there exists $\xi \in [0, b-a]$ such that

$$\int_0^{b-a} f_1 g_1 = f_1(0) \int_0^\xi g_1.$$

That is, $\int_0^{b-a} f(b-x)g(b-x)dx = f(b) \int_0^\xi g(b-x)dx$. Performing the substitution $b - x = t$, we easily obtain

$$\int_a^b fg = f(b) \int_\eta^b g,$$

where $\eta = b - \xi$, as desired. $\qquad\square$

Now, to prove the assertion, let $M = \int_0^p |g|$. If $M = 0$, the assertion is trivial because we would have $\int_a^b fg = 0$ for all $b > a$, from which, the assertion trivially follows. So suppose $M > 0$. As $\lim_{x \to +\infty} f(x) = 0$ and $f$ is decreasing on $[0, +\infty)$, we see that $f(x) \geq 0$ on $[0, +\infty)$ and that for given $\varepsilon > 0$, there exists an $x_0 > 0$ such that

$$f(x) = |f(x)| < \frac{\varepsilon}{2M},$$

whenever $x \geq x_0$. It now follows from Lemma 2 that for all $x_0 < x_1 < x_2$, there exists $\xi \in [x_1, x_2]$ such that

$$\left| \int_{x_1}^{x_2} fg \right| = \left| f(x_1) \int_{x_1}^\xi g \right| = f(x_1) \left| \int_{x_1}^\xi g \right|.$$

Since $g$ is periodic with period $p$, so is $|g|$. And since

$$\int_0^p g = 0,$$

we see that

$$\int_{ip}^{jp} g = 0, \quad \int_{(i-1)p}^{ip} |g| = \int_0^p |g|,$$

for all $i, j \in \mathbb{N}$. Now, set $T = \{kp : k \in \mathbb{N}\} \cap [x_1, \xi]$. It is plain that $T$ is a finite set. For $x_0 < x_1 < x_2$, there are two cases to consider. (i) $T = \emptyset$. (ii) $T \neq \emptyset$.

First, if $T = \emptyset$, then we have $[x_1, \xi] \subseteq [(i-1)p, ip]$ for some $i \in \mathbb{N}$. So we can write

$$\left| \int_{x_1}^{x_2} fg \right| = f(x_1) \left| \int_{x_1}^{\xi} g \right| \leq f(x_1) \int_{x_1}^{\xi} |g|$$

$$\leq \frac{\varepsilon}{2M} \int_{(i-1)p}^{ip} |g| = \frac{\varepsilon}{2M} \int_{0}^{p} |g| = \frac{\varepsilon}{2M} M = \frac{\varepsilon}{2} < \varepsilon.$$

That is,

$$\left| \int_{x_1}^{x_2} fg \right| < \varepsilon.$$

Next, if $T \neq \emptyset$, let $ip$ and $jp$ be the closest points of $T$ to $x_1$ and $\xi$, respectively. Note that $i, j \in \mathbb{N}$ and $i \leq j$. We can write

$$\left| \int_{x_1}^{x_2} fg \right| = f(x_1) \left| \int_{x_1}^{\xi} g \right| = f(x_1) \left| \int_{x_1}^{ip} g + \int_{ip}^{jp} g + \int_{jp}^{\xi} g \right|$$

$$= f(x_1) \left| \int_{x_1}^{ip} g + \int_{jp}^{\xi} g \right| \leq \frac{\varepsilon}{2M} \left( \int_{x_1}^{ip} |g| + \int_{jp}^{\xi} |g| \right).$$

But $[x_1, ip] \subseteq [(i-1)p, ip]$ and $[jp, \xi] \subseteq [jp, (j+1)p]$. So we can write

$$\left| \int_{x_1}^{x_2} fg \right| \leq \frac{\varepsilon}{2M} \left( \int_{(i-1)p}^{ip} |g| + \int_{jp}^{(j+1)p} |g| \right)$$

$$= \frac{\varepsilon}{2M} \left( \int_{0}^{p} |g| + \int_{0}^{p} |g| \right) = \frac{\varepsilon}{2M} (M + M) = \varepsilon.$$

That is,

$$\left| \int_{x_1}^{x_2} fg \right| \leq \varepsilon.$$

Thus, in any case,

$$\left| \int_{x_1}^{x_2} fg \right| < 2\varepsilon,$$

and hence from Lemma 1, it follows that $\int_{0}^{+\infty} fg$ converges, which is what we want.

**Second solution:** Let $G : [0, +\infty) \to \mathbb{R}$ be defined by $G(x) = \int_{0}^{x} g$. It follows from the hypothesis that $\int_{x}^{x+p} g = 0$, and hence

$$G(x) = \int_{np}^{x} g = \int_{0}^{x-np} g,$$

where $n = \lfloor \frac{x}{p} \rfloor$ and $\lfloor . \rfloor$ denotes the integer part function. Consequently, $|G(x)| \leq \int_{0}^{p} |g| = M$ for all $x \geq 0$. Thus, the function $G$ is bounded. Now, we prove that the assertion holds true under this hypothesis as well. That is, the assertion will remain true if in the problem one replaces "$g$ is a $p$-periodic function such that $\int_{0}^{p} g = 0$" by "the function $G : [0, +\infty) \to \mathbb{R}$, defined by $G(x) = \int_{0}^{x} g$, is bounded". It is worth mentioning that the idea of this proof is very much like that of Dirichlet's Test for convergent series. Just as in the first solution, it suffices to show that for given $\varepsilon > 0$ there is an $N > 0$ such that $\left|\int_x^y fg\right| < \varepsilon$ whenever $x, y \in \mathbb{R}$ with $N < x < y$. By the First Fundamental Theorem of Calculus, $G' = g$. So we can write

$$
\begin{aligned}
\left|\int_x^y fg\right| &= \left|\int_x^y fG'\right| = \left|\int_x^y f dG\right| = \left|fG\Big|_x^y - \int_x^y G df\right| \\
&\leq |f(y)G(y) - f(x)G(x)| + \left|\int_x^y G d(-f)\right| \leq 2Mf(x) + \left|\int_x^y G d(-f)\right| \\
&\leq 2Mf(x) + M\int_x^y d(-f) = 2Mf(x) + M\big(f(x) - f(y)\big) \\
&\leq 3Mf(x).
\end{aligned}
$$

But $\lim_{x \to +\infty} f(x) = 0$. Thus, there is an $N > 0$ such that $f(x) < \frac{\varepsilon}{3M}$ whenever $x > N$. Therefore,

$$
\left|\int_x^y fg\right| < \varepsilon,
$$

whenever $N < x < y$, and hence $\int_0^{+\infty} fg$ converges, as desired. ∎

## 2.21. Twenty-First Competition

**2.21.1. Analysis. 1.** (a) Suppose $\alpha$ is a limit point of the set $S$ such that for every $\varepsilon > 0$ the set $\{x \in S : |f(x)| \geq \varepsilon\}$ is finite. For given $\varepsilon > 0$, let

$$
\delta = \min \big\{|\alpha - x| : |f(x)| \geq \varepsilon\big\}.
$$

It is easily verified that $|f(x)| < \varepsilon$ whenever $0 < |x - \alpha| < \delta$. This proves the assertion.

(b) To prove the assertion by contradiction, suppose there exist an $\varepsilon > 0$ and an infinite sequence $(x_i)_{i=1}^{+\infty}$ of $S$ such that $|f(x_i)| \geq \varepsilon$ for all $i \in \mathbb{N}$. Since the set $S$ is compact, the subset $\{x_i\}_{i=1}^{+\infty}$ has a limit point in $S$, say, $\alpha \in S$. It is obvious that $\lim_{x \to \alpha} f(x) \neq 0$, a contradiction. So the assertion follows. ∎

**2.** (a) To prove the assertion by contradiction, suppose that there exists an $\varepsilon_0 > 0$ such that to each $n \in \mathbb{N}$, there corresponds an $x_n \in \mathbb{R}$ with $x_n > n$ and a $t_n \in K$ satisfying

$$
\big|g(x_n + t_n) - g(x_n)\big| \geq \varepsilon_0,
$$

for all $n \in \mathbb{N}$. Since $K$ is compact, if necessary, by passing to a subsequence of $(t_n)_{n=1}^{+\infty}$, we may assume that $\lim_n t_n = t_0$ for some $t_0 \in K$. Define $f : [0, +\infty) \to \mathbb{R}$ by $f(x) = g(x + t_0) - g(x)$. As $\lim_{x \to +\infty} f(x) = 0$ and $g$ is continuous, we easily see that $f$ is uniformly continuous on $[0, +\infty)$. Thus, for given $\varepsilon_0 > 0$, there exists a $\delta_0 > 0$ such that

$$
\big|g(x + t_0) - g(y + t_0)\big| = |f(x) - f(y)| < \frac{\varepsilon_0}{2},
$$

whenever $x, y \in [0, +\infty)$ and $|x - y| < \delta_0$. On the other hand, it follows from the hypothesis that for given $\varepsilon_0 > 0$, there exists an $M > 0$ such that

$$
|g(x + t_0) - g(x)| < \frac{\varepsilon_0}{2},
$$

whenever $x > M$. Now, since $K$ is compact, and hence bounded, and $\lim_n t_n = t_0$, choose $n \in \mathbb{N}$ large enough so that $x_n + t_n - t_0 > 0$, $x_n > M$, and $|t_n - t_0| < \delta_0$, where $M > 0$ and $\delta_0$ are as in the above. We can write

$$
\begin{aligned}
\varepsilon_0 &\leq |g(x_n + t_n) - g(x_n)| \\
&\leq |g((x_n + t_n - t_0) + t_0) - g(x_n + t_0)| + |g(x_n + t_0) - g(x_n)| \\
&< \frac{\varepsilon_0}{2} + \frac{\varepsilon_0}{2} = \varepsilon_0,
\end{aligned}
$$

implying $\varepsilon_0 < \varepsilon_0$. So the assertion follows by way of contradiction.

(b) To prove $\lim_{x \to +\infty} \left( \int_x^{x+1} g(u)du - g(x) \right) = 0$, note first that by the Mean Value Theorem for integrals, for all $x \in \mathbb{R}$, there exists a $t_x \in [0,1]$ such that

$$
\int_x^{x+1} g(u)du = g(x + t_x).
$$

Now, letting $K = [0,1]$ in (a), for given $\varepsilon > 0$, find $M > 0$ from (a). We have

$$
\left| \int_x^{x+1} g(u)du - g(x) \right| = |g(x + t_x) - g(x)| < \varepsilon,
$$

whenever $x > M$. That is, $\lim_{x \to +\infty} \left( \int_x^{x+1} g(u)du - g(x) \right) = 0$, as desired.

To prove $\lim_{x \to +\infty} \frac{g(x)}{x} = 0$, it suffices to show that $\lim_{x \to +\infty} \frac{g(x)}{[x]} = 0$, where [.] denotes the integer part function, for $\lim_{x \to +\infty} \frac{[x]}{x} = 1$. To this end, note that we can write

$$
\frac{g(x)}{[x]} = \frac{g([x] + t_x) - g([x])}{[x]} + \frac{g([x])}{[x]},
$$

where $t_x = x - [x] \in [0,1)$ for all $x \in \mathbb{R}$ with $x > 1$. We need to show that $\lim_{x \to +\infty} \frac{g([x]+t_x)-g([x])}{[x]} = \lim_{x \to +\infty} \frac{g([x])}{[x]} = 0$. To this end, letting $K = [0,1]$ in (a), for given $\varepsilon > 0$, there exists an $M > 0$ such that

$$
|g(x + t) - g(x)| < \varepsilon,
$$

whenever $x > M$ and $t \in [0,1]$. In particular, if $x > M + 1$, then $[x] > M$, implying that

$$
|g([x] + t_x) - g([x])| < \varepsilon,
$$

because $t_x \in [0,1]$. Thus, $\lim_{x \to +\infty} (g(x) - g([x])) = 0$, and hence

$$
\lim_{x \to +\infty} \frac{g([x] + t_x) - g([x])}{[x]} = 0.
$$

To prove $\lim_{x \to +\infty} \frac{g([x])}{[x]} = 0$, it suffices to show that $\lim_{n \to +\infty} \frac{g(n)}{n} = 0$. To see this, note that part (ii) of the lemma presented in Solution 3 of 2.7.1, which is known as Stolz's Second Theorem, together with the hypothesis yields

$$
\lim_{n \to +\infty} \frac{g(n)}{n} = \lim_{n \to +\infty} \frac{g(n) - g(n-1)}{n - (n-1)} = \lim_{n \to +\infty} (g(n) - g(n-1)) = 0,
$$

as desired. Therefore, $\lim_{x \to +\infty} \frac{g(x)}{x} = 0$, which is what we want. ∎

**3.** We need the following lemma which is (essentially) due to Leo M. Levine (1977).

**Lemma.** *Let* $f : (a, b) \to \mathbb{R}$ *be a bounded function. Let* $D$, $L$, *and* $R$ *denote the set of points at which* $f$ *is discontinuous, has a left hand limit, and has a right hand limit, respectively. Then,* $D \cap (L \cup R)$ *is countable.*

**Proof.** Define $\omega : (a, b) \to [0, +\infty)$ by

$$\omega(x) = \lim_{\delta \to 0+} \Big( \sup f\big([x - \delta, x + \delta]\big) - \inf f\big([x - \delta, x + \delta]\big) \Big).$$

It is readily verified that $x \in D$ if and only if $\omega(x) > 0$. This implies $D = \bigcup_{n \in \mathbb{N}} D_n$, where $D_n = \{x \in (a, b) : \omega(x) > \frac{1}{n}\}$, and hence

$$D \cap (L \cup R) = \Big( \bigcup_{n \in \mathbb{N}} (D_n \cap L) \Big) \bigcup \Big( \bigcup_{n \in \mathbb{N}} (D_n \cap R) \Big).$$

Thus, to establish the lemma, it suffices to show that $D_n \cap L$ and $D_n \cap R$ are countable for all $n \in \mathbb{N}$. To this end, suppose $x_0 \in D_n \cap L$ (resp. $x_0 \in D_n \cap R$). As $x_0 \in L$ (resp. $x_0 \in R$), there exists a $\delta > 0$ such that $|f(x) - f(x_0^-)| < \frac{1}{2n}$ (resp. $|f(x) - f(x_0^+)| < \frac{1}{2n}$) whenever $x \in (x_0 - \delta, x_0)$ (resp. $x \in (x_0, x_0 + \delta)$). From this, we obtain

$$\big|f(x_1) - f(x_2)\big| < \frac{1}{n},$$

whenever $x_1, x_2 \in (x_0 - \delta, x_0)$ (resp. $x_1, x_2 \in (x_0, x_0 + \delta)$). Therefore, $\omega(x) \leq \frac{1}{n}$ for all $x \in (x_0 - \delta, x_0)$ (resp. $x \in (x_0, x_0 + \delta)$), implying $x \notin D_n$. It thus follows that any point of $D_n \cap L$ (resp. $D_n \cap R$) is the right (resp. left) end point of an open interval which contains no point of $D_n \cap L$ (resp. $D_n \cap R$). But these intervals are obviously disjoint and hence form a countable set. So $D_n \cap L$ (resp. $D_n \cap R$) is countable for all $n \in \mathbb{N}$, as desired.  $\square$

It is now obvious that the lemma together with Lebesgue's Integrability Criterion for Riemann integrals implies that a bounded function $f : [a, b] \to \mathbb{R}$ is integrable in the Riemann sense if and only if $[a, b] \setminus (L \cup R)$ is a set of measure zero. This clearly proves the assertion. It is also worth mentioning that, in view of the lemma, one can mimic the second proof presented for Solution 4 of 2.9.1 to give a direct proof of the assertion. We omit the details for the sake of brevity.  ∎

**2.21.2. Algebra. 1.** It is easily checked that *if* $I, J, K$ *are ideals of* $R$ *with* $J \subseteq I$, *then* $I \cap (J + K) = J + I \cap K$. Let $A \overset{s}{\subseteq} B$ and $C \overset{s}{\subseteq} D$. We prove that $A + C \overset{s}{\subseteq} B + D$. To see this, suppose $E$ is an ideal of $R$ such that $A + C + E = B + D$. We need to show that $E = B + D$. To this end, we can write

$$B \cap (A + C + E) = B \cap (B + D).$$

But $A \subseteq B$ and $B \subseteq B + D$. So, we have

$$A + B \cap (C + E) = B,$$

implying that $B \cap (C + E) = B$ because $A \overset{s}{\subseteq} B$. Thus, $B \subseteq C + E$. Likewise, from $D \cap (A + C + E) = D \cap (B + D)$, we obtain $D \subseteq A + E$. Note that $A \subseteq B$ and $B \subseteq C + E$. Thus, $A \subseteq C + E$, and hence $B + D = A + C + E = C + E$. Now, from $C + E = B + D$, we obtain $D \cap (C + E) = D \cap (B + D)$, yielding $C + (D \cap E) = D$, which, in turn in view of $C \overset{s}{\subseteq} D$, implies $D \cap E = D$. Therefore, $D \subseteq E$. Analogously, since $C \subseteq D \subseteq A + E$, we have $B + D = C + A + E = A + E$, yielding $B \cap (A + E) = B \cap (B + D)$. Hence, $A + B \cap E = B$, which in view of $A \overset{s}{\subseteq} B$, implies $B \cap E = B$. Therefore, $B \subseteq E$. It follows that $B + D \subseteq E + E = E \subseteq B + D$. In other words, $E = B + D$, as desired. ∎

**2.** First we need to recall that *if $p$ is a prime, then every group $G$ of order $p^2$ is abelian* (for a proof see Solution 1 of 2.16.2).

We now prove the assertion. From $H \leq Z(G) \leq G$, we see that

$$p^2 = [G : H] = [G : Z(G)][Z(G) : H],$$

implying that $[G : Z(G)] = 1$ or $p$ or $p^2$. If $[G : Z(G)] = 1$, then $G = Z(G)$. In other words, $G$ is abelian and hence $G' = \{e\}$, which is cyclic. If $[G : Z(G)] = p$, then $\frac{G}{Z(G)}$ is cyclic, implying that $G$ is abelian. This again yields $G' = \{e\}$, which is cyclic. Finally, if $[G : Z(G)] = p^2$, then $\frac{G}{Z(G)}$ is abelian because every group of order $p^2$ is abelian whenever $p$ is a prime. It thus follows from the Fundamental Theorem of finite abelian groups that $\frac{G}{Z(G)}$ is either cyclic or there are $x, y \in G$ such that $\frac{G}{Z(G)} = \langle xZ(G), yZ(G) \rangle$. Just as we saw in the above, if $\frac{G}{Z(G)}$ is cyclic, then $G$ is abelian, in which case the assertion trivially holds. So assume that $\frac{G}{Z(G)} = \langle xZ(G), yZ(G) \rangle$ for some $x, y \in G$. It follows that there is a $z \in Z(G)$ such that $xy = yxz$, for $\frac{G}{Z(G)}$ is abelian. Note that $z = x^{-1}y^{-1}xy \in Z(G)$. By proving that $G' = \langle x^{-1}y^{-1}xy \rangle = \{z^k : k \in \mathbb{Z}\}$, we complete the proof. A straightforward induction on $i + j$ reveals that

$$x^i y^j = y^j x^i z^{ij},$$

for all $i, j \in \mathbb{N}$. Also, as $yx = xyz^{-1}$ and $y^{-1}x = xy^{-1}z$, inducting on $i + j$, we obtain

$$y^j x^i = x^i y^j z^{-ij}, \quad y^{-j} x^i = x^i y^{-j} z^{ij},$$

for all $i, j \in \mathbb{N}$. Again, as $\frac{G}{Z(G)} = \langle xZ(G), yZ(G) \rangle$ is abelian, for all $g \in G$ there are $i, j \in \mathbb{N}$ and $z_g \in Z(G)$ such that $g = x^i y^j z_g$. Therefore, for given $g, g' \in G$, there are $i, j, i', j' \in \mathbb{N}$ and $z_g, z_{g'} \in Z(G)$ such that $g = x^i y^j z_g$ and

$g' = x^{i'} y^{j'} z_{g'}$. We can write

$$
\begin{aligned}
g^{-1}g'^{-1}gg' &= \left(y^{-j}x^{-i}\right)\left(y^{-j'}x^{-i'}\right)\left(x^{i}y^{j}\right)\left(x^{i'}y^{j'}\right) \\
&= \left(z^{-ij}x^{-i}y^{-j}\right)\left(y^{-j'}x^{-i'}\right)\left(x^{i}y^{j}\right)\left(y^{j'}x^{i'}z^{i'j'}\right) \\
&= x^{-i}y^{-(j+j')}x^{-i'}x^{i}\left(y^{j+j'}x^{i'}\right)z^{i'j'-ij} \\
&= x^{-i}y^{-(j+j')}x^{-i'}x^{i}x^{i'}y^{j+j'}z^{-(j+j')i'}z^{i'j'-ij} \\
&= x^{-i}\left(y^{-(j+j')}x^{i'}\right)y^{j+j'}z^{-j(i+i')} \\
&= x^{-i}x^{i}y^{-(j+j')}y^{j+j'}z^{i(j+j')}z^{-j(i+i')} \\
&= z^{ij'-i'j}.
\end{aligned}
$$

In other words, $g^{-1}g'^{-1}gg' = (x^{-1}y^{-1}xy)^{k}$, where $k = ij' - i'j \in \mathbb{Z}$. Consequently, $G'$ is the cyclic group generated by $z = x^{-1}y^{-1}xy \in Z(G)$, completing the proof. ∎

**3.** (a) Let $A^{t} \in M_{n}(\mathbb{R})$ denote the transpose of the matrix $A \in M_{n}(\mathbb{R})$. For all $1 \leq i,j \leq n$, use $(AA^{t})_{ij}$ to denote the $ij$ entry of the matrix $AA^{t}$. It follows from the hypothesis that

$$(AA^{t})_{ij} = \mathrm{tr}(A)a_{ij},$$

for all $1 \leq i,j \leq n$. It is now obvious that $\mathrm{tr}(A) \neq 0$, for otherwise $(AA^{t})_{ij} = 0$ for all $1 \leq i,j \leq n$, yielding $\mathrm{tr}(AA^{t}) = 0$, which obtains $A = 0$, contradicting the hypothesis. Thus, $\mathrm{tr}(A) \neq 0$, which is what we want.

(b) In view of (a), we have

$$A = \frac{1}{\mathrm{tr}(A)}AA^{t}.$$

Since $AA^{t}$ is symmetric, so is $A$, as desired.

(c) In view of the second proof presented for Solution 7 of 2.9.3, it suffices to show that $\mathrm{rank}(A) = 1$. To this end, suppose $A \in M_{n}(\mathbb{R})$ satisfies $a_{ik}a_{jk} = a_{kk}a_{ij}$ for all $1 \leq i,j,k \leq n$. For $1 \leq j \leq n$, let us use $A_{j}$ to denote the $j$th column of the matrix $A$. As $\mathrm{tr}(A) \neq 0$, there exists a $1 \leq k \leq n$ such that $a_{kk} \neq 0$. It thus follows from the hypothesis that for all $1 \leq j \leq n$, we have $A_{j} = b_{jk}A_{k}$, where $b_{jk} = \frac{a_{jk}}{a_{kk}}$. This obviously proves that $\mathrm{rank}(A) = 1$, which is what we want. ∎

## 2.22. Twenty-Second Competition

**2.22.1. Analysis. 1.** By the Mean Value Theorem, there exists a $c \in (0,1)$ such that $f'(c) = \frac{f(1)-f(0)}{1-0} = 1$. If $f'(x) \geq 1$ (resp. $f'(x) \leq 1$) for all $x \in (0,1)$, we see that $f(x) = x$ for all $x \in [0,1]$, in which case the assertion is trivial. To see this, suppose to the contrary that there exists $c \in (0,1)$ such that $f(c) > c$ or $f(c) < c$. Suppose $f(c) > c$. It follows from the Mean Value Theorem that there exists $d \in (c,1)$ (resp. $d \in (0,c)$) such that $f'(d) = \frac{f(1)-f(c)}{1-c} < 1$ (resp. $f'(d) = \frac{f(c)-f(0)}{c-0} > 1$), a contradiction in any event. Likewise, if $f(c) < c$, we obtain a contradiction. Thus, there exist $a, b \in (0, 1)$ with $a < c < b$ such that $(f'(a) - 1)(f'(b) - 1) < 0$. It follows from Darboux's Theorem that the range of $f'$, denoted by $R$, includes the closed interval $[m, M]$, where $m = \min\left(f'(a), f'(b)\right) < 1 < M = \max\left(f'(a), f'(b)\right)$. It thus follows that $[1 - \varepsilon, 1 + \varepsilon] \subseteq R$, where $\varepsilon = \min(1 - m, M - 1)$. It is obvious that for all $y_1 \in [1 - \varepsilon, 1 + \varepsilon]$, with $y_1 = f'(x_1) < 1$ for some $x_1 \in [0, 1]$, there exists $y_2 \in [1 - \varepsilon, 1 + \varepsilon]$, with $y_2 = f'(x_2) > 1$ for some $x_2 \in [0, 1]$, such that

$$\frac{1}{f'(x_1)} + \frac{1}{f'(x_2)} = 2. \qquad (*)$$

This obviously proves the assertion when $n \in \mathbb{N}$ is even. To prove the assertion when $n \in \mathbb{N}$ is odd, use $(*)$ and the fact that $\frac{1}{f'(c)} = 1$, where $c$ is as in the above. ∎

**2.** (a) We have $f_n \leq \sum_{i=1}^{+\infty} f_i = f^2$ for all $n \in \mathbb{N}$ because $f_n$'s are all nonnegative. Define $g_n : X \longrightarrow \mathbb{R}$ by

$$g_n(x) = \begin{cases} 0 & x \in f^{-1}(\{0\}), \\ \frac{f_n(x)}{f(x)} & x \in X \setminus f^{-1}(\{0\}). \end{cases}$$

It is obvious that $f_n = g_n f$ for all $n \in \mathbb{N}$. It remains to show that $g_n$ is continuous for all $n \in \mathbb{N}$. To this end, note that $g_n$ is continuous on the open set $X \setminus f^{-1}(\{0\})$ because so are $f$ and $f_n$ on $X \setminus f^{-1}(\{0\})$ $(n \in \mathbb{N})$. Suppose that $x_0 \in X \setminus f^{-1}(\{0\})$ is arbitrary. It follows that for given $\varepsilon > 0$, there exists $\delta > 0$ such that

$$\left| \frac{f_n(x)}{f(x)} - \frac{f_n(x_0)}{f(x_0)} \right| < \varepsilon,$$

whenever $x \in X \setminus f^{-1}(\{0\})$ and $|x - x_0| < \delta$. Since $X \setminus f^{-1}(\{0\})$ is an open set and $x_0 \in X \setminus f^{-1}(\{0\})$, choose $\delta$ small enough such that $x \in X \setminus f^{-1}(\{0\})$ whenever $|x - x_0| < \delta$. So we have

$$\left| g_n(x) - g_n(x_0) \right| < \varepsilon,$$

whenever $|x - x_0| < \delta$. That is, $g_n$ is continuous at $x_0$. Next, suppose $x_0 \in f^{-1}(\{0\})$ is arbitrary. As $f$ is continuous, we see that for given $\varepsilon > 0$, there exists a $\delta > 0$ such that

$$\left| f(x) \right| = \left| f(x) - f(x_0) \right| < \varepsilon,$$

whenever $|x - x_0| < \delta$. Now if $|x - x_0| < \delta$, we can write

$$\left| g_n(x) - g(x_0) \right| = \left| g_n(x) \right| = \begin{cases} 0 < \varepsilon & x \in f^{-1}(\{0\}) \\ \left| \frac{f_n(x)}{f(x)} \right| \leq \left| f(x) \right| < \varepsilon & x \in X \setminus f^{-1}(\{0\}) \end{cases}.$$

In other words, $\left| g_n(x) - g(x_0) \right| < \varepsilon$ whenever $|x - x_0| < \delta$. That is, $g_n$ is continuous at $x_0$, as desired.

(b) Suppose that $\sum_{n=1}^{+\infty} g_n$ converges uniformly on $X$ and that the interior of $f^{-1}(\{0\})$ is empty. It follows that $\sum_{n=1}^{+\infty} g_n$ is a continuous function on $X$ and that $X \setminus f^{-1}(\{0\})$ is a dense subset of $X$. We see from (a) that $\sum_{n=1}^{+\infty} g_n f = f^2$, yielding $f\left(f - \sum_{n=1}^{+\infty} g_n\right) = 0$. This easily implies $f(x) - \sum_{n=1}^{+\infty} g_n(x) = 0$ whenever $x \in X \setminus f^{-1}(\{0\})$. That is, the continuous function $f - \sum_{n=1}^{+\infty} g_n$

vanishes on the dense subset $X \setminus f^{-1}(\{0\})$. Thus, $f - \sum_{n=1}^{+\infty} g_n$ vanishes on $X$, and hence $f = \sum_{n=1}^{+\infty} g_n$, which is what we want. ∎

**3.** Note first that $f$ is continuous on $D$ because the sequence $(f_n)_{n=1}^{+\infty}$ uniformly converges to $f$ on $D$ as $n$ tends to infinity and that $f_n$'s are all analytic, and hence continuous. Thus, in view of Morera's Theorem, it suffices to show that $\int_{\gamma_0} f(z)dz = 0$, where $\gamma_0$ is any simple closed curve inside $\gamma$. We have

$$\int_{\gamma_0} f(z)dz = \lim_n \int_{\gamma_0} f_n(z)dz = 0,$$

because $(f_n)_{n=1}^{+\infty}$ uniformly converges to $f$ on $\gamma_0$ and $f_n$'s are analytic. Assuming that $z_0$ is inside $\gamma$, we see from Cauchy's Integral Formula that

$$f_n(z_0) = \frac{1}{2\pi i} \int_\gamma \frac{f_n(z)}{z - z_0}dz, \quad f(z_0) = \frac{1}{2\pi i} \int_\gamma \frac{f(z)}{z - z_0}dz,$$

proving the assertion because $f_n(z_0) \to f(z_0)$ as $n \to +\infty$. ∎

**2.22.2. Algebra. 1.** Suppose $[G : K] = 2n - 1$ for some $n \in \mathbb{N}$. We have $[G : H] = [G : K][K : H] = 2(2n-1)$. The group $G$ acts on $\Omega = \{Hg : g \in G\}$, the set of all right cosets of $H$ in $G$, by multiplication from the right. That is, the action is given by the map $\varphi : G \times \Omega \to \Omega$ defined by $\varphi(a, Hg) = Hga$. Recall that $\varphi$ gives rise to a group homomorphism $\psi : G \to S(\Omega)$ defined by $\psi(a) = \varphi(a, .) : \Omega \to \Omega$, where $S(\Omega)$ denotes the group of the permutations of $\Omega$, i.e., the set of all one-to-one maps from $\Omega$ onto $\Omega$, which forms a group under composition of maps. The kernel of the action, by definition, is $\ker \psi = \bigcap_{g \in G} g^{-1}Hg$. It follows from the First Isomorphism Theorem for groups that $\frac{G}{\ker \psi} \cong \text{im}(\psi) \le S(\Omega) \cong S_{2(2n-1)}$. That is, $\frac{G}{\ker \psi}$ is isomorphic to a subgroup of the symmetric group of degree $2(2n - 1)$. Since $k \in K$ has order 2, we see that for $k \ker \psi \in \frac{G}{\ker \psi}$, we have $(k \ker \psi)^2 = k^2 \ker \psi = \ker \psi$. Thus, the order of $k \ker \psi$, as an element of $\frac{G}{\ker \psi}$, is equal to 1 or 2. But $k \notin \ker \psi$ because otherwise, in particular, $k \in H$, which is impossible. Consequently, $\text{ord}(k \ker \psi) = 2$. Note that $\frac{G}{\ker \psi}$ acts on $\Omega$ via $(a \ker \psi, Hg) \to Hga$ and that the element $(k \ker \psi, .) \in S(\Omega) \cong S_{2(2n-1)}$ has no fixed point, for otherwise $Hgk = Hg$ for some $g \in G$, yielding $gkg^{-1} \in H$, contradicting the hypothesis. Therefore, the element $(k \ker \psi, .) \in S(\Omega)$ corresponds to an odd permutation of $S(\Omega)$ because, in view of $(k \ker \psi, .)^2 = id \in S(\Omega)$, it decomposes into the product of $2n - 1$ transpositions or 2-cycles. Thus, the subgroup of $\frac{G}{\ker \psi}$, say, $\frac{J}{\ker \psi}$ for some $J \supseteq \ker \psi$, which consists of all elements that correspond to even permutations of $S(\Omega)$ forms a subgroup of $\frac{G}{\ker \psi}$ of index 2. Since $[G : \ker \psi] = [G : J][J : \ker \psi]$, we can write

$$[G : J] = \left[\frac{G}{\ker \psi} : \frac{J}{\ker \psi}\right] = 2,$$

which is what we want. ∎

**2.** If $J = R$ or $J = \{0\}$, the assertion is trivial. So suppose that $J$ is a nontrivial ideal of $R$ and $J$ has a maximal ideal, say, $M$. By obtaining a contradiction, we settle the proof. Let $f : J \to \frac{J}{M}$, defined by $f(x) = x + M$, be the natural homomorphism from $J$ onto $\frac{J}{M}$. As $M$ is a maximal ideal of $J$, it follows that $\frac{J}{M}$ has no nontrivial ideal. This implies that $\frac{J}{M}$ is a field or it is a finite ring with zero multiplication. The latter is impossible because it would mean $(x + M)(y + M) = M$ for all $x, y \in J$, yielding $J^2 \subseteq M$, which is impossible. Thus, $\frac{J}{M}$ is a field. Let $e + M$ denote the identity element of the field $\frac{J}{M}$, where $e \in J \setminus M$. Define $f^* : R \to \frac{J}{M}$ by $f^*(x) = f(xe + M)$. It is readily seen that $f^*$ is a homomorphism of rings. So, we see from the First Isomorphism Theorem for rings that $\frac{R}{\ker f^*} \cong \frac{J}{M}$. Consequently, $\ker f^*$ is a maximal ideal of $R$, for $\frac{J}{M}$ is field. Thus, $J \subseteq \ker f^*$. This, in particular, implies $f^*(e) = M$. On the other hand, $f^*(e) = f(e^2 + M) = (e + M)^2 = e + M$. So, we must have $M = e + M$, yielding $e \in M$, which is impossible. The assertion thus follows by contradiction. ∎

**3.** To prove the assertion by contradiction, suppose that $\mathrm{rank}(L) = 1$ so that $LV = \langle \alpha \rangle$ for some $\alpha \in V$. With no loss of generality, we may assume that the characteristic polynomial of $T$ is irreducible over $F$. We claim that

$$T^k \alpha \in \ker L,$$

for each $k = 0, 1, 2, \ldots$. As the rank of $LT^k$ is at most one and

$$\mathrm{tr}(LT^k) = \mathrm{tr}(TST^k - ST^{k+1}) = 0,$$

for each $k = 0, 1, 2, \ldots$, it follows that $LT^k$ is nilpotent. And hence its eigenvalues are all zero. But $LT^k \alpha \in LV = \langle \alpha \rangle$ for all nonnegative integers $k$. That is, $\alpha$ is an eigenvector of $LT^k$, and hence $LT^k \alpha = 0$ for all nonnegative integers $k$. Now, set

$$W = \langle \{ T^k \alpha : k = 0, 1, 2, \ldots \} \rangle.$$

It is plain that $W$ is invariant under $T$ and that $W$ is a nontrivial subspace of $V$ because $\alpha$ is nonzero and $W \subseteq \ker L$. It thus follows that the characteristic polynomial of $T|_W$, the restriction of $T$ to the invariant subspace $W$, divides the minimal polynomial of $T$, which is equal to the characteristic polynomial of $T$, for the characteristic polynomial of $T$ is irreducible over $F$. This contradicts the hypothesis that the characteristic polynomial of $T$ is irreducible. Thus, the assertion follows by contradiction. ∎

### 2.22.3. General.

| | a | b | c | d | | a | b | c | d | | a | b | c | d |
|---|---|---|---|---|---|---|---|---|---|---|---|---|---|---|
| 1. | | | ★ | | 15. | | | ★ | | 29. | ★ | | | |
| 2. | ★ | | | | 16. | | | ★ | | 30. | ★ | | | |
| 3. | | ★ | | | 17. | ★ | | | | 31. | | | ★ | |
| 4. | | | | ★ | 18. | | ★ | | | 32. | | | | ★ |
| 5. | ★ | | | | 19. | | | ★ | | 33. | ★ | | | |
| 6. | ★ | | | | 20. | | ★ | | | 34. | | | ★ | |
| 7. | | | | ★ | 21. | | | ★ | | 35. | | | | ★ |
| 8. | | ★ | | | 22. | | | ★ | | 36. | | | | ★ |
| 9. | | ★ | | | 23. | | ★ | | | 37. | ★ | | | |
| 10. | | ★ | | | 24. | | | | ★ | 38. | | | | ★ |
| 11. | | | ★ | | 25. | | | | ★ | 39. | | ★ | | |
| 12. | ★ | | | | 26. | ★ | | | | 40. | | | ★ | |
| 13. | | | ★ | | 27. | | ★ | | | | | | | |
| 14. | ★ | | | | 28. | | ★ | | | | | | | |

## 2.23. Twenty-Third Competition

**2.23.1. Analysis. 1.** (a) We prove the assertion by way of contradiction. Suppose that the set $\{f_\alpha : \alpha \in \mathbb{R}\}$ is compact with respect to the uniform norm of $\mathcal{C}_b(\mathbb{R})$ and yet $f$ is not uniformly continuous on $\mathbb{R}$. It follows that there are $\epsilon > 0$ and sequences $(x_n)_{n=1}^{+\infty}$ and $(y_n)_{n=1}^{+\infty}$ in $\mathbb{R}$ such that $\lim_n x_n = \lim_n y_n = \pm\infty$, $\lim_n(x_n - y_n) = 0$, and $|f(x_n) - f(y_n)| \geq \epsilon$ for all $n \in \mathbb{N}$. By the hypothesis, the sequence $(g_n)_{n=1}^{+\infty}$, where $g_n = f_{x_n}$, has a subsequence, say $(g_{n_k})_{k=1}^{+\infty}$, which converges uniformly to $f_\alpha$ on $\mathbb{R}$ for some $\alpha \in \mathbb{R}$. It follows that $\lim_k g_{n_k}(y_{n_k} - x_{n_k}) = f_\alpha(0)$. On the other hand, $\lim_k g_{n_k}(0) = f_\alpha(0)$. Consequently, $\lim_k \big(g_{n_k}(y_{n_k} - x_{n_k}) - g_{n_k}(0)\big) = 0$. But

$$\big|g_{n_k}(y_{n_k} - x_{n_k}) - g_{n_k}(0)\big| = \big|f(y_{n_k}) - f(x_{n_k})\big| \geq \epsilon,$$

for all $k \in \mathbb{N}$, which is in contradiction with

$$\lim_k \big(g_{n_k}(y_{n_k} - x_{n_k}) - g_{n_k}(0)\big) = 0.$$

Thus, $f$ is uniformly continuous on $\mathbb{R}$, as desired.

(b) We disprove the proposition by showing that for the function $f : \mathbb{R} \longrightarrow \mathbb{R}$ defined by $f(x) = e^{-x^2}$, the sequence $\{f_\alpha : \alpha \in \mathbb{R}\}$ is not compact with respect to the uniform norm of $\mathcal{C}_b(\mathbb{R})$. Note first that $f$ is bounded and uniformly continuous on $\mathbb{R}$ because $\lim_{x \to \pm\infty} f(x) = 0$. To disprove the proposition by contradiction, suppose that $\{f_\alpha : \alpha \in \mathbb{R}\}$ is compact with respect to the uniform norm of $\mathcal{C}_b(\mathbb{R})$. It follows that the sequence $(f_n)_{n=1}^{+\infty}$ has a subsequence, say $(f_{n_k})_{k=1}^{+\infty}$, which converges uniformly on $\mathbb{R}$ to $f_\alpha$ for some $\alpha \in \mathbb{R}$. Thus, $0 = \lim_k e^{-(x+n_k)^2} = e^{-(x+\alpha)^2}$, implying that $e^{-(x+\alpha)^2} = 0$ for all $x \in \mathbb{R}$, which is obviously impossible. This completes the proof by way of contradiction. ∎

**2. First solution:** First we need to recall Abel's Continuity Theorem which asserts that *if $\sum_{n=0}^{+\infty} a_n$ is a convergent series of complex numbers, then*

$$\lim_{r \to 1^-} \sum_{n=0}^{+\infty} a_n r^n = \sum_{n=0}^{+\infty} a_n.$$

Without loss of generality, if necessary by adding or subtracting multiples of $2\pi$, we may assume that $0 < \alpha < 2\pi$. Set

$$S_n(\alpha) = \sum_{k=1}^{n} \frac{e^{ik\alpha}}{k}.$$

We have

$$S_n'(\alpha) = i \sum_{k=1}^{n} e^{ik\alpha} = i \frac{e^{i(n+1)\alpha} - e^{i\alpha}}{e^{i\alpha} - 1}.$$

Consequently,

$$
\begin{aligned}
S_n(\alpha) - S_n(\pi) &= i \int_{\pi}^{\alpha} \frac{e^{i(n+1)t}}{e^{it} - 1} dt - i \int_{\pi}^{\alpha} \left(1 + \frac{1}{e^{it} - 1}\right) dt \\
&= \int_{\pi}^{\alpha} \frac{1}{e^{it} - 1} d\left(\frac{e^{i(n+1)t}}{n+1}\right) - i(\alpha - \pi) - \\
&\quad i \int_{\pi}^{\alpha} \frac{(\cos t - 1) - i \sin t}{2(1 - \cos t)} dt \\
&= \frac{1}{n+1} \frac{e^{i(n+1)t}}{e^{it} - 1}\Big|_{\pi}^{\alpha} + \frac{i}{n+1} \int_{\pi}^{\alpha} \frac{e^{i(n+2)t}}{(e^{it} - 1)^2} dt - i(\alpha - \pi) + \\
&\quad \frac{i}{2} \int_{\pi}^{\alpha} dt - \int_{\pi}^{\alpha} \frac{1}{\sin \frac{t}{2}} d\left(\sin \frac{t}{2}\right) \\
&= \frac{1}{n+1} \frac{e^{i(n+1)\alpha}}{e^{i\alpha} - 1} + \frac{(-1)^n}{2(n+1)} + \frac{i}{n+1} \int_{\pi}^{\alpha} \frac{e^{i(n+2)t}}{(e^{it} - 1)^2} dt - \\
&\quad \frac{i(\alpha - \pi)}{2} - \ln \sin \frac{\alpha}{2}.
\end{aligned}
$$

So we can write

$$
\begin{aligned}
S_n(\alpha) - \left(-\ln 2 \sin \frac{\alpha}{2} + i \frac{\pi - \alpha}{2}\right) &= \left(\ln 2 + S_n(\pi)\right) + \frac{1}{n+1} \frac{e^{i(n+1)\alpha}}{e^{i\alpha} - 1} \\
&\quad + \frac{(-1)^n}{2(n+1)} + \frac{i}{n+1} \int_{\pi}^{\alpha} \frac{e^{i(n+2)t}}{(e^{it} - 1)^2} dt.
\end{aligned}
$$

On the other hand, we have

$$\lim_n S_n(\pi) = -\sum_{k=1}^{+\infty} \frac{(-1)^{k-1}}{k} = -\ln 2.$$

To see this, just apply Abel's Continuity Theorem to the power series $\sum_{n=1}^{+\infty} \frac{(-1)^{n-1}}{n} x^n = \ln(1+x)$ and note that the alternating series $\sum_{n=1}^{+\infty} \frac{(-1)^n}{n}$ converges by Leibniz's Theorem. Also, it is plain that

$$\lim_n \left( \frac{1}{n+1} \frac{e^{i(n+1)\alpha}}{e^{i\alpha} - 1} + \frac{(-1)^n}{2(n+1)} \right) = 0.$$

Finally,

$$\left| \frac{i}{n+1} \int_\pi^\alpha \frac{e^{i(n+2)t}}{(e^{it} - 1)^2} dt \right| \leq \frac{1}{n+1} \left| \int_\pi^\alpha \frac{\left| e^{i(n+2)t} \right|}{\left| e^{it} - 1 \right|^2} dt \right|$$

$$= \frac{1}{n+1} \left| \int_\pi^\alpha \frac{1}{\left| e^{it} - 1 \right|^2} dt \right|,$$

for all $n \in \mathbb{N}$ and $\lim_n \frac{1}{n+1} = 0$, implying that

$$\lim_n \frac{i}{n+1} \int_\pi^\alpha \frac{e^{i(n+2)t}}{(e^{it} - 1)^2} dt = 0.$$

Consequently,

$$\lim_n \left( S_n(\alpha) - \left( -\ln 2 \sin \frac{\alpha}{2} + i \frac{\pi - \alpha}{2} \right) \right) = 0.$$

In other words,

$$\sum_{n=1}^{+\infty} \frac{e^{in\alpha}}{n} = -\ln 2 \sin \frac{\alpha}{2} + i \frac{\pi - \alpha}{2}.$$

Now, equating the real and imaginary parts of the above equality obtains

$$\sum_{n=1}^{+\infty} \frac{\cos(n\alpha)}{n} = -\ln 2 \sin \frac{\alpha}{2},$$

$$\sum_{n=1}^{+\infty} \frac{\sin(n\alpha)}{n} = \frac{\pi - \alpha}{2},$$

which is what we want.

**Second solution:** For this solution, first we need to recall Dirichlet's Theorem which asserts that *if* $(a_n)_{n=1}^{+\infty}$ *and* $(b_n)_{n=1}^{+\infty}$ *are sequences of complex and real numbers, respectively, such that the sequence of partial sums of* $(a_n)_{n=1}^{+\infty}$ *is bounded and that* $(b_n)_{n=1}^{+\infty}$ *is monotonic and tends to zero as* $n \to +\infty$, *then* $\sum_{n=1}^{+\infty} a_n b_n$ *converges.* Note that, as we saw in the first solution, we have

$$\sum_{k=1}^n e^{ik\alpha} = \frac{e^{i(n+1)\alpha} - e^{i\alpha}}{e^{i\alpha} - 1},$$

implying that

$$\left| \sum_{k=1}^n e^{ik\alpha} \right| \leq \frac{2}{\left| e^{i\alpha} - 1 \right|},$$

for all $n \in \mathbb{N}$. On the other hand, the sequence $(\frac{1}{n})_{n=1}^{+\infty}$ is decreasing and tends to zero as $n \to +\infty$. Thus, by Dirichlet's Theorem, the series

$$\sum_{n=1}^{+\infty} \frac{e^{in\alpha}}{n}$$

converges whenever $\alpha \in \mathbb{R} \setminus \{2k\pi : k \in \mathbb{Z}\}$.

Again, without loss of generality, we may assume that $0 < \alpha < 2\pi$. Let $z = re^{i\alpha}$, where $0 \le r < 1$. Note that $1 - z = (1 - r\cos\alpha) + i(-r\sin\alpha)$. We can write

$$-\ln(1 - z) = \sum_{n=1}^{+\infty} \frac{z^n}{n} = \sum_{n=1}^{+\infty} \frac{e^{in\alpha}}{n} r^n,$$

where $\ln$ denotes the principal value of the natural logarithm, which is defined by $\ln z = \ln |z| + i \arg z$, where $\arg z = \arctan \frac{\operatorname{Im}(z)}{\operatorname{Re}(z)} \in (-\frac{\pi}{2}, \frac{\pi}{2})$. On the other hand,

$$\begin{aligned}
-\ln(1 - z) &= -\big(\ln|1 - z| + i\arg(1 - z)\big) \\
&= -\left(\ln \sqrt{1 - 2r\cos\alpha + r^2} + i\arctan \frac{-r\sin\alpha}{1 - r\cos\alpha}\right).
\end{aligned}$$

Letting $r \to 1^-$, in view of Abel's Continuity Theorem, we obtain

$$\begin{aligned}
\sum_{n=1}^{+\infty} \frac{e^{in\alpha}}{n} &= -\ln\sqrt{2 - 2\cos\alpha} - i\arctan\frac{-\sin\alpha}{1 - \cos\alpha} \\
&= -\ln\sqrt{4\sin^2\frac{\alpha}{2}} + i\arctan\cot\left(\frac{\alpha}{2}\right) \\
&= -\ln 2\sin\frac{\alpha}{2} + i\frac{\pi - \alpha}{2}.
\end{aligned}$$

Thus,

$$\sum_{n=1}^{+\infty} \frac{e^{in\alpha}}{n} = -\ln 2\sin\frac{\alpha}{2} + i\frac{\pi - \alpha}{2}.$$

Equating the real and imaginary parts of the above equality, we obtain

$$\sum_{n=1}^{+\infty} \frac{\cos(n\alpha)}{n} = -\ln 2\sin\frac{\alpha}{2},$$

$$\sum_{n=1}^{+\infty} \frac{\sin(n\alpha)}{n} = \frac{\pi - \alpha}{2},$$

which is what we want.                                                 ∎

**3.** It is worth mentioning that the hypothesis that $f'$ and $f_n'$'s are continuous is redundant. Let $x \in [a, b)$ be arbitrary. In view of the definition of the derivative, the Mean Value Theorem, and the continuity of $g$, we can write

$$\begin{aligned}
f'(x) &= \lim_{n\to+\infty} n\left(f\left(x + \frac{1}{n}\right) - f(x)\right) \\
&= \lim_{n\to+\infty}\lim_{k\to+\infty} n\left(f_k\left(x + \frac{1}{n}\right) - f_k(x)\right) = \lim_{n\to+\infty}\lim_{k\to+\infty} f_k'(x_n) \\
&= \lim_{n\to+\infty} g(x_n) = g(x),
\end{aligned}$$

where $x < x_n < x + \frac{1}{n}$. Thus, $f'(x) = g(x)$ for all $x \in [a, b)$. If $x = b$, just write $f'(b) = \lim_{n\to+\infty} -n\left(f(b - \frac{1}{n}) - f(b)\right)$ and repeat the above argument to get $f'(b) = g(b)$. So the proof is complete.                    ∎

**2.23.2. Algebra. 1.** This problem is wrong! For $n \in \mathbb{N}$, use $D_n$ to denote the dihedral group of order $2n$. A presentation of $D_n$ is

$$\langle a, b : a^n = e, b^2 = e, a^k \neq e \ (0 < k < n), ab = ba^{-1} \rangle,$$

and hence $D_n = \{a^i b^j : 0 \leq i < n, \ j = 0, 1\}$. We show that $D_8$ is a counterexample. First, recall that $\mathrm{Inn}(G) \cong \frac{G}{Z(G)}$, where $Z(G)$ denotes the center of the group $G$ (see Solution 1 of 2.16.2). Therefore, $\mathrm{Inn}(D_8) \cong \frac{D_8}{Z(D_8)}$. To show that $D_8$ is a counterexample, it suffices to prove that $\frac{D_{2n}}{Z(D_{2n})} \cong D_n$ for all $n \in \mathbb{N}$. To this end, just note that $Z(D_{2n}) = \{e, a^n\}$, where $a$ is the element of order $2n$ in the above presentation of $D_{2n}$, and that

$$\frac{D_{2n}}{Z(D_{2n})} = \{A^i B^j : A = aZ(D_{2n}), \ B = bZ(D_{2n}), 0 \leq i < n, \ j = 0, 1\}$$

$$\cong D_n,$$

for one can easily verify that $A^n = Z(D_{2n}) = B^2$, $AB = BA^{-1}$, and that $A^k \neq Z(D_{2n})$ for all $0 < k < n$. ∎

**2.** First, we prove that $H$ is a two-sided ideal of $R$ by showing that for every $r \in R$ and $x \in H$, there is a $t \in R$ such that $rx = xt$. To this end, note first that $rR$ is a right ideal of $R$. If $rR = \{0\}$, there is nothing to prove. If $rR \neq \{0\}$, then $x \in rR$, and hence there is a $y \in R$ such that $x = ry$. Likewise, considering the right ideal $yR$, we obtain a $t \in R$ such that $x = yt$. So we can write $rx = (ry)t = xt$. That is, for $r \in R$ and $x \in H$, there is a $t \in R$ such that $rx = xt$, as desired. So far, we have actually shown that the intersection of all nonzero right ideals of $R$ is the same of those of the nonzero left ideals of $R$.

Next, suppose that $H^2 \neq 0$. We prove that $R$ is a division ring. To this end, we first show that $R$ has no zero divisors. Suppose to the contrary that there are nonzero elements $x, y \in R$ such that $xy = 0$. As $H^2 \neq 0$, there are $a, b \in H$ such that $ab \neq 0$. For $a, b \in R$, just as we saw in the above, there are $s, t \in R$ such that $xs = a$ and $yt = b$ because $a, b \in H$. Again, for $s \in R$ and $b \in H$, there is a $u \in R$ such that $sb = bu$. We can write

$$xy = 0 \implies xyt = 0 \implies xb = 0 \implies xbu = 0 \implies xsb = 0 \implies ab = 0,$$

a contradiction. Thus, $R$ has no zero divisors. To show that $R$ has a multiplicative identity, choose a nonzero $a \in H$ and note that $a \in aR$, implying that $a = ae$ for some $e \in R$. Since $R$ has no zero divisors and $a(ea - a) = 0$, we obtain $ea = a$. Now for an arbitrary $x \in R$, we have $(xe - x)a = a(ex - x) = 0$, from which, we obtain $xe = ex = x$. That is, $e$ is the multiplicative identity of $R$. For the rest, since $H$ is an ideal of $R$, we need to prove that every nonzero element of $H$ is invertible. To this end, for a nonzero $a \in H$, note that $a^2 \neq 0$, and hence $a = a^2 t$ for some $t \in R$. It follows that $a(e - at) = 0$, which, in turn, yields $at = e$. This implies that $ta(e - ta) = 0$, from which, we obtain $e = ta$. Thus $e = ta = at$. That is, $a$ is invertible, which is what we want. ∎

**3.** Let $A = (a_{ij}) \in M_n(F)$. Use $E_{ij}$ and $I_n$ to denote the matrix whose $ij$ entry is 1 and zero elsewhere and the identity matrix, respectively. If $1 \leq i, j \leq n$ with $i \neq j$, then $\mathrm{tr}(E_{ij}) = 0$. By the hypothesis, $a_{ji} = \mathrm{tr}(AE_{ij}) = 0$ for all $1 \leq i, j \leq n$ with $i \neq j$. Thus, $A$ is a diagonal matrix. For $1 \leq i \leq n$, let

$B_i = \text{diag}(1, 0, \ldots, 0) - \text{diag}(\delta_{1i}, \delta_{2i}, \ldots, \delta_{ni})$, where $\delta_{ji}$ denotes the Kronecker delta. Since $\text{tr}(B_i) = 0$, the hypothesis implies $a_{ii} - a_{11} = \text{tr}(AB_i) = 0$ for all $1 \leq i \leq n$. Therefore, $A = a_{11} I_n$, as desired. ∎

## 2.24. Twenty-Fourth Competition

**2.24.1. First Day. 1.** Define $F : [a, b] \longrightarrow \mathbb{R}$ by $F(x) = \int_a^x f(t)dt$. By the First Fundamental Theorem of Calculus, $F'(x) = f(x)$ for all $x \in [a, b]$. Let $h = F - g$. Obviously, $h$ is differentiable on $[a, b]$ and we have $h' = f - g'$. It follows from the hypothesis that $h'(a)h'(b) < 0$. This together with Darboux's Theorem implies that there exists a $c \in (a, b)$ such that $h'(c) = 0$. In other words, $f(c) = g'(c)$, which is what we want. ∎

**2.** Note first that all real harmonic functions satisfy the Mean Value Property. That is, if $D$ is a domain such that $\overline{B_{r_0}(a_0)} \subseteq D$ for some $a_0 = (x_0, y_0) \in D$ and $r_0 > 0$ and $u : D \longrightarrow \mathbb{R}$ is a harmonic function, then

$$u(a_0) = \frac{1}{2\pi} \int_0^{2\pi} u(a_0 + r_0 e^{i\theta}) d\theta.$$

To see this, let $B$ be an open ball such that $\overline{B_{r_0}(a_0)} \subseteq B \subseteq D$ and $f$ a holomorphic function on the simply connected domain $B$ such that $u|_B = \text{Re} f$. It easily follows from Cauchy's Integral Formula that

$$f(a_0) = \frac{1}{2\pi} \int_0^{2\pi} f(a_0 + r_0 e^{i\theta}) d\theta.$$

Taking the real part of the both sides of the above equation proves the counterpart of it with $f$ replaced by $u$, as desired. Next, we need the following version of the Maximum Modulus Theorem for harmonic functions, which in fact holds true for all functions satisfying the Mean Value Property.

**Theorem.** *Let $D$ be a domain and $u : D \longrightarrow \mathbb{R}$ be a harmonic function. If there is a point $a_0 = (x_0, y_0) \in D$ such that $u(z) \leq u(a_0)$ for all $z \in D$, then $u(z) = u(a_0)$ for all $z \in D$.*

**Proof.** Set

$$L = \{z \in D : u(z) = u(a_0)\}.$$

It is plain that $L$ is a nonempty closed subset of $D$. By proving that $L$ is an open subset of $D$ as well, in view of the connectedness of $D$, we see that $L = D$, proving the assertion. To this end, suppose to the contrary that there is a $z_0 \in L$ which is not an interior point of $L$. Choose $r > 0$ such that $\overline{B_r(z_0)} \subseteq D$. It follows that there is a $b_0 \in B_r(z_0)$ such that $u(b_0) < u(a_0) = u(z_0)$. And hence, from the continuity of $u$, we see that $u(z) < u(z_0)$ for all $z$ in a small enough neighborhood of $b_0$. Letting $r_0 = |z_0 - b_0|$ and $b_0 = z_0 + r_0 e^{i\theta_0}$ for some $0 \leq \theta_0 \leq 2\pi$, there thus exists a proper subinterval $I$ of $[0, 2\pi]$ such that

$\theta_0 \in I$ and $u(z_0 + r_0 e^{i\theta}) < u(z_0)$ for all $\theta \in I$. This, in view of the Mean Value Property of harmonic functions, yields

$$u(z_0) = \frac{1}{2\pi} \int_0^{2\pi} u(z_0 + r_0 e^{i\theta})d\theta < u(z_0),$$

which is a contradiction. Thus, $L$ is open, completing the proof. $\qquad\square$

To prove the assertion, let the set $A$ be as in the statement of the problem. By the Maximum Modulus Theorem for harmonic functions, the set $A$ is open. The assertion follows as soon as we prove that $A$ is closed because $D$ is connected and $A$ is nonempty by the hypothesis. To prove that $A$ is closed, suppose that $(a_n)_{n=1}^{+\infty}$ is a sequence of the elements of $A$ such that $\lim_n a_n = a$ for some $a \in D$. We need to show that $a \in A$. To this end, choose $r > 0$ such that $B_r(a) \subseteq D$. It follows that there is an $n \in \mathbb{N}$ such that $a_n \in B_r(z_0)$. As $a_n \in A$, there is an $r_n > 0$ such that $B_{r_n}(a_n) \subseteq B_r(a) \subseteq D$ and that $u(z) \le u(a_n)$ for all $z \in B_{r_n}(a_n)$. Since $B_r(a)$ is simply connected, there is a holomorphic function $f$ on $B_r(a)$ such that $u|_{B_r(a)} = \mathrm{Re} f$. By the Maximum Modulus Theorem for harmonic functions, $\mathrm{Re} f|_{B_{r_n}(a_n)} = \mathrm{Re} f(a_n)$, which, in view of the Cauchy-Riemann Equations, yields $f|_{B_{r_n}(a_n)} = f(a_n)$. Since $f$ is holomorphic on $B_r(a)$, it follows that $f|_{B_r(a)} = f(a_n) = f(a)$, which, in turn, implies $u(z) = u(a)$ for all $z \in B_r(a)$. Therefore, $a \in A$, which is what we want, completing the proof. $\qquad\blacksquare$

**3. First solution:** Let $p = t(3^n - 2^n)$. In particular, $p|3^n - 2^n$. In other words, $3^n \overset{p}{\equiv} 2^n$. Obviously, $\gcd(6, p) = 1$. And hence, $2$ has an inverse modulo $p$. That is, there is an $a \in \mathbb{N}$ with $1 < a < p$ such that $2a \overset{p}{\equiv} 1$. So we can write

$$3^n a^n \overset{p}{\equiv} 2^n a^n \overset{p}{\equiv} (2a)^n \overset{p}{\equiv} 1,$$

implying $(3a)^n \overset{p}{\equiv} 1$. Use $m$ to denote the order of $3a$ modulo $p$, i.e., $m$ is the least positive integer satisfying $(3a)^m \overset{p}{\equiv} 1$. It follows that $m|n$. By Fermat's Little Theorem, $(3a)^{p-1} \overset{p}{\equiv} 1$, which obtains $m|p-1$. Consequently, $m|\gcd(n, p-1)$. Note that $m > 1$, for otherwise $3a \overset{p}{\equiv} 1$, yielding $3 \overset{p}{\equiv} 2$, which is impossible. This implies that any prime divisor of $m$, say $q$, is a divisor of $n$ as well and that $q \le p - 1 < p$. Thus, $t(n) \le q < p = t(3^n - 2^n)$, as desired.

**Second solution:** Let $m$ be the least positive integer such that $3^m \overset{p}{\equiv} 2^m$. Obviously, $2^m 3^m \overset{p}{\not\equiv} 0$. Note that if $3^k \overset{p}{\equiv} 2^k$ for some $k \in \mathbb{N}$, then $m|k$. To see this, letting $k = qm + r$, where $r + 1, q \in \mathbb{N}$ and $0 \le r < k$, we can write

$$3^r \overset{p}{\equiv} 3^{k-qm} \overset{p}{\equiv} 3^k (3^m)^{-q} \overset{p}{\equiv} 2^k (2^m)^{-q} \overset{p}{\equiv} 2^{k-mq} = 2^r,$$

implying that $3^r \overset{p}{\equiv} 2^r$. Thus, $r = 0$, yielding $k = qm$, i.e., $m|k$. This together with the hypothesis implies that $m|n$. On the other hand, as $6 \overset{p}{\not\equiv} 0$, in view of Fermat's Little Theorem, we can write

$$3^{p-1} \overset{p}{\equiv} 1 \overset{p}{\equiv} 2^{p-1},$$

yielding $m|p-1$. Consequently, $m|\gcd(n, p-1)$, which, as we saw in the first solution, entails $t(n) < t(3^n - 2^n)$, as desired. ∎

**4.** Use $Z$ to denote the center of $G$. It follows from the hypothesis that there exist $n \in \mathbb{N}$ and $g_i \in G$ $(1 \leq i \leq n)$ such that

$$\frac{G}{Z} = \{g_1 Z, \ldots, g_n Z\}.$$

It is obvious that $g_i Z$'s $(1 \leq i \leq n)$ are all infinite subsets of $G$. It suffices to show that $g_i g_j = g_j g_i$ for all $1 \leq i, j \leq n$. To this end, for given $1 \leq i, j \leq n$ with $i \neq j$, it follows from the hypothesis that there exist $z_i, z_j \in Z$ such that $(g_i z_i)(g_j z_j) = (g_j z_j)(g_i z_i)$. This easily yields $(g_i g_j)(z_i z_j) = (g_j g_i)(z_i z_j)$, which, in turn, implies $g_i g_j = g_j g_i$ for all $1 \leq i, j \leq n$, completing the proof. ∎

**5. First solution:** First, we show that *if there are $m$ $n$-tuples from the set $\{0, 1\}$ in such a way that every two of which differ at least in $d$ components, then*

$$\frac{mn}{2} \geq (m - 1)d.$$

To this end, the given $n$-tuples form the following $m \times n$ matrix

$$\begin{pmatrix} a_{11} & \cdots & a_{1n} \\ \vdots & \cdots & \vdots \\ a_{m1} & \cdots & a_{mn} \end{pmatrix}.$$

whose row $i$ is the $i$th $n$-tuple $(1 \leq i \leq m)$. Denote by $A$ the total number of differences between any two rows of these $m$ rows. In other words, $A = \sum_{1 \leq i < j \leq m} d_{ij}$, where $d_{ij}$ is the number of components in which row $i$ and row $j$ differ. As $d_{ij} \geq d$ for all $1 \leq i < j \leq m$, we obtain $A \geq \binom{m}{2}d$. To find an upper bound for $A$, fix a column, say column $j$, and use $c_j$ to denote the number of differences that these $m$ vector can mutually have in their $j$th component. It is obvious that $A = \sum_{1 \leq j \leq n} c_j$. If there are $x_j$ zeros in column $j$, then $c_j = x_j(m - x_j) \leq \frac{m}{2}(m - \frac{m}{2}) = \frac{m^2}{4}$ because the polynomial function $f(x) = x(m - x)$ assumes its maximum at $x = \frac{m}{2}$. It follows that $A = \sum_{1 \leq j \leq n} c_j \leq \sum_{1 \leq j \leq n} \frac{m^2}{4} = \frac{m^2 n}{4}$. Consequently,

$$\frac{m^2 n}{4} \geq A \geq \binom{m}{2}d = \frac{m(m-1)}{2}d,$$

from which, we obtain

$$\frac{mn}{2} \geq (m - 1)d,$$

as desired.

Now, to prove the assertion, let $m = M(2d - 1, d)$ and note that $n = 2d - 1$. In view of the above inequality, we have

$$\frac{m(2d - 1)}{2} \geq (m - 1)d,$$

yielding

$$2md - m \geq 2md - 2d \Longrightarrow m = M(2d - 1, d) \leq 2d,$$

which is what we want.

**Second solution:** First we need to recall Turán's Theorem  from Graph Theory. For the sake of completeness, we quote a simple proof of it which is due to William Staton.

**Theorem (Turán, 1941).** *Graphs with $n$ vertices containing no $K_r$ have no more than $\frac{(r-2)n^2}{2r-2}$ edges, for $r \geq 2$.*

**Proof (William Staton).** Induct on $r$. If $r = 2$, the result is obvious. Now if the statement is true for $K_r$-free graphs it must be shown that $K_{r+1}$-free graphs have no more than $\frac{(r-1)n^2}{2r}$ edges. Let $G$ be such a graph, and let $x$ be the number of vertices in a largest $K_r$-free induced subgraph of $G$. Since the neighbors of any vertex induce a $K_r$-free subgraph, no vertex of $G$ has degree exceeding $x$. Let $A$ be a largest induced $K_r$-free subgraph of $G$. By induction, there are at most $\frac{(r-2)x^2}{(2r-2)}$ edges in $A$. Each edge of $G$ not in $A$ is incident with at least one of the $n - x$ vertices not in $A$, so summing the degrees of these vertices counts each such edge at least once. Hence there are at most $x(n - x)$ such edges and so $G$ has at most $\frac{(r-2)x^2}{(2r-2)} + x(n - x)$ edges. Since

$$\frac{r - 2}{2r - 2}(x^2) + x(n - x) = \frac{r - 1}{2r}n^2 - \frac{r}{2r - 2}\left(x - \frac{(r - 1)n}{r}\right)^2,$$

the result follows.                                                                  □

To prove the assertion by contradiction, suppose that $M(2d - 1, d) > 2d$ so that there are $2d + 1$ $(2d - 1)$-tuples any two of which differ in at least $d$ components. Let $G$ be an edge-labeled graph as follows. The set of vertices is the $2d + 1$ $(2d - 1)$-tuples. For any two vertices, if the two vertices differ in their $k$ component ($1 \leq k \leq 2d-1$), then connect the two vertices with an edge with label $k$. Consequently, there are at least $d$ labeled edges connecting any two vertices of the graph $G$ and hence the graph $G$ has at least $\binom{2d+1}{2} \times d = (2d + 1)d^2$ edges. Since $2d - 1$ numbers are assigned to the edges and there are at least $(2d + 1)d^2$ edges, we see that there is a label $1 \leq k_0 \leq 2d - 1$ which is assigned to at least $\left[\frac{(2d+1)d^2}{2d-1}\right] + 1$ edges of the graph, where $[.]$ denotes the integer part function. Now consider the induced graph on these edges to each of which $k_0$ is assigned. This induced graph is indeed a simple graph having $2d + 1$ vertices and $\left[\frac{(2d+1)d^2}{2d-1}\right] + 1$ edges. Note that

$$\left[\frac{(2d + 1)d^2}{2d - 1}\right] + 1 > \frac{(2d + 1)d^2}{2d - 1} > d^2 + d = \left[\frac{(2d + 1)^2}{4}\right].$$

It thus follows from Turán's Theorem  that the induced graph has a $K_3$, i.e., a triangle. This means that there are three vertices any two of which differ in their $k_0$ component. This is a contradiction because the components are either 0 or 1, completing the proof.                                    ■

**6.** First we need the following simple lemma.

**Lemma.** *Let two lookalike boxes contain $w_1$ white marbles and $b_1$ black marbles, and $w_2$ white marbles and $b_2$ black marbles, respectively. The probability of picking a white marble from one of the boxes is equal to $\frac{1}{2}\left(\frac{w_1}{w_1+b_1} + \frac{w_2}{w_2+b_2}\right)$.*

**Proof.** Use $A$ to denote the event that one picks a marble from the box that contains $w_1$ white marbles and $b_1$ black marbles, and use $B$ to denote the event that one picks a marble from the other box. If $W$ denotes the event of picking a white marble from one of the boxes, as $\{A, B\}$ is a partition of the probability space, we can write

$$
\begin{aligned}
P(W) &= P\big((A \cap W) \cup (B \cap W)\big) = P(A)P(W|A) + P(B)P(W|B) \\
&= \frac{1}{2} \times \frac{w_1}{w_1 + b_1} + \frac{1}{2} \times \frac{w_2}{w_2 + b_2} = \frac{1}{2}\left(\frac{w_1}{w_1 + b_1} + \frac{w_2}{w_2 + b_2}\right),
\end{aligned}
$$

proving the lemma. $\qquad\square$

Suppose that the first person has picked a marble from one of the boxes, say, box $b$. Use $b'$ to denote the other box. For each $i = 1, 2, 3$, define the following events

$W_i$ (resp. $W_i'$): the event that the $i$th person picks a white marble from $b$ (resp. $b'$).

$B_i$ (resp. $B_i'$): the event that the $i$th person picks a black marble from $b$ (resp. $b'$).

By the lemma above, we have

$$
P(W_1) = \frac{1}{2}\left(\frac{1}{1+2} + \frac{2}{1+2}\right) = \frac{1}{2}.
$$

Likewise, $P(B_1) = \frac{1}{2}$.

Use $B_{011}$ and $B_{001}$ (resp. $B_{011}'$ and $B_{001}'$) to denote the events that box $b$ (resp. $b'$) contains one black marble and two white marbles and two black marbles and one white marble, respectively.

First, suppose that the first person has picked a white marble from $b$. Using Bayes' Theorem, we can write

$$
\begin{aligned}
P(B_{011}|W_1) &= \frac{P(B_{011})P(W_1|B_{011})}{P(W_1)} = \frac{\frac{1}{2} \times \frac{2}{3}}{\frac{1}{2}} = \frac{2}{3}, \\
P(B_{001}|W_1) &= \frac{P(B_{001})P(W_1|B_{001})}{P(W_1)} = \frac{\frac{1}{2} \times \frac{1}{3}}{\frac{1}{2}} = \frac{1}{3}.
\end{aligned}
$$

And hence, $P(B_{011}'|W_1) = 1 - P(B_{011}|W_1) = 1 - \frac{2}{3} = \frac{1}{3}$ and $P(B_{001}'|W_1) = 1 - P(B_{001}|W_1) = 1 - \frac{1}{3} = \frac{2}{3}$. Let $U = B_{011}|W_1$, $V = B_{001}|W_1$, $U' = B_{011}'|W_1$, and $V' = B_{001}'|W_1$. We have $P(U) = P(V') = \frac{2}{3}$ and $P(V) = P(U') = \frac{1}{3}$. Noting that $\{U, V\}$ is a partition of the probability space, we have

$$
P(W_2) = P(U)P(W_2|U) + P(V)P(W_2|V) = \frac{2}{3} \times \frac{1}{2} + \frac{1}{3} \times 0 = \frac{1}{3}.
$$

That is, assuming that the first person has picked a white marble from $b$, the probability that the second person picks a white marble from $b$ is $\frac{1}{3}$. Analogously, as $\{U', V'\}$ is a partition of the probability space, we can write

$$P(W_2') = P(U')P(W_2'|U') + P(V')P(W_2'|V') = \frac{1}{3} \times \frac{2}{3} + \frac{2}{3} \times \frac{1}{3} = \frac{4}{9}.$$

In other words, assuming that the first person has picked a white marble from $b$, the probability that the second person picks a white marble from $b'$ is $\frac{4}{9}$, which is greater than $\frac{1}{3}$. So, in this case, the second person should pick a marble from the other box, i.e., $b'$.

Now, assuming that the first person has picked a black marble from $b$, again using Bayes' Theorem, we will have $P(B_{011}|B_1) = \frac{1}{3}$ and $P(B_{001}|B_1) = \frac{2}{3}$, and hence $P(B_{011}'|B_1) = 1 - P(B_{011}|B_1) = 1 - \frac{1}{3} = \frac{2}{3}$ and $P(B_{001}'|B_1) = 1 - P(B_{001}|B_1) = 1 - \frac{2}{3} = \frac{1}{3}$. So, letting $S = B_{011}|B_1$, $T = B_{001}|B_1$, $S' = B_{011}'|B_1$, and $T' = B_{001}'|B_1$, we can write

$$P(W_2) \quad = \quad P(S)P(W_2|S) + P(T)P(W_2|T) = \frac{1}{3} \times 1 + \frac{2}{3} \times \frac{1}{2} = \frac{2}{3},$$

$$P(W_2') \quad = \quad P(S')P(W_2'|S') + P(T')P(W_2'|T') = \frac{2}{3} \times \frac{2}{3} + \frac{1}{3} \times \frac{1}{3} = \frac{5}{9}.$$

Since $P(W_2) = \frac{2}{3}$ is greater than $P(W_2') = \frac{5}{9}$, the second person should pick a marble from box $b$. Therefore, the probability of survival for the second person is

$$\frac{1}{2} \times \frac{4}{9} + \frac{1}{2} \times \frac{2}{3} = \frac{5}{9},$$

for, by the above lemma, the first person picks a white (resp. black) marble with the probability of $\frac{1}{2}$.

To investigate the probability of survival for the third person, there are two cases to consider.

(i) the first person survives.

Without loss of generality, we may assume that the first person has picked a white marble from box $b$. In view of what we showed in the above, there are two subcases to consider. (a) The second person picks a white marble from box $b'$; and (b) The second person picks a black marble from box $b'$. In case (a), since the first and second person have picked white marbles from $b$ and $b'$, respectively, the boxes do not make any difference for the third person to pick a marble from. Thus, by the lemma above, the third person picks a white marble from one of the boxes with the probability of $\frac{1}{4}$. So if we use $S_1$ to denote the event that the third person survives in this case, we will have

$$P(S_1) = \frac{1}{2} \times \frac{4}{9} \times \frac{1}{4} = \frac{1}{18}.$$

In case (b), letting $C = W_1 \cap B_2'$, in view of Bayes' Theorem, we can write

$$P(B_{011}|C) \quad = \quad \frac{P(C|B_{011})P(B_{011})}{P(B_{011})P(C|B_{011}) + P(B_{001})P(C|B_{001})}$$

$$= \quad \frac{\frac{2}{3} \times \frac{2}{3} \times \frac{1}{2}}{\frac{1}{2} \times \frac{2}{3} \times \frac{2}{3} + \frac{1}{2} \times \frac{1}{3} \times \frac{1}{3}} = \frac{4}{5}.$$

Thus, $P(B_{001}|C) = 1 - P(B_{011}|C) = \frac{1}{5}$. Consequently, the probability that the third person picks a white marble from $b$ is equal to

$$P(B_{001}|C)P(W_3|B_{001}|C) + P(B_{011}|C)P(W_3|B_{011}|C) = \frac{1}{5} \times 0 + \frac{4}{5} \times \frac{1}{2} = \frac{2}{5}.$$

Likewise, the probability that the third person picks a white marble from $b'$ is equal to

$$P(B_{001}|C)P(W_3'|B_{001}|C) + P(B_{011}|C)P(W_3'|B_{011}|C) = \frac{1}{5} \times 1 + \frac{4}{5} \times \frac{1}{2} = \frac{3}{5}.$$

Thus, the third person should pick a marble from $b'$, which is the box from which the second person picked a marble. So, the probability of survival for the third person in this case, denoted by $P(S_2)$, is equal to

$$P(S_2) = \frac{1}{2} \times \frac{5}{9} \times \frac{3}{5} = \frac{1}{6}.$$

(ii) the first person does not survive.

Again, we may assume that the first person has picked a black marble from box $b$. There are two subcases to consider. (a) The second person picks a white marble from box $b$; and (b) The second person picks a black marble from box $b$. In case (a), letting $C = B_1 \cap W_2$, in view of Bayes' Theorem, we can write

$$P(B_{001}|C) = \frac{P(C|B_{001})P(B_{001})}{P(B_{001})P(C|B_{001}) + P(B_{011})P(C|B_{011})}$$

$$= \frac{\frac{2}{3} \times \frac{1}{2} \times \frac{1}{2}}{\frac{1}{2} \times \frac{2}{3} \times \frac{1}{2} + \frac{1}{2} \times \frac{1}{3} \times 1} = \frac{1}{2}.$$

Thus, the boxes do not make any difference for the third person to pick a marble from. So the probability of picking a white marble becomes $\frac{2}{4} = \frac{1}{2}$. And hence, if $S_3$ denotes the event that the third person survives in this case, we will have

$$P(S_3) = \frac{1}{2} \times \frac{2}{3} \times \frac{1}{2} = \frac{1}{6}.$$

Finally, if the first and second person both pick black marbles from $b$, then the third person can survive by picking the remaining white marble from box $b$. So the probability of survival in this case, which is denoted by $P(S_4)$, is equal to

$$P(S_4) = \frac{1}{2} \times \frac{1}{3} \times 1 = \frac{1}{6}.$$

Consequently, the probability of survival for this person, denoted by $P(S)$, is equal to

$$P(S) = P(S_1) + \cdots + P(S_4) = \frac{1}{18} + \frac{1}{6} + \frac{1}{6} + \frac{1}{6} = \frac{5}{9}.$$

That is, the probability of survival for the third person is the same as that of the second person. ∎

**2.24.2. Second Day. 1.** (a) Use $\overline{B}$ to denote the closure of $B$ with respect to the Euclidean norm of $\mathbb{R}^n$. As $\overline{B}$ is compact because it is bounded and closed, there is a positive real $r$ such that

$$\mathrm{diam}(\overline{B}) := \sup\left\{||a - b|| : a, b \in \overline{B}\right\} = 2r, .$$

As $\mathrm{diam}(\overline{B}) = \mathrm{diam}(B)$, in view of the hypothesis, we see that for all $n \in \mathbb{N}$ there are $x_n, y_n \in B$ and an open ball $B_{r_n}(z_n)$, the ball centered at $z_n$ with radius $r_n$, such that $x_n, y_n \in B_{r_n}(z_n) \subseteq B$ and $||x_n - y_n|| > 2r - \frac{1}{n}$. Since $z_n$'s are in the compact set $\overline{B}$, if necessary by passing to a subsequence of $(z_n)_{n=1}^{+\infty}$, we may assume that there is a $z \in \overline{B}$ such that $\lim_n z_n = z$. We claim that $B = B_r(z)$, the open ball with radius $r$ centered at $z$, proving the assertion. First, we prove that $B_r(z) \subseteq B$. To this end, let $x \in B_r(z)$ be arbitrary. We can write $||x - z|| = r - \epsilon$ for some $\epsilon > 0$. As $\lim_n \frac{1}{n} = 0$, there is an $N \in \mathbb{N}$ such that $\frac{1}{N} < \epsilon$ and $||z - z_N|| < \frac{\epsilon}{2}$. We have

$$||x - z_N|| \leq ||x - z|| + ||z - z_N|| < r - \epsilon + \frac{\epsilon}{2}$$
$$= r - \frac{\epsilon}{2} < r - \frac{1}{2N} < \frac{||x_N - y_N||}{2} \leq r_N,$$

implying that $x \in B_{r_N}(z_N) \subseteq B$. In other words, $x \in B$, and hence $B_r(z) \subseteq B$. To prove that $B \subseteq B_r(z)$, we argue by contradiction. Suppose that there is a point $x \in B \setminus B_r(z)$ so that $||x - z|| > r$. Join the two points $x, z$ to intersect the closed ball $\overline{B}_r(z)$ at two antipodal points $y, y' \in \overline{B}_r(z)$ so that $y$ and $y'$ are, respectively, on and off the line segment that joins $x$ and $z$. It is obvious that

$$||x - y'|| = ||x - z|| + ||z - y'|| = ||x - z|| + r > 2r = \mathrm{diam}(\overline{B}) = \mathrm{diam}(B),$$

which is a contradiction because $x, y' \in \overline{B}$. Therefore, $B \subseteq B_r(z)$, and hence $B = B_r(z)$, completing the proof.

(b) **First solution:** Let $X = (\mathcal{C}[0, 1], ||.||_\infty)$ be the normed space of all real valued continuous functions on the closed interval $[0, 1]$ equipped with the uniform norm. As is well-known, $X$ is a Banach space, and hence in particular a complete metric space. Let $Y = (\mathcal{C}[0, 1], d)$ be the metric space equipped with the metric $d$ which is defined on $\mathcal{C}[0, 1]$ by

$$d(f, g) := \frac{||f - g||_\infty}{1 + ||f - g||_\infty},$$

where $||f - g||_\infty = \sup_{x \in [0,1]} |f(x) - g(x)|$. It is readily verified that $Y$ is a complete metric space. We claim that the set $B = \{f \in \mathcal{C}[0, 1] : f(x) > 0\}$ is a bounded subset of $Y$ with the property that for each pair of points $x, y$ in $B$, there exists an open ball $U$ such that $U \subseteq B$ and $x, y \in U$ and yet $B$ is not an open ball in $Y$. Plainly, $Y$ is bounded, and hence so is $B$. Now, suppose $f, g \in B$ are arbitrary. Set $M = \sup_{x \in [0,1]} \max\left(f(x), g(x)\right)$ and $m = \inf_{x \in [0,1]} \min\left(f(x), g(x)\right)$. We have $0 < m, M \in \mathbb{R}$ because $f$ and $g$ are continuous on the compact interval $[0, 1]$. Define $t : [0, +\infty) \to [0, +\infty)$ by $t(x) = \frac{x}{1+x}$. First, we show that $f, g \in B_r(h) \subseteq B$, where $r = t(\frac{M - m}{2})$ and $h \in B$ is defined by $h(x) = \frac{M+m}{2}$ for all $x \in [0, 1]$. To see $f \in B_r(h)$, we need to show that $d(f,h) < r$. To this end, note first that $m \leq f(x) \leq M$ implying that $\frac{3m}{4} < f(x) < M + \frac{m}{4}$ for all $x \in [0,1]$, from which, we easily obtain $||f - h||_\infty < \frac{M - \frac{m}{2}}{2}$. But $t$ is strictly increasing on $[0, +\infty)$. This yields

$$d(f,h) = t(||f - h||_\infty) < t\left(\frac{M - \frac{m}{2}}{2}\right) = r,$$

as desired. Likewise, we see that $g \in B_r(h)$. Next, we need to show that $B_r(h) \subseteq B$. To this end, let $k \in B_r(h)$ be arbitrary. It follows that

$$d(k,h) = t(||k - h||_\infty) < r = t\left(\frac{M - \frac{m}{2}}{2}\right),$$

which yields

$$-\frac{M - \frac{m}{2}}{2} < k(x) - \frac{M + m}{2} < \frac{M - \frac{m}{2}}{2},$$

for all $x \in [0,1]$ because $t$ is strictly increasing on $[0, +\infty)$. This obtains

$$0 < \frac{3m}{4} < k(x) < M + \frac{m}{4},$$

for all $x \in [0,1]$, implying that $k \in B$, which is what we want. It remains to prove that $B$ is not a an open ball. Suppose to the contrary that $B = B_{t(s)}(l)$, for some $s > 0$ and $l \in B$. Now, let $k \in B = B_{t(s)}(l)$ be arbitrary. We see that

$$d(k,l) = t(||k - l||_\infty) < t(s),$$

which obtains

$$-s < k(x) - l(x) < s,$$

for all $x \in [0,1]$. Consequently, $\sup_{x \in [0,1]} k(x) < s + M_0$, where $M_0 = \sup_{x \in [0,1]} l(x)$. This is a contradiction because $k \in B$ is arbitrary. Therefore, $B$ is a bounded subset of the complete metric space $Y = (\mathcal{C}[0,1], d)$ with the property that for each pair of points $f, g$ in $B$ there exists an open ball $B_r(h)$ such that $f, g \in B_r(h) \subseteq B$ and yet $B$ is not an open ball in $Y$, proving the claim. $\qquad \cdot$

**Second solution:** A simpler proof similar to that of first solution shows that the open interval $(0, +\infty)$ of the complete metric space $(\mathbb{R}, d)$, where $d(x,y) := \frac{|x-y|}{1+|x-y|}$ and $|.|$ denotes the absolute value function, is a counterexample to (a) when $\mathbb{R}^n$ is replaced by the complete metric space $(\mathbb{R}, d)$. We omit the details for the sake of brevity.

**Third solution:** Let $X = \{A_1, \ldots, A_5\}$, where $A_1 = (1,0,0)$, $A_2 = (0,1,0)$, $A_3 = -A_1$, $A_4 = -A_2$, and $A_5 = (0,0,\frac{4}{3})$. Note that if we use $d$ and $s$ to, respectively, denote the diameter and side length of the square whose vertices are $A_1, A_2, A_3, A_4$, we have $d = 2$ and $s = \sqrt{2}$. Also note that $s < A_5 A_i = \frac{5}{3} < d$ for each $i = 1,2,3,4$. It is readily checked that $X$ with respect to the Euclidean metric of $\mathbb{R}^3$ is a complete metric space. Set $B = \{A_1, \ldots, A_4\}$. It is now easily verified that $B$ is a bounded subset of the complete metric space $X$ with the property that for each pair of points $x, y$ in $B$, there exists an open ball $B_r(z)$, where $r = \frac{s + A_5 A_1}{2}$, such that $x, y \in B_r(z) \subseteq B$ and yet $B$ is not an open ball in $X$, which is what we want. $\blacksquare$

**2. First solution:** It might be worth noting that the hypothesis that $f$ is continuous is redundant. To prove the assertion, it is enough to assume that $f$ is integrable on $[a, b]$. Without loss of generality, we may assume that $\int_a^b f(t)dt > 0$. For given $k \in (0, 1)$, define the continuous function $g_k : [a, b] \longrightarrow \mathbb{R}$ by

$$g_k(x) = \int_a^x f(t)dt - k \int_a^b f(t)dt.$$

We have $g_k(a) = -k \int_a^b f(t)dt < 0$ and $g_k(b) = (1 - k) \int_a^b f(t)dt > 0$. Thus, by the Intermediate Value Theorem, there is a $c_k \in (a, b)$ such that $g(c_k) = 0$, proving the assertion.

**Second solution:** Define $g : [a, b] \longrightarrow \mathbb{R}$ by

$$g(x) = \frac{\int_a^x f(t)dt}{\int_a^b f(t)dt}.$$

Plainly, $g$ is continuous on $[a, b]$ and $g(a) = 0 < 1 = g(b)$. Thus, by the Intermediate Value Theorem, there exists a $c_k \in (a, b)$ such that $g(c_k) = k$, implying $\int_a^{c_k} f(t)dt - k \int_a^b f(t)dt = 0$, as desired. ∎

**3.** (a) The inequality is known as the Frobenius Inequality. We prove the assertion over division rings. Let $D$ be a division ring and $A \in M_{m \times n}(D), B \in M_{n \times p}(D), C \in M_{p \times q}(D)$, where $m, n, p, q \in \mathbb{N}$. View $A$ (resp. $B, C$) as a linear transformations acting on the left of $D^n$ (resp. $D^p, D^q$), the right vector space of all $n \times 1$ (resp. $p \times 1$, $q \times 1$) column vectors over $D$. Let $A_1 = A|_{B(D^p)}$. We can write

$$\dim \ker A_1|_{BC(D^q)} \leq \dim \ker A_1 = \dim B(D^p) - \dim A_1(B(D^p)),$$

which, in view of $A_1(B(D^p)) = AB(D^p)$, yields

$$\dim \ker A_1|_{BC(D^q)} \leq r(B) - r(AB).$$

On the other hand,

$$\begin{aligned} \dim \ker A_1|_{BC(D^q)} &= \dim BC(D^q) - \dim A_1(BC(D^q)) \\ &= r(BC) - \dim ABC(D^q) = r(BC) - r(ABC). \end{aligned}$$

Therefore,

$$r(BC) - r(ABC) \leq r(B) - r(AB),$$

implying

$$r(BC) + r(AB) \leq r(ABC) + r(B),$$

as desired.

(b) To prove the assertion which is known as the Sylvester Inequality, just set $A = A$, $B = I_n$, and $C = B$ in (a). ∎

**4.** Let $I$ and $K$ be a two-sided ideal and a left ideal of the ring $R$, respectively. It is plain that $KI \subseteq I \cap K$. To prove $I \cap K \subseteq KI$, choose an arbitrary element $x \in I \cap K$. It suffices to show that $x \in KI$. To this end, by the hypothesis, we have $(I \cap K)^2 = I \cap K$ because $I \cap K$ is a left ideal of $R$. It follows that $x \in (I \cap K)^2$. And hence there are $n \in \mathbb{N}$ and $x_i, y_i \in I \cap K$ ($1 \leq i \leq n$) such that $x = \sum_{i=1}^{n} x_i y_i$. But $x_i y_i \in KI$ for all $1 \leq i \leq n$. Therefore, $x \in KI$, which is what we want, completing the proof. ∎

**5. First solution:** First, suppose that the number of wins of no two teams are equal. As there are $n$ teams and any team has played against all of the other $n-1$ teams, we see that for each $i = 0, \ldots, n-1$, there is exactly one team, which we call the $i$th team, whose wins is equal to $i$. It is obvious that the $(n-1)$st team has won all the other $n-1$ teams, the $(n-2)$nd team losses to the $(n-1)$st team and wins all the remaining $n-2$ teams, and so on and so forth. For the $i$th team, where $i \in \{0, 1, \ldots, n-1\}$, let's call $i$ to be the label of the team. Thus, for any three teams, the team with the smallest label has lost to the other two teams. Therefore, there are no three teams $A, B, C$ such that $A$ wins $B$, $B$ wins $C$, and $C$ wins $A$.

Next, suppose that there are no such three teams. Define the relation $\geq$ on the teams as follows. For two teams $A$ and $B$, we write $A \geq B$ if and only if $A = B$ or $A$ wins $B$. Since there are no such three teams, in view of the hypothesis, it is easily verified that the relation $\geq$ is a linear order on the set of the participating teams. So if we put the teams in the decreasing order, say $T_1 > T_2 > \cdots > T_n$, then $T_i$, the $i$th team, has won exactly $n - i + 1$ teams, namely, $T_{i+1} > \cdots > T_n$. Consequently, no two teams have scored the same number of wins, which is what we want.

**Second solution:** First, suppose that there are two teams $A$ and $B$ that have scored $k$ wins, where $k \in \{0, \ldots, n-1\}$. Without loss of generality, we may assume that $A$ wins $B$. We claim that there is a team $C$ such that $A$ wins $B$, $B$ wins $C$, and $C$ wins $A$. To see this, as $B$ has won $k$ teams, one of them, say $C$, must have won $A$, for otherwise $A$ must have won $k+1$ teams, the $k$ teams lost to $B$ plus $B$ itself, which is impossible. Thus, such team $C$ exists, settling the implication.

Next, suppose that there are three teams $A$, $B$, and $C$ such that $A$ has won $B$, $B$ has won $C$, and $C$ has won $A$. We prove the assertion by induction on $n$, the number of participating teams. If $n = 3$, the assertion is trivial. Assuming that the assertion holds for $n-1$, to prove the assertion for $n$, argue by contradiction. So no two teams have scored the same number of wins. Consequently, there is exactly one team, say $D$, that has scored $n-1$ wins, for $n$ teams have participated in the game. Now, $D$ is not one of $A$, $B$, or $C$ because $D$ has won them all. Exclude $D$ and consider the game between the remaining $n-1$ teams which include $A, B, C$. It follows from the induction hypothesis that there are two teams in the remaining $n-1$ teams whose wins are equal. But the two teams have both lost to $D$. Thus, the two teams have scored the same number of wins in the original game with $n$ teams, which is a contradiction. So the assertion follows. ∎

**6.** We convert this problem into the following coin-flipping game of which the problem is a special case. We then present a proof of the counterpart of the assertion for the coin-flipping game, which we have taken from "Concrete Mathematics", a book by Ronald L. Graham, Donald E. Knuth, and Oren Patashnik. Here is the counterpart of the problem.

*Two persons, A and B, are playing the following coin-flipping game.*

*First, A chooses a pattern of length $\ell$ ($\ell \geq 3$) of heads and tails (for instance, HTH, where H stands for heads and T for tails). Then B, who knows the pattern chosen by A, chooses a different pattern of the same length of heads and tails. Then a fair coin is flipped until one of the patterns is first obtained, in which case the the player whose pattern occurs first is to win the game. Show that no matter what choice is made by A, there is a choice for B so that the probability of winning the game by B is greater than $\frac{1}{2}$.*

Denote, respectively, by $\mathcal{S}_A$ and $\mathcal{S}_B$ the sum of $A$'s and $B$'s winning positions. Use $\mathcal{N}$ to denote the sum of the patterns each of which does not contain any occurrences of the patterns $\mathcal{A}$ and $\mathcal{B}$ chosen by $A$ and $B$, respectively. For instance, if $A$ chooses $\mathcal{A} = HTH$ and $B$ chooses $\mathcal{B} = TTH$, we have

$$\mathcal{S}_A =$$

$$HTH + HHTH + THTH + HHHTH + TTHTH + THHTH + \cdots,$$

$$\mathcal{S}_B =$$

$$TTH + HTTH + TTTH + HHTTH + TTTTH + THTTH + HTTTH + \cdots,$$

$$\mathcal{N} =$$

$$1 + H + T + HH + TT + TH + HT + HHH + HHT + THT + TTT + \cdots.$$

Obviously, if we set $H = T = \frac{1}{2}$, the resulting values for $\mathcal{S}_A$ and $\mathcal{S}_B$, respectively, become the probability that $A$ and $B$ wins the game. We have

$$1 + \mathcal{N}(H + T) = \mathcal{N} + \mathcal{S}_A + \mathcal{S}_B,$$

$$\mathcal{N}\mathcal{A} = \mathcal{S}_A \sum_{k=1}^{\ell} \mathcal{A}^{(\ell-k)} \delta_{\mathcal{A}^{(k)},\mathcal{A}_{(k)}} + \mathcal{S}_B \sum_{k=1}^{\ell} \mathcal{A}^{(\ell-k)} \delta_{\mathcal{B}^{(k)},\mathcal{A}_{(k)}},$$

$$\mathcal{N}\mathcal{B} = \mathcal{S}_A \sum_{k=1}^{\ell} \mathcal{B}^{(\ell-k)} \delta_{\mathcal{A}^{(k)},\mathcal{B}_{(k)}} + \mathcal{S}_B \sum_{k=1}^{\ell} \mathcal{B}^{(k-\ell)} \delta_{\mathcal{B}^{(k)},\mathcal{B}_{(k)}},$$

where $\delta_{\alpha,\beta}$ denotes the Kronecker delta, $\mathcal{A}^{(k)}$ and $\mathcal{A}_{(k)}$ (resp. $\mathcal{B}^{(k)}$ and $\mathcal{B}_{(k)}$) denote the last and the first $k$ characters of $\mathcal{A}$ (resp. $\mathcal{B}$). To see the first equality, just note that every term on the left side of it either ends with $\mathcal{A}$, or $\mathcal{B}$, or it does not end with either of $\mathcal{A}$ and $\mathcal{B}$ meaning that the term belongs to $\mathcal{N}$; conversely, every term on the right of the first equality, is either empty or it belongs to $\mathcal{N}H$ or $\mathcal{N}T$. The second equality holds because every term on the left either completes a term of $\mathcal{S}_A$ in such a way that the last $k$ characters of $\mathcal{A}$ coincides with its first $k$ characters for some $1 \leq k \leq \ell$, or a term of $\mathcal{S}_B$ in such a way that the last $k$ characters of $\mathcal{B}$ equals the first $k$ characters of $\mathcal{A}$ for some $1 \leq k \leq \ell$; and conversely because every term on the right belongs to the left. Analogously, the third equality holds. As noted in the above, by setting $H = T = \frac{1}{2}$, we obtain the wining probabilities for $A$ and $B$, which we denote by $P(A)$ and $P(B)$, respectively. It follows from the first equality above that $P(A) + P(B) = 1$. Let

$$\mathcal{A}{:}\mathcal{A} = \sum_{k=1}^{\ell} 2^{k-1} \delta_{\mathcal{A}^{(k)},\mathcal{A}_{(k)}}, \mathcal{B}{:}\mathcal{A} = \sum_{k=1}^{\ell} 2^{k-1} \delta_{\mathcal{B}^{(k)},\mathcal{A}_{(k)}},$$

$$A{:}B = \sum_{k=1}^{\ell} 2^{k-1} \delta_{A^{(k)}, B_{(k)}}, \, B{:}B = \sum_{k=1}^{\ell} 2^{k-1} \delta_{B^{(k)}, B_{(k)}}.$$

Using the second and third equalities, we can write

$$\mathcal{N} = 2\Big(P(A)(A{:}A) + P(B)(B{:}A)\Big),$$

$$\mathcal{N} = 2\Big(P(A)(A{:}B) + P(B)(B{:}B)\Big),$$

which obtains

$$\frac{P(A)}{P(B)} = \frac{B{:}B - B{:}A}{A{:}A - A{:}B}.$$

We now claim that if $A$ chooses the pattern $\mathcal{A} = \tau_1 \tau_2 \cdots \tau_\ell$, then $B$ has a better chance of winning the game by choosing $\mathcal{B} = \tau_2' \tau_1 \cdots \tau_{\ell-1}$, where $\tau_2'$ is the heads/tails opposite of $\tau_2$. It suffices to show that $P(A) < P(B)$. Suppose the contrary, implying that

$$B{:}B - B{:}A \geq A{:}A - A{:}B. \qquad (*)$$

Note that $A{:}A \geq 2^{\ell-1}$ and $B{:}B < 2^{\ell-1} + 2^{\ell-3}$, and $B{:}A \geq 2^{\ell-2}$, for $A^{(\ell)} = A_{(\ell)}$, $B^{(\ell)} = B_{(\ell)}$ but $B^{(\ell-2)} \neq B_{(\ell-2)}$, and $B^{(\ell)} \neq A_{(\ell)}$ but $B^{(\ell-1)} = B_{(\ell-1)}$. It follows that $B{:}B - B{:}A < 2^{\ell-1} + 2^{\ell-3} - 2^{\ell-2}$. Since $A^{(\ell)} \neq B_{(\ell)}$ and $A^{(\ell-1)} \neq B_{(\ell-1)}$, we conclude $A^{(\ell-2)} = B_{(\ell-2)}$, for otherwise $A{:}B \leq 2^{\ell-3}$, from which, we obtain

$$A{:}A - A{:}B \geq 2^{\ell-1} - 2^{\ell-3} \geq 2^{\ell-1} + 2^{\ell-3} - 2^{\ell-2} > B{:}B - B{:}A,$$

which is impossible. Consequently, $A^{(\ell-2)} = B_{(\ell-2)}$, yielding $\tau_2' = \tau_3, \tau_1 = \tau_4, \tau_2 = \tau_5, \tau_3 = \tau_6, \ldots, \tau_{\ell-3} = \tau_\ell$. But then, $A{:}A \geq 2^{\ell-1} + 2^{\ell-4} + \cdots$, $A{:}B \leq 2^{\ell-3} + 2^{\ell-6} + \cdots$, $B{:}A \geq 2^{\ell-2} + 2^{\ell-5} + \cdots$, and $B{:}B < 2^{\ell-1} + 2^{\ell-4} + \cdots$, implying that

$$\begin{aligned} A{:}A - A{:}B &\geq (2^{\ell-1} - 2^{\ell-3}) + (2^{\ell-4} - 2^{\ell-6}) + \cdots \\ &> (2^{\ell-1} - 2^{\ell-2}) + (2^{\ell-4} - 2^{\ell-5}) + \cdots \\ &> B{:}B - B{:}A. \end{aligned}$$

In other words, $A{:}A - A{:}B > B{:}B - B{:}A$, which is in contradiction with $(*)$. Therefore, $P(A) < P(B)$, proving the assertion. ∎

## 2.25. Twenty-Fifth Competition

### 2.25.1. First Day. 1. As $|G| = n$ and $[G : Z(G)] = 4$, we have

$$Z(G) \subsetneq C_G(x) \subsetneq G,$$

for all $x \in G \setminus Z(G)$, where $C_G(x) := \{g \in G : xg = gx\}$ denotes the centralizer of the element $x$ in $G$. We can write

$$4 = [G : Z(G)] = [G : C_G(x)][C_G(x) : Z(G)],$$

which easily implies $[G : C_G(x)] = 2$. It follows that the size of any conjugacy class of the elements of $G \setminus Z(G)$ is equal to 2, and hence $2 \mid \left|G \setminus Z(G)\right|$. On the other hand, $\left|G \setminus Z(G)\right| = \frac{3n}{4}$. Thus, $2 \mid \frac{3n}{4}$, implying $8 \mid 3n$, which, in turn, yields $8 \mid n$. For a given $n \in \mathbb{N}$ satisfying $8 \mid n$, set $G := \mathbb{Q}_8 \times \mathbb{Z}_{\frac{n}{8}}$, where $\mathbb{Q}_8$ denotes the quaternionic group with 8 elements. It is obvious that $Z(G) = \{-1,1\} \times \mathbb{Z}_{\frac{n}{8}}$, yielding $[G : Z(G)] = 4$, as desired. ∎

**2.** From $T^2 = T$, we see that $V = \ker T \oplus \operatorname{im} T$. Now, let $\alpha \in V$ be arbitrary. It follows that there are $\beta \in \ker T$ and $\gamma \in V$ such that $\alpha = \beta + T\gamma$. In view of the hypotheses, we can write

$$(T + S)\alpha = (T + S)(\beta + T\gamma) = T\beta + T^2\gamma + S\beta + ST\gamma = T\gamma + S\beta.$$

Since $\beta \in \ker T$, by the hypothesis, there is a $\delta \in V$ such that $\beta = S\delta$. So we can write

$$(T + S)\alpha = T\gamma + S\beta = T\gamma + S^2\delta = T\gamma + S\delta = \beta + T\gamma = \alpha.$$

Thus, $T + S$ is the identity transformation, for $\alpha \in V$ was arbitrary, finishing the proof. ∎

**3.** By the Mean Value Theorem, there is a $c \in (y, y + 1)$ such that

$$f(y + 1) - f(y) = f'(c).$$

As $f''(t) < 0$ for all $t \in \mathbb{R}$, $f'$ is strictly decreasing on $\mathbb{R}$. Thus, $f'(y) > f'(c)$ because $y < c$. So we can write

$$f(y + 1) - f(y) = f'(c) < f'(y) < f(y + 1) - x,$$

yielding $x < f(y)$, as desired. ∎

**4.** Define $g : (a, b) \longrightarrow \mathbb{R}$ by $g(x) = \ln\left(1 + f^2(x)\right) - x$. We have

$$g'(x) = \frac{2f(x)f'(x)}{1 + f^2(x)} - 1 = \frac{2f(x)f'(x) - f^2(x) - 1}{1 + f^2(x)} \geq 0,$$

for all $x \in (a, b)$. Thus, $g$ is nondecreasing on $(a, b)$, implying that $-a = \lim_{x \to a^+} g(x) \leq \lim_{y \to b^-} g(y) = 1 - b$. Therefore, $-a \leq 1 - b$, yielding $b - a \leq 1$. As for an example for which $b - a = 1$, just let $a = 0 = b - 1$ and $f(x) = \sqrt{e^x - 1}$ on $(0, 1)$. ∎

**5.** Note first that the game stops exactly when the 7th numbered one marble is drawn from a box. To see this, it is obvious that if the 7th numbered one marble is drawn from a box, the the game stops because one needs to draw a marble from box one which is empty. Conversely, if the game stops when a marble numbered $i$ is drawn from a box, then $i = 1$. Because otherwise the box numbered $i$ must be empty for some $i \geq 2$, which means seven marbles numbered $i$ must have already been drawn from the $i$th box, implying that there are 8 marbles numbered $i$, a contradiction.

Now consider the following extended game. Assume that, after a stop in the original game, the game is continued by choosing the box which is not empty yet and whose number is minimal among all nonempty boxes, and that the game is continued in this manner until all the marbles are drawn from all boxes. In this extended game, to each permutation of the 49 marbles, there corresponds a round of the game. Conversely, to each round of the game, there corresponds a permutation of the 49 marbles. Also, a round of the original game continues until all the marbles are drawn from all boxes only when the number on the last drawn marble, i.e., the 49th drawn marble, reads one, in which case the original game and the extended game are the same. It is now obvious that the probability of the event that all the marbles are drawn from all boxes in the original game is equal to that of that same event in the extended game which is equal to

$$\frac{\frac{48!}{6!7!^6}}{\frac{49!}{7!^7}} = \frac{1}{7},$$

which is what we want. ∎

**6.** Let $0 < c \le b \le a$ with $a, b, c \in \mathbb{N}$ and $a + b + c = n$ for some $n \in \mathbb{N}$, be the side lengths of a desired triangle. The following gives an algorithm for finding a partition of the number $n$ into the summands 2, 3, and 4 in which the summand 3 appears at least once. As the following algorithm is reversible, the assertion follows. We have $a < b + c$, e.g., $7 < 6 + 3$. We explain the algorithm in three stages.

Stage (i): To create summands of 3, subtract one from the three side lengths, i.e., $a, b, c$, one at a time and continue this for $c + b - a$ times so that the inequality $a < b + c$ becomes an equality, e.g.,

$$7 < 6 + 3 \quad \rightarrow \quad 6 < 5 + 2 \quad \rightarrow \quad 5 = 4 + 1.$$

Stage (ii): To create summands of 4, in the equality obtained in stage (i), subtract two units and one unit, one at a time, from the left hand side and right hand side of the equality, respectively. Continue in this manner for $a - b$ times till a zero appears in the equality , e.g.,

$$5 = 4 + 1 \quad \rightarrow \quad 3 = 3 + 0.$$

Stage (iii): To create summands of 2, in the last equality obtained in stage (ii), subtract one unit at a time from each nonzero number of the equality and continue in this way for $b - c$ times till all numbers become zero, e.g.,

$$3 = 3 + 0 \quad \rightarrow \quad 2 = 2 + 0 \quad \rightarrow \quad 1 = 1 + 0 \quad \rightarrow \quad 0 = 0 + 0.$$

So, the corresponding partition for a triangle with side lengths $0 < a < b < c$ and $a + b + c = n$ is given by $n = 4(a - b) + 3(b + c - a) + 2(b - c)$, e.g., $16 = 4 + 3 + 3 + 2 + 2 + 2$. ∎

**2.25.2. Second Day. 1.** The assertion is a quick consequence of the following proposition.

*Let $R$ be a unital ring without zero divisors, i.e., $\forall a, b \in R : ab = 0 \Rightarrow a = 0$ or $b = 0$, and $U(R)$ denote the multiplicative group of the units of $R$. Then, every finite abelian subgroup of $U(R)$ is cyclic.*

We present two proofs for this proposition.

**First proof.** Let $G$ be an abelian subgroup of $U(R)$ and $Z$ denote the prime ring of $R$, i.e., $Z = \{k1 : k \in \mathbb{Z}\}$. It follows from the hypothesis that

$$Z[G] := \{\sum_{i=1}^{n} k_i g_i : n \in \mathbb{N}, k_i \in Z, g_i \in G \ (1 \le i \le n)\}$$

is an integral domain and that $G$ is a subgroup of the multiplicative group of the units of $Z[G]$. So without loss of generality, we may assume that $R$ is an integral domain. To prove the assertion, we need the following lemma.

**Lemma.** *(i) Let $G$ be a group and $g \in G$ with $\mathrm{ord}(g) = n_1 \cdots n_k$, where $k \in \mathbb{N}$ and $n_j$'s $(1 \le j \le k)$ are pairwise relatively prime. Then, there exist unique $g_1, \ldots, g_k \in G$ such that $g = g_1 \cdots g_k$, $g_i g_j = g_j g_i$, and $\mathrm{ord}(g_i) = n_i$ for each $i, j = 1, \ldots, k$. Conversely, if for $g \in G$, there are $g_1, \ldots, g_k \in G$ satisfying $g = g_1 \cdots g_k$, $g_i g_j = g_j g_i$ and $\mathrm{ord}(g_i) = n_i$ for each $i, j = 1, \ldots, k$ such that $n_j$'s $(1 \le j \le k)$ are pairwise relatively prime, then $\mathrm{ord}(g) = n_1 \cdots n_k$.*

*(ii) Let $G$ be a group and $g_1, \ldots, g_k \in G$ be of finite order such that $g_i g_j = g_j g_i$ for each $i, j = 1, \ldots, k$. Then there is a $g \in G$ such that $\mathrm{ord}(g) = \mathrm{lcm}(\mathrm{ord}(g_1), \ldots, \mathrm{ord}(g_k))$.*

**Proof.** (i) First, recall that *if $g \in G$ is such that $\mathrm{ord}(g) = n$ for some $n \in \mathbb{N}$, then $\mathrm{ord}(g^i) = \frac{n}{\gcd(i,n)}$ for all $i \in \mathbb{Z}$.* Now, set $m_i = \prod_{i \ne j=1}^{k} n_j$. As $n_j$'s $(1 \le j \le k)$ are pairwise relatively prime, we see that $\gcd(m_1, \ldots, m_k) = 1$ and hence there are $c_j$'s $(1 \le j \le k)$ in $\mathbb{Z}$ such that $\sum_{j=1}^{k} c_j m_j = 1$. We note that $\gcd(c_i, n_i) = 1$ for each $i = 1, \ldots, k$. To see this, it suffices to show that no prime $p$ divides $\gcd(c_i, n_i)$. Suppose to the contrary that there is a prime $p$ that divides $\gcd(c_i, n_i)$ for some $1 \le i \le k$. It follows that $p | c_i$ and $p | n_i$ and hence $p | c_i$ and $p | m_j$ for all $j \in \{1, \ldots, k\} \setminus \{i\}$. Consequently, $p | \sum_{j=1}^{k} c_j m_j = 1$, which is impossible. Thus, $\gcd(c_i, n_i) = 1$ for each $i = 1, \ldots, k$. Now, letting $g_i = g^{c_i m_i}$, we can write

$$
\mathrm{ord}(g_i) = \mathrm{ord}(g^{c_i m_i}) = \frac{\mathrm{ord}(g)}{\gcd(c_i m_i, \mathrm{ord}(g))} = \frac{m_i n_i}{\gcd(c_i m_i, m_i n_i)}
$$

$$
= \frac{n_i}{\gcd(c_i, n_i)} = n_i.
$$

Also, we can write

$$
g_1 \cdots g_k = g^{c_1 m_1} \cdots g^{c_k m_k} = g^{\sum_{j=1}^{k} c_j m_j} = g.
$$

Finally, it is obvious that $g_j$'s $(1 \le j \le k)$ commute. To see that $g_j$'s $(1 \le j \le k)$ are unique, suppose that there are $g_1', \ldots, g_k' \in G$ such that $g = g_1' \cdots g_k'$, $g_i' g_j' = g_j' g_i'$, and $\mathrm{ord}(g_i') = n_i$ for each $i, j = 1, \ldots, k$. We can write

$$
g^{c_j m_j} = (g_1' \cdots g_k')^{c_j m_j} = g_1'^{c_j m_j} \cdots g_k'^{c_j m_j}.
$$

But $g_i'^{m_j} = e$ whenever $i \ne j$ because $\mathrm{ord}(g_i') = n_i$ and $m_j = \prod_{j \ne i=1}^{k} n_i$. Thus,

$$
g^{c_j m_j} = g_j'^{c_j m_j} = g_j'^{1 - \sum_{j \ne i=1}^{k} c_i m_i} = g_j',
$$

for all $1 \le j \le k$, as desired.

As for the converse, suppose that $\mathrm{ord}(g) = m$. We need to show that $m = n_1 \cdots n_k$. Note first that $m | n_1 \cdots n_k$ because $g^{n_1 \cdots n_k} = (g_1 \cdots g_k)^{n_1 \cdots n_k} = e$. It thus suffices to show that $n_1 \cdots n_k | m$. To this end, letting $1 \le i \le k$ be arbitrary but fixed, we can write

$$
g^m = e \implies (g_1 \cdots g_k)^m = e \implies g_i^m = \prod_{i \ne j=1}^{k} g_j^{-m}.
$$

As $g_i^{n_i} = e$, we get $(g_i^m)^{n_i} = e$, from which, we obtain $\operatorname{ord}(g_i^m)\big|n_i$. On the other hand,

$$(g_i^m)^{\prod_{i\neq j=1}^k n_j} = \Big( \prod_{\substack{i\neq l=1}}^k g_l^{-m} \Big)^{\prod_{i\neq j=1}^k n_j} = \prod_{\substack{i\neq l=1}}^k \big(g_l^{\prod_{i\neq j=1}^k n_j}\big)^{-m} = e,$$

implying that $\operatorname{ord}(g_i^m)\big| \prod_{i\neq j=1}^k n_j$. Thus, $\operatorname{ord}(g_i^m)\big|\gcd\big(n_i, \prod_{i\neq j=1}^k n_j\big) = 1$, which yields $\operatorname{ord}(g_i^m) = 1$. This obtains $g_i^m = e$ for all $1 \leq i \leq k$. Hence, $n_i\big|m$ for all $1 \leq i \leq k$. It thus follows that $n_1 \cdots n_k\big|m$ because $n_j$'s $(1 \leq j \leq k)$ are pairwise relatively prime. Therefore, $m = \operatorname{ord}(g) = n_1 \cdots n_k$, which is what we want.

(ii) It is plain that there exists an $l \in \mathbb{N}$ and primes $p_1, \ldots, p_l$ such that

$$\operatorname{ord}(g_i) = p_1^{m_{i1}} \cdots p_l^{m_{il}},$$

where $m_{ij}$'s are nonnegative integers $(1 \leq i \leq k, 1 \leq j \leq l)$. For $1 \leq i \leq k$, we see from (i) that there are $g_{i1}, \ldots, g_{il} \in G$ such that $g_i = g_{i1} \cdots g_{il}$, that $g_{ij}$'s $(1 \leq i \leq k, 1 \leq j \leq l)$ all commute, and that $\operatorname{ord}(g_{ij}) = p_j^{m_{ij}}$. It is obvious that for each $1 \leq j \leq l$, there is an $1 \leq i_j \leq k$ such that $\operatorname{ord}(g_{i_j j}) = \max\big\{\operatorname{ord}(g_{1j}), \ldots, \operatorname{ord}(g_{kj})\big\}$. Let $g' = \prod_{j=1}^l g_{i_j j}$. It thus follows from (i) that

$$\operatorname{ord}(g') = \prod_{j=1}^l p_j^{\max\{m_{1j}, \ldots, m_{kj}\}} = \operatorname{lcm}\big(\operatorname{ord}(g_1), \ldots, \operatorname{ord}(g_k)\big),$$

which is what we want. $\qquad\square$

Note that $U(R)$, and hence $G$, is commutative. Let $a \in G$ be such that its order, say $m$, is maximal among the elements of $G$. Let $b$ be an arbitrary element of $G$. Since $G$ is commutative, by part (ii) of the above lemma, there is a $c \in G$ whose order is equal to the least common multiple of $\operatorname{ord}(a)$ and $\operatorname{ord}(b)$. It follows that $\operatorname{ord}(c) = \operatorname{ord}(a)$ because $\operatorname{ord}(a)$ is maximal. Thus, $\operatorname{ord}(b)\big|\operatorname{ord}(c) = \operatorname{ord}(a) = m$, yielding $b^m = 1$. In other words, every element of $G$ is a root of $f = x^m - 1 \in R[x]$. As $R$ is an integral domain, $f$ has at most $m$ roots, and hence $|G| \leq m$. Therefore, $|G| = m = \operatorname{ord}(a)$. That is, $G$ is cyclic, as desired.

**Second proof.** Just as we saw in the first proof, we may assume that $R$ is an integral domain. Consequently, the equation $x^n = 1$ has at most $n$ solutions in $G$ for all $n \in \mathbb{N}$. It thus follows from the lemma presented in Solution 1 of 2.2.2 that $G$ is cyclic, as desired. $\qquad\blacksquare$

**2.** Let $q$ be a prime factor of $2^p + 1$ which is different from 3. We have $q\big|2^p+1$, and hence $q\big|2^{2p} - 1$, yielding $2^{2p} \overset{q}{\equiv} 1$. Use $t = \operatorname{ord}_q(2)$ to denote the order of 2 modulo $q$. It follows that $t$ divides $2p$. By showing that $t \neq 1, 2, p$, we conclude that $t = 2p$. Firstly, $t \neq 1$ because $2^1 \overset{q}{\not\equiv} 1$. Secondly, $t \neq 2$ because $2^2 \overset{q}{\not\equiv} 1$, for $q \neq 3$. Thirdly, $t \neq p$ because otherwise $2^p \overset{q}{\equiv} 1$, which, in view of $q\big|2^p + 1$, obtains $q\big|2$, which is impossible, for $q$ is odd. Thus, $t = 2p$. Now, since $t = \text{ord}_q(2) \mid |\mathbb{Z}_q^*| = q - 1$, where $\mathbb{Z}_q^* = \mathbb{Z}_q \setminus \{0\}$, we have $2p | q - 1$ or $q - 1 = 2kp$ for some $k \in \mathbb{N}$, which is what we want. ∎

**3.** It is plain that $f$ is a one-to-one function from $X$ onto $X$. Thus, $f$ is invertible. Use $f^{-1}$ to denote the inverse of $f$. It follows from the hypothesis that

$$d(f^{-1}(x), f^{-1}(y)) \leq d(x, y),$$

for all $x, y \in X$. Thus, the function $f^{-1} : X \longrightarrow X$ is a Lipschitz function, and hence it is (uniformly) continuous. Now, since $f^{-1}$ is continuous and $X$ is compact, we see that $f^{-1}$ is a closed map, implying that $f = (f^{-1})^{-1}$ is continuous, which is what we want. ∎

**4.** Let $\alpha = f(1)$ and $\beta = f(i)$. By the hypothesis, $|f(z)| = |z|$, $|f(z) - \alpha| = |z - 1|$, and $|f(z) - \beta| = |z - i|$ for all $z \in \mathbb{C}$. In particular, by substituting $z = 1, i$ in the above equalities, we obtain

$$|\alpha| = |\beta| = 1, |\alpha - \beta| = \sqrt{2}.$$

We can write

$$
\begin{aligned}
\alpha^2 + \beta^2 &= \alpha^2 \beta\bar{\beta} + \beta^2 \alpha\bar{\alpha} = \alpha\beta(\alpha\bar{\beta} + \beta\bar{\alpha}) \\
&= \alpha\beta(\alpha\bar{\alpha} + \beta\bar{\beta} - (\alpha - \beta)(\bar{\alpha} - \bar{\beta})) = \alpha\beta(|\alpha|^2 + |\beta|^2 - |\alpha - \beta|^2) \\
&= \alpha\beta(1 + 1 - 2) = 0,
\end{aligned}
$$

yielding $\beta = \varepsilon\alpha$, where $\varepsilon = \pm i$. Simplify $|f(z) - \alpha| = |z - 1|$ and $|f(z) - \beta| = |z - i|$ to, respectively, obtain $\bar{\alpha}f(z) + \alpha\overline{f(z)} = z + \bar{z}$ and $\bar{\alpha}f(z) - \alpha\overline{f(z)} = -\varepsilon iz + \varepsilon i\bar{z}$ for all $z \in \mathbb{C}$. Adding up these two equalities, we obtain

$$2\bar{\alpha}f(z) = (1 - \varepsilon i)z + (1 + \varepsilon i)\bar{z},$$

which, in view of $\alpha = f(1)$ and $\varepsilon = \pm i$, yields $f(z) = f(1)z$ for all $z \in \mathbb{C}$ or $f(z) = f(1)\bar{z}$ for all $z \in \mathbb{C}$, as desired. ∎

**5.** Let $n = 2k - 1$ for some $k \in \mathbb{N}$ and $K_n$ denote the $n \times n$ matrix whose elements are all 1. For a given $n \times n$ matrix $A$ with entries in $\{-1, 1\}$, let's call

$$C(A) := \sum_{i=1}^{n} a_i + \sum_{j=1}^{n} b_j$$

*the content* of the matrix $A$, where $a_i$'s and $b_j$'s are as in the statement of the problem. We have

$$C(K_n) = n + n = 4k - 2.$$

It is plain that any $n \times n$ matrix $A$, whose entries are in $\{-1, 1\}$, can be obtained from $K_n$ within a finite number of steps so that at each step a fixed entry, say, $a_{ij} = 1$, of $K_n$ is changed to $-a_{ij} = -1$. After each step, the number two will be added to or will be subtracted from each of the two sums in the definition of the content. Thus, the content remains intact or it will be increased or decreased by four at each step. Therefore, the content of any $n \times n$ matrix $A$ will always be of the form $4k' - 2$, for some $k' \in \mathbb{Z}$, which is never zero.

This proves the first part of the assertion. As for the case when $n$ is even, the assertion does not hold because for the matrix $A$ defined by

$$A = \begin{pmatrix} K_m & -K_m \\ K_m & K_m \end{pmatrix},$$

where $m = \frac{n}{2}$, we have $C(A) = 0$.                ∎

**6.** For a member $a \in S$, use $A(a)$ to denote the set of all of the acquaintances of $a$. It follows from the hypothesis that $A(a) \neq A(b)$ whenever $a \neq b$. Proceed by contradiction. Let $x \in S$ be a member for which $A(x)$ is maximal among all members of $S$. Since $S$ is ideal but $S \setminus \{x\}$ is not ideal, there are two distinct members $p, q \in S \setminus \{x\}$ such that

$$A(p) = A(q) \setminus \{x\}, \quad x \in A(q). \tag{$*$}$$

Likewise, since $S$ is ideal but $S \setminus \{p\}$ is not ideal, there are two distinct members $r, s \in S \setminus \{p\}$ such that

$$A(r) = A(s) \setminus \{p\}, \quad p \in A(s). \tag{$**$}$$

We claim that $r = x$. By $(**)$, $p$ and $s$ are acquainted. This, in view of $(*)$, implies that $s \neq x$ and that $s$ and $q$ are acquainted. But since, by $(**)$, $r$ is acquainted with all of the acquaintances of $s$ but $p$, we see that $r$ and $q$ are acquainted. This implies $r = x$ because $r$ is not acquainted with $p$. Thus, $A(x) = A(s) \setminus \{p\}$. That is, the number of the acquaintances of $s$ is greater than those of $x$ by one, which is a contradiction, proving the assertion.                ∎

## 2.26. Twenty-Sixth Competition

**2.26.1. First Day. 1. First solution:** Define $g : [a, b] \longrightarrow \mathbb{R}$ by $g(x) = f(x) + x - (a + b)$. The function $g$ is continuous on $[a, b]$ and differentiable on $(a, b)$. Moreover, $g(a) = a - b < 0$ and $g(b) = b - a > 0$. So, by the Intermediate Value Theorem, there is a $c \in (a, b)$ such that $g(c) = 0$, which means $f(c) = a + b - c$. It now follows from the Mean Value Theorem that there are $x_1, x_2 \in (a, b)$ with $x_1 \neq x_2$ such that

$$f'(x_1) = \frac{f(a) - f(c)}{a - c} = \frac{c - b}{a - c}, \quad f'(x_2) = \frac{f(c) - f(b)}{c - b} = \frac{a - c}{c - b},$$

yielding $f'(x_1) f'(x_2) = 1$, proving the assertion.

**Second solution:** Let $g = f \circ f : [a, b] \longrightarrow \mathbb{R}$. We have $g(a) = a$ and $g(b) = b$. By the Mean Value Theorem, there is a $c \in (a, b)$ such that

$$1 = \frac{g(b) - g(a)}{b - a} = f'(c) f'(f(c)),$$

yielding $f'(c) f'(f(c)) = 1$. There are two cases to consider. If $f(c) \neq c$, then the assertion is proved by letting $x_1 = c$ and $x_2 = f(c)$. If $f(c) = c$, then using the Mean Value Theorem for $f$ on the intervals $[a, c]$ and $[c, b]$, respectively, we see that there are $x_1 \in (a, c)$ and $x_2 \in (c, b)$ such that

$$f'(x_1) = \frac{f(c) - f(a)}{c - a} = 1, \quad f'(x_2) = \frac{f(b) - f(c)}{b - c} = 1,$$

yielding $f'(x_1) = f'(x_2) = 1$, and hence $f'(x_1)f'(x_2) = 1$, as desired.    ∎

**2.** Suppose to the contrary that $f$ has no zero in $U$. Define $g : U \longrightarrow \mathbb{C}$ by $g(z) = \frac{1}{f(z)}$. It follows that $g$ is analytic on $U$. Moreover, $g(0) = 1$ and $|g(z)| \leq \frac{1}{2}$ on $|z| = 1$. This contradicts the Maximum Modulus Principle, and hence proves the assertion.    ∎

**3.** Since $\gcd(p, 2p - 2) = 1$, it suffices to prove the congruence modulo $p$ and modulo $2p - 2$. Note that, by Fermat's Little Theorem, $(2p - 2)^{p-1} \stackrel{p}{\equiv} 1$. This, together with $p^{p-1} \stackrel{p}{\equiv} 0$, yields $p^{p-1} + (2p - 2)^{p-1} \stackrel{p}{\equiv} 1$. Also, note that

$$p^{p-1} - 1 = (p - 1)(p^{p-2} + \cdots + p + 1) = 2k(p - 1),$$

which implies $p^{p-1} \stackrel{2p-2}{\equiv} 1$. This, together with $(2p - 2)^{p-1} \stackrel{2p-2}{\equiv} 0$, yields $p^{p-1} + (2p - 2)^{p-1} \stackrel{2p-2}{\equiv} 1$. Therefore, $p^{p-1} + (2p - 2)^{p-1} \stackrel{p(2p-2)}{\equiv} 1$, proving the assertion.    ∎

**4.** We need to show that for all $x, y \in R$, $x$ divides $y$ or $y$ divides $x$. To prove this by contradiction, suppose that there are $x, y \in R$ such that $x \nmid y$ and $y \nmid x$. It follows from the hypothesis that there is a nonzero $z \in R$ such that $Rx + Ry = Rz$. Thus, there are $r, s \in R$ such that $x = rz$ and $y = sz$. The elements $r, s \in R$ are not units because otherwise $x|y$ or $y|x$, which is not possible. Therefore, the elements $r$ and $s$ belong to $M$, the only maximal ideal of $R$. From $Rx + Ry = Rz$, we see that there are $a, b \in R$ such that $z = ax + by$. This, together with $x = rz$ and $y = sz$, yields $z = (ar + bs)z$, implying $(1 - (ar + bs))z = 0$. On the other hand, $1 - (ar + bs)$ is a unit in $R$, for otherwise $1 - (ar + bs) \in M$, yielding $1 \in M$, which is a contradiction. Therefore, $z = 0$, a contradiction. So the assertion follows.    ∎

**5.** For the sake of brevity, we say a $k$-cover for the set $S$ to mean a $k$-element cover for the set $S$. It is plain that any element of a minimal cover of $S$ is nonempty and that it includes exactly one element of $S$ by itself. Therefore, a minimal $(n - 1)$-cover for the set $S$ is either a partition of the set $S$, or there exists exactly one element $x \in S$ which belongs to two or more elements of the $(n - 1)$-cover. In the first case, the number of such $(n - 1)$-covers of $S$, i.e., such partitions of $S$ into $(n - 1)$ subsets, is equal to $\binom{n}{2}$. In the latter case, there are $\binom{n}{1}$ ways to choose $x \in S$ and $2^{n-1} - \binom{n}{1} = 2^{n-1} - n$ ways to put $x$ in the elements of the $(n - 1)$-cover so that $x$ is put in at least two elements of the cover. So, we must have

$$M(n, n - 1) = \binom{n}{2} + \binom{n}{1}(2^{n-1} - n) = \frac{n}{2}(2^n - n - 1),$$

which is what we want.    ∎

**6.** Let $g(R, S) = \sum \{f(Y) : Y \in X^{\{m\}}, R \subseteq Y \subseteq S\}$. We prove the following more general proposition. *Under the hypotheses of the problem, $g(R, S) = 0$ whenever $R \subseteq S$ and $|S \setminus R| \geq k$.* The assertion is a consequence of the above proposition. To see this, note that for every $T \in X^{\{m\}}$, there is an $S \in X^{\{m+k\}}$ such that $T \subseteq S$. This implies $f(T) = g(S, T) = 0$, as desired. We prove the above proposition by induction on $|R|$. If $|R| = 0$, assuming that $|S| \geq k$, we have

$$\binom{|S|}{k} g(\emptyset, S) = \sum_{T \in X^{\{k\}}, T \subseteq S} g(\emptyset, T) = 0,$$

for, by the hypothesis, $g(\emptyset, T) = \sum_{Y \in X^{\{m\}}, Y \subseteq T} f(Y) = 0$ for all $T \in X^{\{k\}}$ such that $T \subseteq S$. On the other hand, the following recurrence equation holds.

$$g(R, S) = g\big(R \setminus \{\alpha\}, S\big) - g\big(R \setminus \{\alpha\}, S \setminus \{\alpha\}\big),$$

for all $\alpha \in R$. Therefore, the assertion follows by induction. ∎

**2.26.2. Second Day. 1.** Let $a \in [0, 1] \setminus \mathbb{Q}$ be arbitrary. We show that $f$ is continuous at $a$. To this end, for a given $\varepsilon > 0$, choose $k \in \mathbb{N}$ such that $\sum_{i=k}^{+\infty} \frac{1}{2^i} < \varepsilon$ and let

$$\delta = \min \left\{ \big|a - r_1\big|, \ldots, \big|a - r_k\big| \right\}.$$

Note that $\delta > 0$. If $a - \delta < x \leq a$, we can write

$$\big|f(x) - f(a)\big| = \sum_{i \in \{n \in \mathbb{N}: x < r_n \leq a\}} \frac{1}{2^i} \leq \sum_{i=k+1}^{+\infty} \frac{1}{2^i} < \varepsilon.$$

Analogously, if $a \leq x < a + \delta$, then

$$\big|f(x) - f(a)\big| = \sum_{i \in \{n \in \mathbb{N}: a < r_n \leq x\}} \frac{1}{2^i} \leq \sum_{i=k+1}^{+\infty} \frac{1}{2^i} < \varepsilon.$$

That is, $\lim_{x \to a} \big(f(x) - f(a)\big) = 0$. Thus, $f$ is continuous at $a$, which is what we want. ∎

**2.** It easily follows from the hypothesis that $f$ is one-to-one. Since $f$ is continuous, we conclude that $f$ is strictly increasing or decreasing on $\mathbb{R}$. But $f$ is not strictly decreasing, for otherwise $f \circ f$ would be strictly increasing and hence $f \circ f \circ f = I$ would be strictly decreasing, which is impossible. Thus, $f$ is strictly increasing on $\mathbb{R}$. Now, to prove the assertion by contradiction, suppose that there is an $a \in \mathbb{R}$ such that $f(a) \neq a$. Two cases to consider. If $a < f(a)$, we see that $f(a) < f(f(a))$, yielding $f(a) < f\big(f(a)\big) < f\big(f(f(a))\big) = a$, a contradiction. Likewise, if $a > f(a)$, we obtain a contradiction. So, the assertion follows by way of contradiction. ∎

**3.** Suppose by way of contradiction that there is an $x \in H \cap K \setminus \{e\}$, where $e$ denotes the identity element of $G$. Since $H$ is abelian and $\mathrm{Z}(G) = \{e\}$, we have $H \subseteq \mathrm{C}_G(x) \subsetneqq G$, where $\mathrm{C}_G(x)$ denotes the centralizer of the element $x$ in $G$. Thus, $H = \mathrm{C}_G(x)$ because $H$ is maximal. Likewise, we see that $K = \mathrm{C}_G(x)$. Therefore, $H = K$, which is a contradiction. So the assertion follows. ∎

**4.** We prove the assertion by showing that $A^n = \lambda I$, where $I$ denotes the identity matrix of size $n$. Let $F^n$ denote the $n$-dimensional vector space of all $n \times 1$ column vectors with entries in $F$. By the hypothesis, the vectors

$x, Ax, \ldots, A^{n-1}x$ are linearly independent and hence they form a basis for $F^n$. We can write

$$A^n(A^i x) = A^i(A^n x) = \lambda A^i x,$$

implying

$$(A^n - \lambda I)(A^i x) = 0,$$

for all $0 \leq i \leq n-1$. This yields $A^n - \lambda I = 0$, and hence $A^n = \lambda I$, because $\{x, Ax, \ldots, A^{n-1}x\}$ forms a basis for $F^n$. ∎

**5.** Let $(a_1, \ldots, a_6)$ and $(b_1, \ldots, b_6)$ be the sequences of the numbers appearing on two extended dice. Let $f(x) = x^{a_1} + \cdots + x^{a_6}$ and $g(x) = x^{b_1} + \cdots + x^{b_6}$. It follows from the hypothesis that

$$f(x)g(x) = (x^1 + \cdots + x^6)^2 = x^2(1+x)^2(1+x+x^2)^2(1-x+x^2)^2.$$

As $f(0) = g(0) = 0$, the factor $x$ must appear in the factorizations of both of $f$ and $g$. Since $f(1) = g(1) = 6$, the factors $1+x$ and $1+x+x^2$ must appear in the factorizations of both of $f$ and $g$ as well. Therefore, the only choices which do not lead to two ordinary dice are as follows

$$\begin{aligned} f(x) &= x(1+x)(1+x+x^2) = x + 2x^2 + 2x^3 + x^4, \\ g(x) &= x(1+x)(1+x+x^2)(1-x+x^2)^2 = x + x^3 + x^4 + x^5 + x^6 + x^8. \end{aligned}$$

That is, aside from the ordinary dice, the sequences of the numbers on the desired extended dice are $(1, 2, 2, 3, 3, 4)$ and $(1, 3, 4, 5, 6, 8)$. ∎

**6.** In view of the definition of a balanced matrix and the hypothesis that the balanceness remains intact under interchanging the rows, it follows that for all $j_1 < j_2$ and $i_1 < i < i_2$, we have

$$\left. \begin{array}{c} c_{i_1 j_1} + c_{i j_2} \leq c_{i_1 j_2} + c_{i j_1} \\ c_{i j_1} + c_{i_1 j_2} \leq c_{i j_2} + c_{i_1 j_1} \end{array} \right\} \iff c_{i_1 j_1} + c_{i j_2} = c_{i j_2} + c_{i j_1}$$

$$\iff c_{i_1 j_1} + c_{i j_1} = c_{i_1 j_2} + c_{i j_2}.$$

That is, the difference of two rows $i$ and $i'$ is a constant vector whenever $i_1 \leq i \leq i' \leq i_2$. Conversely, it follows from the above relations that if the difference of two rows is a constant vector, then the balanceness remains intact under interchanging the two rows. ∎

## 2.27. Twenty-Seventh Competition

### 2.27.1. First Day. 1. First solution: See Problem 2 of 2.1.1.

**Second solution:** Since all norms on the $n^2$-dimensional normed space $M_n(\mathbb{R})$ are equivalent, it suffices to show that $GL_n(\mathbb{R})$ is open and disconnected with respect to the operator norm of $M_n(\mathbb{R})$. To this end, let $A \in GL_n(\mathbb{R})$ be arbitrary. By showing that $B \in GL_n(\mathbb{R})$ whenever $\|B - A\| < \|A^{-1}\|^{-1}$, where $\|.\|$ denotes the operator norm of $M_n(\mathbb{R})$, we prove that $GL_n(\mathbb{R})$ is open. It is plain that $B \in GL_n(\mathbb{R})$ if and only if $BA^{-1} \in GL_n(\mathbb{R})$. But $\|I - BA^{-1}\| \leq \|A - B\| \, \|A^{-1}\| < 1$. Thus,

$$(BA^{-1})^{-1} = \left(I - (I - BA^{-1})\right)^{-1} = \sum_{i=1}^{+\infty} (I - BA^{-1})^n.$$

Consequently, $BA^{-1} \in GL_n(\mathbb{R})$, implying $B \in GL_n(\mathbb{R})$, as desired. To show that $GL_n(\mathbb{R})$ is disconnected, suppose to the contrary that it is connected. Since $GL_n(\mathbb{R})$ is open and connected, it is path connected. Choose $A, B \in GL_n(\mathbb{R})$ with $\det(A) = -\det(B) = 1$. It follows that there is a continuous function $\phi : [0, 1] \longrightarrow GL_n(\mathbb{R})$ with $\phi(0) = A$ and $\phi(1) = B$. Since $\det : M_n(\mathbb{R}) \longrightarrow \mathbb{R}$ is continuous, so is $\det(\phi) : [0, 1] \longrightarrow \mathbb{R}$. But $\det\left(\phi(0)\right) = \det(A) = 1$ and $\det\left(\phi(1)\right) = \det(B) = -1$. It thus follows from the Intermediate Value Theorem that there is a $c \in (0, 1)$ such that $\det\left(\phi(c)\right) = 0$. This implies $\phi(c) \notin GL_n(\mathbb{R})$, which is a contradiction. Therefore, $GL_n(\mathbb{R})$ is disconnected, proving the assertion. ∎

**2.** Let $0 \leq \theta < \pi$ be the angle of inclination of the line $L$, i.e., $\theta$ is the angle between $L$ and the $x$-axis. Define the function $g : \mathbb{C} \longrightarrow \mathbb{C}$ by $g(z) = \left(f(e^{i\theta} z + A) - A\right)e^{-i\theta}$. It is plain that $g$ is analytic on $\mathbb{C}$, $g$ takes the $x$-axis to the $x$-axis and the $y$-axis to the $y$-axis, and that $g(0) = 0$. We can write

$$g(z) = \sum_{n=1}^{+\infty} a_n z^n,$$

where $a_n = \frac{g^{(n)}(0)}{n!}$. Set

$$\widetilde{g}(z) = \overline{g(\bar{z})} = \sum_{n=1}^{+\infty} \bar{a}_n z^n,$$

Since $g$ maps the $x$-axis into the $x$-axis, we obtain

$$(g - \widetilde{g})(z) = \sum_{n=1}^{+\infty} (a_n - \bar{a}_n) z^n = 0,$$

whenever $z \in \mathbb{R}$. It thus follows from the Uniqueness Theorem for analytic functions that $g = \widetilde{g}$ because the zeros of the analytic function $g - \widetilde{g}$ has a limit point in $\mathbb{C}$. Therefore, $a_n = \bar{a}_n$ which, in turn, implies $a_n \in \mathbb{R}$ for all $n \in \mathbb{N}$. Define $h : \mathbb{C} \longrightarrow \mathbb{C}$ by

$$h(z) = ig(iz) = \sum_{n=1}^{+\infty} a_n i^{n+1} z^n.$$

Since $g$ maps the $x$-axis into the $x$-axis and the $y$-axis into the $y$-axis, it follows that the analytic function $h$ maps the $x$-axis into the $x$-axis. Hence, just as we saw in the above, we see that $a_n i^{n+1} \in \mathbb{R}$ for all $n \in \mathbb{N}$. This yields $a_{2n} = 0$ for all $n \in \mathbb{N}$. Therefore,

$$g(z) = \sum_{n=1}^{+\infty} a_{2n-1} z^{2n-1},$$

where $a_{2n-1} \in \mathbb{R}$ for all $n \in \mathbb{N}$. Obviously, $g(-z) = -g(z)$ for all $z \in \mathbb{C}$. That is, if $z_1, z_2 \in \mathbb{C}$ are symmetric with respect to $0$, then so are $g(z_1)$ and $g(z_2)$ with respect to $0$. Now, suppose $z_1$ and $z_2$ are symmetric with respect to $A$. So, $z_1 + z_2 = 2A$. We can write

$$
\begin{aligned}
f(z_2) &= e^{i\theta}g(e^{-i\theta}(z_2 - A)) + A \\
&= e^{i\theta}g(-e^{-i\theta}(z_1 - A)) + A = -e^{i\theta}g(e^{-i\theta}(z_1 - A)) + A \\
&= -f(z_1) + 2A,
\end{aligned}
$$

implying $f(z_1) + f(z_2) = 2A$. Thus, $f(z_1)$ and $f(z_2)$ are symmetric with respect to $A$, finishing the proof. ∎

**3.** Note first that $\sinh : \mathbb{R} \longrightarrow \mathbb{R}$ defined by $\sinh(x) = \frac{e^x - e^{-x}}{2}$ is one-to-one and onto. Thus, for each $n \in \mathbb{N}$, there is a unique $\alpha_n \in \mathbb{R}$ such that $a_n = \sinh(\alpha_n)$. We can write

$$
\begin{aligned}
\sinh(\alpha_{n+1}) &= \sinh(\alpha_n)\sqrt{1 + \sinh^2(\alpha_{n-1})} + \sinh(\alpha_{n-1})\sqrt{1 + \sinh^2(\alpha_n)} \\
&= \sinh(\alpha_n)\cosh(\alpha_{n-1}) + \sinh(\alpha_{n-1})\cosh(\alpha_n) \\
&= \sinh(\alpha_{n-1} + \alpha_n),
\end{aligned}
$$

implying $\alpha_{n+1} = \alpha_{n-1} + \alpha_n$ for each $n \in \mathbb{N}$. The characteristic polynomial of this recurrence equation is $x^2 - x - 1 = 0$, yielding $x = \frac{1 \pm \sqrt{5}}{2}$. Thus, $\alpha_n = A\left(\frac{1+\sqrt{5}}{2}\right)^n + B\left(\frac{1-\sqrt{5}}{2}\right)^n$ for some $A, B \in \mathbb{R}$, where $n \in \mathbb{N} \cup \{0\}$. To determine $A$ and $B$, we have

$$
\begin{aligned}
0 &= \sinh^{-1}(0) = \alpha_0 = A + B, \\
\sinh^{-1}(b) &= A\left(\frac{1+\sqrt{5}}{2}\right) + B\left(\frac{1-\sqrt{5}}{2}\right),
\end{aligned}
$$

implying $A = -B = \frac{\sinh^{-1}(b)}{\sqrt{5}}$. So we have

$$
\sinh^{-1}(a_n) = \alpha_n = \frac{\sinh^{-1}(b)}{\sqrt{5}}\left(\left(\frac{1+\sqrt{5}}{2}\right)^n - \left(\frac{1-\sqrt{5}}{2}\right)^n\right),
$$

yielding

$$
a_n = \sinh\left(\frac{\sinh^{-1}(b)}{\sqrt{5}}\left(\left(\frac{1+\sqrt{5}}{2}\right)^n - \left(\frac{1-\sqrt{5}}{2}\right)^n\right)\right),
$$

for all $n \in \mathbb{N}$, which is what we want. ∎

**4.** To prove the assertion by contradiction, suppose that there is an $i_0 \in I$ such that $i_0 \notin A_{i_0}$. It follows that $\{i : i \notin A_i\} \neq \emptyset$, from which, in view of the hypothesis, we see that $\{i : i \notin A_i\} = A_{i_1}$ for some $i_1 \in I$. Two cases to consider. If $i_1 \in A_{i_1}$, then $i_1 \notin \{i : i \notin A_i\} = A_{i_1}$, a contradiction. If $i_1 \notin A_{i_1}$, then $i_1 \in \{i : i \notin A_i\} = A_{i_1}$, a contradiction again. Therefore, $i \in A_i$ for all $i \in I$, as desired. ∎

**5.** It is easily verified that $(AB)^2 = 9(AB)$. We can write

$$
(BA)^3 = B(AB)^2A = 9B(AB)A = 9(BA)^2,
$$

yielding $(BA)^2(BA - 9I_2) = 0$. If we use $f$ to denote the minimal polynomial of $BA$, we see that $f$ divides $x^2(x - 9)$. As the degree of $f$ is less than or equal to 2, there are four cases to consider. If $f = x$, then $BA = 0$, yielding $AB = 0$, for $A(BA)B = (AB)^2 = 9(AB)$, which is impossible. If $f = x^2$, then $(BA)^2 = 0$, implying $\text{tr}(AB) = \text{tr}(BA) = 0$, which is again impossible. If $f = x(x - 9)$, then $\text{tr}(AB) = \text{tr}(BA) = 9$, which is impossible. If $f = x - 9$, we obtain $BA - 9I_2 = 0$. Thus, $BA = 9I_2$, which is what we want.  ∎

**6.** We first show that $H$ is normal in $G$. To prove this by contradiction, suppose that there exist $g \in G$ and $h \in H$ such that $g^{-1}hg \notin H$. Since $g^{-1}hg \in G \setminus H$, it follows from the hypothesis that there is a $u \in H$ such that $g(g^{-1}hg)g^{-1} = u^{-1}(g^{-1}hg)u$, from which, we obtain $g^{-1}hg = uhu^{-1} \in H$, a contradiction. Thus, $H$ is normal in $G$. To prove that $G/H$ is abelian, it suffices to show that $g_1 g_2 H = g_2 g_1 H$, or equivalently, $g_2^{-1} g_1^{-1} g_2 g_1 \in H$ for all $g_1, g_2 \in G$. If $g_2 \in H$, we see that $g_2^{-1}(g_1^{-1} g_2 g_1) \in H$ because $H$ is normal in $G$. So we may, with no loss of generality, assume that $g_2 \notin H$. Hence, for $g_1 \in G$, there is a $u \in H$ such that $g_1^{-1} g_2 g_1 = u^{-1} g_2 u$. This implies $g_2^{-1} g_1^{-1} g_2 g_1 = (g_2^{-1} u^{-1} g_2)u \in H$, for $H$ is normal in $G$. It thus follows that $G/H$ is abelian, as desired.  ∎

### 2.27.2. Second Day. 1. First solution:

Since $f'(a) > 0$, $f'(b) > 0$, and $f(a) = f(b) = 0$, there is a $\delta > 0$ such that $a + \delta < b - \delta$, $f(x) > 0$ whenever $a < x < a + \delta$, and that $f(x) < 0$ whenever $b - \delta < x < b$. It follows that $f(a + \delta) \geq 0$ and $f(b - \delta) \leq 0$. So by the Intermediate Value Theorem, there is a $c \in [a + \delta, b - \delta]$ such that $f(c) = 0$. We have $f(a) = f(c) = f(b) = 0$. Applying the Rolle's Theorem, we obtain $c_1 \in (a, c)$ and $c_2 \in (c, b)$ such that $f'(c_1) = f'(c_2) = 0$, as desired.

**Second solution:** Recall that by Darboux's Theorem, the derivative function has the intermediate value property. If there is a $c \in (a, b)$ such that $f'(c) < 0$, then $f'(a)f(c) < 0$ and $f'(c)f'(b) < 0$. Hence, by Darboux's Theorem, there are $c_1 \in (a, c)$ and $c_2 \in (c, b)$ such that $f'(c_1) = f'(c_2) = 0$, proving the assertion in this case. If $f'(x) \geq 0$ for all $x \in (a, b)$, then $f$ is increasing on $[a, b]$. This implies $0 = f(a) \leq f(x) \leq f(b) = 0$, whence $f(x) = 0$ for all $x \in [a, b]$, in which case the assertion is trivial. So the proof is complete.  ∎

**2.** First we need to recall the following lemma from theory of metric spaces. Also recall that a subset of a metric space is said to be *perfect* if it is closed and has no isolated points.

**Lemma.** *Every nonempty perfect subset of a complete metric space is uncountable.*

**Proof.** Let $P$ be a nonempty perfect subset of a complete metric space $X$. As $P$ is closed, $P$ equipped with the metric $X$ induces on $P$ forms a complete metric space itself. Now to prove the assertion by contradiction, suppose that $P$ is countable so that $P = \{x_i\}_{i=1}^{+\infty}$, where $x_i \in X$ for all $i \in \mathbb{N}$. As $P$ is perfect, each singleton $\{x_i\}$ ($i \in \mathbb{N}$) is a nowhere dense subset of $P$, implying that $P$ is a countable union of nowhere dense subsets, namely $\{x_i\}$'s where $i \in \mathbb{N}$. This is in contradiction with the Baire Category Theorem which asserts that every complete metric space is of second category in itself, i.e., the space cannot be written as a countable union of nowhere dense subsets of it. Thus, $P$ is uncountable, which is what we want. $\qquad\qquad\qquad\qquad\qquad\qquad\qquad\qquad\square$

To prove the assertion by contradiction, suppose that $[0,1]$ is written as a union of mutually disjoint closed intervals each of which having a positive length less than one. There are a countable number of the intervals participating in the union because they are mutually disjoint. So we may assume that

$$[0,1] = \bigcup_{n=1}^{+\infty} I_n,$$

where $I_n = [a_n, b_n]$ for some $a_n, b_n \in \mathbb{R}$ with $0 < b_n - a_n < 1$ and $I_m \cap I_n = \emptyset$ whenever $m, n \in \mathbb{N}$ and $m \neq n$. Set

$$P = \{a_n\}_{n=1}^{+\infty} \cup \{b_n\}_{n=1}^{+\infty}.$$

The set $P$ is closed. Because if $x \in [0,1] \setminus P$, then $x \in (a_n, b_n)$ for some $n \in \mathbb{N}$, and hence $x$ is an interior point of $[0,1] \setminus P$. We now show that any point $x \in P \setminus \{0,1\}$ is a limit point of $P$. Note that 0 and 1 are isolated points of $P$. If $x = a_n > 0$ for some $n \in \mathbb{N}$, then for every $0 < \delta < a_n$, there exists an $m \in \mathbb{N}$ such that $(a_n - \delta, a_n) \cap I_m \neq \emptyset$. Since $I_n \cap I_m = \emptyset$, we see that $b_m \in (a_n - \delta, a_n)$. Thus, $x = a_n$ is a limit point of $P$, as claimed. Likewise, if $x = b_n < 1$ for $n \in \mathbb{N}$, then $x$ is a limit point of $P$. Therefore, $P \setminus \{0,1\}$ is a nonempty perfect subset of $\mathbb{R}$, which is a complete metric space endowed with the ordinary metric of $\mathbb{R}$. By the lemma above, the set $P \setminus \{0,1\}$ is uncountable, and hence so is $P$, which is obviously a contradiction. This completes the proof. $\qquad\blacksquare$

**3.** As $D$ is countable, there are a countable number of lines which are all parallel to the $x$-axis so that each of which interests $D$. Use $\{\ell_{x_i}\}_{i=1}^{N_x}$ to denote these lines, where $\ell_{x_i} : x = x_i$ and $N_x \in \mathbb{N} \cup \{+\infty\}$. Likewise, there are a countable number of lines, denoted by $\{\ell_{y_j}\}_{j=1}^{N_y}$, where $\ell_{y_j} : y = y_j$ and $N_y \in \mathbb{N} \cup \{+\infty\}$, which are all parallel to the $y$-axis so that each of which intersects $D$. It is obvious that $D \subseteq A \times B$, where $A = \{x_i\}_{i=1}^{N_x}$ and $B = \{y_j\}_{j=1}^{N_y}$. Set

$$\begin{aligned} \Gamma &= D \cap \{(x_i, y_j) : j \leq i\}, \\ \Delta &= D \cap \{(x_i, y_j) : j > i\}. \end{aligned}$$

As $D \subseteq A \times B$, $\{\Gamma, \Delta\}$ is a partition of $D$ into two subsets. Now, any line parallel to the $x$-axis whose equation is given by $x = a$, where $a \in \mathbb{R}$, intersects $\Gamma$ at no point or at most at the $i$ points $(x_i, y_1), \ldots, (x_i, y_i)$ depending on whether $a \notin A$ or $a = x_i$ for some $1 \leq i \leq N_x < +\infty$ or for some $i \in \mathbb{N}$ if $N_x = +\infty$. Analogously, any line parallel the $y$-axis whose equation is given by $y = b$, where $b \in \mathbb{R}$, intersects $\Delta$ at no point or at most at the $j - 1$ points $(x_1, y_j), \ldots, (x_{j-1}, y_j)$ depending on whether $b \notin B$ or $b = y_j$ for some $1 \leq j \leq N_y < +\infty$ or for some $j \in \mathbb{N}$ if $N_y = +\infty$. So the assertion follows. $\qquad\blacksquare$

**4.** For a subset $I \subseteq N_n := \{1, \ldots, n\}$, set $B_I = (\bigcup_{i \in I} A_i)^c = S \setminus (\bigcup_{i \in I} A_i)$. Also, let $\mathcal{K} := \{I \subseteq N_n : |I| = k - 1\}$. We claim that if $I, J \in \mathcal{K}$, then

$B_I \cap B_J = \emptyset$ whenever $I \neq J$. To see this, note first that $B_I \cap B_J = B_{I \cup J}$. Since $I, J \in \mathcal{K}$ and $I \neq J$, we see that $|I \cup J| \geq k$, which, in view of the hypothesis, yields $\bigcup_{i \in I \cup J} A_i = S$, and hence $B_I \cap B_J = B_{I \cup J} = S \setminus S = \emptyset$. It is now plain that the number of $B_I$'s, where $I \in \mathcal{K}$, is equal to $\binom{n}{k-1}$ and that such $B_I$'s form a partition of $S$ into $\binom{n}{k-1}$ subsets. In other words,

$$S = \dot{\bigcup} \{ B_I : I \in \mathcal{K} \}, \qquad (*)$$

where $\dot{\bigcup}$ stands for the disjoint union. Thus, $|B_I| = 1$ for all $I \in \mathcal{K}$, because the number of such $B_I$'s is equal to $\binom{n}{k-1} = |S|$ and that they are all nonempty. Now, fix an arbitrary $i_0 \in I$. Note that $\mathcal{K}$ is a disjoint union of $\mathcal{K}_0 := \{ I \in \mathcal{K} : i_0 \in I \}$ and $\mathcal{K}'_0 := \{ I \in \mathcal{K} : i_0 \notin I \}$. It is plain that $|\mathcal{K}_0| = \binom{n-1}{k-2}$ and that if $x \in B_I$ for some $I \in \mathcal{K}_0$ (resp. $I \in \mathcal{K}'_0$), then $x \notin A_{i_0}$ (resp. $x \in A_{i_0}$). This together with $(*)$, in view of the fact that $|B_I| = 1$ for all $I \in \mathcal{K}$, implies that

$$|A_{i_0}| = \binom{n}{k-1} - \binom{n-1}{k-2} = \binom{n-1}{k-1}.$$

In other words, $A_i = \binom{n-1}{k-1}$ for all $i \in N_n$, which is what we want. ∎

**5.** Let $k = \mathrm{ord}_{ab}(a+b)$, $m = \mathrm{ord}_b(a)$, $n = \mathrm{ord}_a(b)$, and $[m,n] = \mathrm{lcm}(m,n)$. We have $a^m \overset{b}{\equiv} 1$ and $b^n \overset{a}{\equiv} 1$. Also, $[m,n] = mq = nq'$ for some $q, q' \in \mathbb{N}$. So we can write

$$(a+b)^{[m,n]} \overset{a}{\equiv} b^{[m,n]} \overset{a}{\equiv} b^{nq'} \overset{a}{\equiv} 1.$$

Analogously,

$$(a+b)^{[m,n]} \overset{b}{\equiv} a^{[m,n]} \overset{b}{\equiv} a^{mq} \overset{b}{\equiv} 1.$$

Since $\gcd(a,b) = 1$, it follows that $(a+b)^{[m,n]} \overset{ab}{\equiv} 1$, which implies $\mathrm{ord}_{ab}(a+b) \leq [m,n] = \mathrm{lcm}[\mathrm{ord}_b(a), \mathrm{ord}_a(b)]$. On the other hand, $(a+b)^k \overset{ab}{\equiv} 1$. So, in particular, $(a+b)^k \overset{a}{\equiv} 1$ and $(a+b)^k \overset{b}{\equiv} 1$, which, in turn, yields $b^k \overset{a}{\equiv} (a+b)^k \overset{a}{\equiv} 1$ and $a^k \overset{b}{\equiv} (a+b)^k \overset{b}{\equiv} 1$. Thus, $m|k$ and $n|k$, and hence $\mathrm{lcm}[\mathrm{ord}_b(a), \mathrm{ord}_a(b)] = [m,n] \big| k = \mathrm{ord}_{ab}(a+b)$. Therefore,

$$\mathrm{ord}_{ab}(a+b) = \mathrm{lcm}[\mathrm{ord}_b(a), \mathrm{ord}_a(b)],$$

as desired. ∎

**6.** The rings $R$ and $R'$ are both abelian and have characteristic 2 because they are Boolean rings. Since $f$ is onto and $0 \in R'$, there is an $x \in R$ such that $f(x) = 0$. We can write

$$f(0) = f(0x) = f(0)f(x) = f(0)0 = 0.$$

That is, $f(0) = 0$. Now, let $a, b \in R$ be arbitrary. By the surjectivity of $f$, there is an $x \in R$ such that $f(a+b) - f(a) - f(b) = f(x)$. We need to show that $x = 0$. We can write

$$f(ax)\big(f(a+b) - f(a) - f(b)\big) = f(ax)\dot{f}(x) = f(ax),$$

implying that

$$f(ax + abx) - f(ax) - f(abx) = f(ax),$$

and hence

$$f(ax + abx) = f(abx).$$

Since $f$ is one-to-one, we obtain $ax+abx = abx$, implying that $ax = 0$. Likewise, we see that $bx = 0$. Now, in view of $ax = bx = 0$, we can write

$$f(x)\big(f(a+b) - f(a) - f(b)\big) = f(x)f(x) = f(x),$$

which yields

$$f(ax + bx) - f(ax) - f(bx) = f(x).$$

Hence, $f(x) = 0 = f(0)$, from which, in view of the injectivity of $f$, we see that $x = 0$, which is what we want. ∎

## 2.28. Twenty-Eighth Competition

**2.28.1. First Day. 1.** Let $f(z) = \sum_{n=0}^{+\infty} c_n z^n$. If $f$ is a polynomial, then, by the Fundamental Theorem of Algebra, we have $f(z) = c_0 + c_1 z$ because $f$ is one-to-one. If not, then $f(\frac{1}{z})$ would have an essential singularity at zero. Thus, by Picard's Great Theorem, $f(\frac{1}{z})$ is not one-to-one in any neighborhood of zero, and hence neither is $f$, a contradiction. This proves the assertion. ∎

**2.** We prove the assertion by showing that $f^{-1}(\{(0,0)\})$ is not bounded, and hence has infinitely many points. As $f$ is surjective, $f^{-1}(\{(0,0)\})$ is nonempty. Suppose to the contrary that $f^{-1}(\{(0,0)\})$ is bounded. Thus, $f^{-1}(\{(0,0)\})$ is a nonvoid closed bounded subset of $\mathbb{R}$. Consequently, there exist $a, b \in \mathbb{R}$ such that $a = \min f^{-1}(\{(0,0)\}) < b = \max f^{-1}(\{(0,0)\})$. Let $A_1 = \{0\} \times (0, +\infty)$, $A_2 = \{0\} \times (-\infty, 0)$, $A_2 = (0, +\infty) \times \{0\}$, and $A_4 = (-\infty, 0) \times \{0\}$. As $f$ is continuous and $[a, b]$ is a compact subset of $\mathbb{R}$, we see that $f([a,b])$ is a bounded subset of $A$. Now, since $f$ is continuous and $f(x) \neq 0$ for all $x \in (-\infty, a) \cup (b, +\infty)$, there are $i_1, i_2 \in \{1, 2, 3, 4\}$, not necessarily distinct, such that $f(-\infty, a) \subseteq A_{i_1}$ and $f(b, +\infty) \subset A_{i_2}$. Therefore, $f(\mathbb{R}) \cap A_i$ is a bounded subset $A_i$ for each $i \in \{1, 2, 3, 4\} \setminus \{i_1, i_2\}$, which is a contradiction because $f$ is surjective. So the assertion follows. ∎

**3.** By the hypothesis and the Möbius Inversion Formula, we have

$$f(n) = \sum_{d|n} \mu(d)\Big(\frac{n}{d}\Big)^2 = n^2 \sum_{d|n} \frac{\mu(d)}{d^2} = n^2 g(n),$$

for all $n \in \mathbb{N}$, where $g(n) = \sum_{d|n} \frac{\mu(d)}{d^2}$. The arithmetic function $g$ is multiplicative because $\frac{\mu(d)}{d^2}$ is a multiplicative function of $d$. If $p$ is prime and $\alpha$ is a positive integer, we can write

$$g(p^\alpha) = \sum_{d|p^\alpha} \frac{\mu(d)}{d^2} = \frac{\mu(1)}{1} + \frac{\mu(p)}{p^2} = 1 - \frac{1}{p^2}.$$

Thus, if $n = \prod_{p|n} p^{\alpha_p}$, then

$$g(n) = g(\prod_{p|n} p^{\alpha_p}) = \prod_{p|n} g(p^{\alpha_p}) = \prod_{p|n} \left(1 - \frac{1}{p^2}\right)$$

$$= \prod_{p|n} \left(1 - \frac{1}{p}\right) \prod_{p|n} \left(1 + \frac{1}{p}\right) = \frac{\phi(n)}{n} \prod_{p|n} \left(1 + \frac{1}{p}\right).$$

Consequently,

$$\frac{f(n)}{\phi(n)} = \frac{n^2 g(n)}{\phi(n)} = \frac{n^2}{\phi(n)} \frac{\phi(n)}{n} \prod_{p|n} \left(1 + \frac{1}{p}\right) = n \prod_{p|n} \left(1 + \frac{1}{p}\right),$$

for all $n \in \mathbb{N}$ with $n > 1$, finishing the proof. ∎

**4.** Let $1 \le i, k \le n$ be arbitrary and $\tau \in G_k \cap Z(G)$. By the hypothesis, there exists a $\sigma \in G$ such that $\sigma(k) = i$. We can write

$$\lambda \in G_i \iff \lambda(i) = i \iff \lambda(\sigma(k)) = \sigma(k)$$
$$\iff \sigma^{-1}\lambda\sigma(k) = k \iff \sigma^{-1}\lambda\sigma \in G_k$$
$$\iff \lambda \in \sigma G_k \sigma^{-1}.$$

Thus, $G_i = \sigma G_k \sigma^{-1}$, and hence $\sigma\tau\sigma^{-1} \in G_i$ because $\tau \in G_k$. On the other hand, $\tau \in Z(G)$ implies $\tau = \sigma\tau\sigma^{-1} \in G_i$. It follows that $\tau \in \cap_{i=1}^n G_i = \{e\}$ because $i$ was arbitrary. This proves $G_k \cap Z(G) = \{e\}$, as desired. ∎

**5.** We prove that the maximum number of such vectors is $n$. To this end, let the vectors $v_1, \ldots, v_k \in A$ be such that any two of which share an even number of one entries. View the elements of $A$ as vectors in the $n$-dimensional vector space $\mathbb{Z}_2^n$, where $\mathbb{Z}_2 = \{0, 1\}$ is the field of integers modulo 2. We claim that the set $\{v_1, \ldots, v_k\}$ is a linearly independent subset of $\mathbb{Z}_2^n$. To see this, suppose $c_1 v_1 + \cdots + c_k v_k = 0$ for some $c_1, \ldots, c_k \in \mathbb{Z}_2$. If "." denotes the usual inner product on $\mathbb{Z}_2^n$, for all $1 \le i \le n$, we have

$$c_i = v_i.(c_1 v_1 + \cdots + c_k v_k) = 0,$$

proving the claim. Thus, $k \le n$. On the other hand, if $e_i$ is the vector with 1 in the $i$th place and zero elsewhere, then any two elements of the set $\{e_1, \ldots, e_n\}$, the standard basis of $\mathbb{Z}_2^n$, share an even number of one entries. Therefore, the maximum number of such vectors is $n$, as desired. ∎

**6.** Let $P = \{p_1, \ldots, p_n\}$. As is usual, use $\text{Sym}(P)$ to denote the set of all permutations on (the poset) $P$. Let's view $\text{Sym}(P)$ as a probability space. Choose a permutation $\sigma$ on $\{1, \ldots, n\}$ at random which, in turn, gives rise to a permutation $s_\sigma \in \text{Sym}(P)$ at random so that $s_\sigma(p_i) = p_{\sigma i}$ for all $1 \le i \le n$. For any $1 \le i \le n$, there exists a $1 \le j \le n$ such that $\sigma j = i$, which is equivalent to $p_i = s_\sigma(p_j) = p_{\sigma j}$. With all that in mind, for an element $p_i \in P$ for which $p_i = s_\sigma(p_j) = p_{\sigma j}$ for some $1 \le j \le n$, let $F_{p_i}$ be the event that the permutation $s_\sigma \in \text{Sym}(P)$ satisfies the following

$$\forall p \in P \setminus U_{p_i} : p = s_\sigma(p_k) = p_{\sigma k} \implies \sigma j \le \sigma k.$$

In other words, if

$$s_\sigma = \begin{pmatrix} p_1 & p_2 & \cdots & p_n \\ p_{\sigma 1} & p_{\sigma 2} & \cdots & p_{\sigma n} \end{pmatrix},$$

then $s_\sigma \in F_{p_i}$ if and only if the element $p_i$ precedes all of the elements of $P \setminus U_{p_i}$ but $p_i$ in the second row of the above representation of $s_\sigma$. Set $E_{p_i} := F_{p_i} \cap (\bigcap_{p \in L_{p_i}} F_p^c)$, where $F_p^c := \mathrm{Sym}(P) \setminus F_p$.

We note that if $p_i$ and $p_{i'}$ are distinct, then the events $E_{p_i}$ and $E_{p_{i'}}$ cannot occur simultaneously, and hence they are disjoint. Let's prove this. First, assume that for $p_i, p_{i'} \in P$, $E_{p_i}$, and hence $F_{p_i}$, occurs, and that $p_i$ and $p_{i'}$ are not comparable. Then, in particular $p_{i'} \not\succ p_i$ yielding $p_{i'} \in P \setminus U_{p_i}$, which, in turn, implies $\sigma j < \sigma j'$, where $i = \sigma j$ and $i' = \sigma j'$. Consequently, $F_{p_{i'}}$, and hence $E_{p_{i'}}$, cannot occur, for otherwise, we must have $\sigma j' < \sigma j$ because $p_i \not\succ p_{i'}$, which is a contradiction. That is, $F_{p_i}$ and $F_{p_{i'}}$, and hence $E_{p_i}$ and $E_{p_{i'}}$, cannot occur simultaneously whenever $p_i$ and $p_{i'}$ are not comparable. Next, if the elements $p_i, p_{i'} \in P$ are comparable, then it is plain that the events $E_{p_i}$ and $E_{p_{i'}}$ cannot occur simultaneously. Thus, the events $E_{p_i}$ and $E_{p_{i'}}$ are disjoint whenever $p_i, p_{i'} \in P$ are distinct.

We now use induction on $|L_{p_i}|$ to prove $X_{p_i} = P(E_{p_i})$. If $|L_{p_i}| = 0$, i.e., if $p_i$ is a minimal element of $P$, then $E_{p_i} = F_{p_i}$ and $s_\sigma \in F_{p_i}$ with $s_\sigma(p_j) = p_i$ if and only if $\sigma j \leq \sigma k$ whenever $p = s(p_k) = p_{\sigma k} \in P \setminus U_{p_i}$ for some $1 \leq k \leq n$. As $|P \setminus U_{p_i}| = n - |U_{p_i}|$, in this case, we can write

$$P(E_{p_i}) = P(F_{p_i}) = \frac{(|U_{p_i}|)! \binom{n}{n - |U_{p_i}|}(n - |U_{p_i}| - 1)!}{n!} = \frac{1}{n - |U_{p_i}|}.$$

Now, assuming that $X_p = P(E_p)$ for all $p \in L_{p_i}$, we prove $X_{p_i} = P(E_{p_i})$. We claim that

$$\bigcap_{p \in L_{p_i}} F_p^c = \bigcap_{p \in L_{p_i}} E_p^c,$$

for all $p_i \in P$. To see this, first, let $s_\sigma \in \bigcap_{p \in L_{p_i}} F_p^c$ be arbitrary. We show $s_\sigma \in \bigcap_{p \in L_{p_i}} E_p^c$. Suppose the contrary, implying that there is a $q \in L_{p_i}$ such that $s_\sigma \notin E_q^c$. It follows that $s_\sigma \in E_q$, and hence $s_\sigma \in F_q$. This is in contradiction with $s_\sigma \in \bigcap_{p \in L_{p_i}} F_p^c$. Thus, $\bigcap_{p \in L_{p_i}} F_p^c \subseteq \bigcap_{p \in L_{p_i}} E_p^c$. Next, letting $s_\sigma \in \bigcap_{p \in L_{p_i}} E_p^c$ be arbitrary, we show $s_\sigma \in \bigcap_{p \in L_{p_i}} F_p^c$. Again suppose the contrary, implying that there is a minimal $q \in L_{p_i}$ so that $s_\sigma \notin F_q^c$ but $s_\sigma \in F_p^c$ for all $p \in L_q$. It follows that $s_\sigma \in F_q$ but $s_\sigma \notin F_p$ for all $p \in L_q$. Consequently, $s_\sigma \in E_q = F_q \cap (\bigcap_{p \in L_q} F_p^c)$, which is a contradiction. Therefore, $\bigcap_{p \in L_{p_i}} E_p^c \subseteq \bigcap_{p \in L_{p_i}} F_p^c$, proving the claim. Now, as $E_p$'s ($p \in L_{p_i}$) are disjoint, in view of the claim, we can write

$$P\left( \bigcap_{p \in L_{p_i}} F_p^c \right) = P\left( \bigcap_{p \in L_{p_i}} E_p^c \right) = 1 - P\left( \bigcup_{p \in L_{p_i}} E_p \right) = 1 - \sum_{p \in L_{p_i}} X_p.$$

On the other hand, the events $F_{p_i}$ and $\bigcap_{p \in L_{p_i}} F_p^c$ are independent because the position of $p_i$ in the permutation has no affect on $F_p$'s ($p \in L_{p_i}$). Thus,

$$P(E_{p_i}) = P\left( \left( \bigcap_{p \in L_{p_i}} F_p^c \right) \cap F_{p_i} \right) = P\left( \bigcap_{p \in L_{p_i}} F_p^c \right) P(F_{p_i}) = \frac{1 - \sum_{p \in L_{p_i}} X_p}{n - |U_{p_i}|}.$$

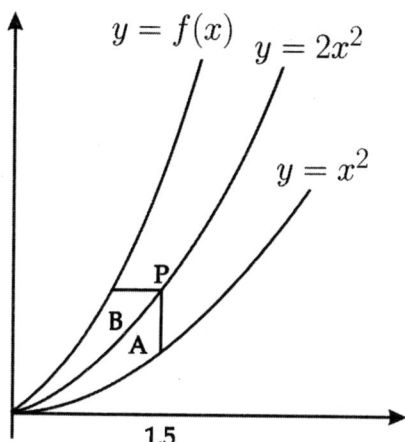

Figure 13

Therefore, $X_{p_i} = P(E_{p_i})$ for all $p_i \in P$, and hence $0 \leq X_{p_i} \leq 1$ for all $p_i \in P$, which is what we want. ∎

**2.28.2. Second Day. 1.** Let $g = f^{-1}$. It follows from the hypothesis that for all $x, y$ with $y = 2x^2$, we have

$$A = \int_0^x (2t^2 - t^2)dt = \frac{1}{3}x^3 = \frac{\sqrt{2}}{12}y^{\frac{3}{2}} =$$

$$B = \int_0^y (\sqrt{\frac{t}{2}} - g(t))dt = \frac{\sqrt{2}}{3}y^{\frac{3}{2}} - \int_0^y g(t)dt,$$

implying

$$\int_0^y g(t)dt = \frac{\sqrt{2}}{4}y^{\frac{3}{2}}.$$

Taking derivative of both sides yields

$$x = g(y) = \frac{3\sqrt{2}}{8}y^{\frac{1}{2}} \implies x^2 = \frac{9}{32}y \implies y = \frac{32}{9}x^2.$$

Thus, $y = f(x) = \frac{32}{9}x^2$, as desired. ∎

**2.** Set $h(x) = f(x) - x$. We have $h(a) > 0$ and $h(b) < 0$. Since $h$ is continuous, there exists a $\delta > 0$ such that $h(a+x) > 0$ and $h(b-x) < 0$ whenever $x \in (0, \delta)$. Set $\alpha = \frac{\delta}{n+1}$ and

$$g(x) = h(x) + h(x + \alpha) + \cdots + h(x + n\alpha)$$
$$= \big(f(x) + \cdots + f(x + n\alpha)\big) - \big(x + \cdots + (x + n\alpha)\big).$$

It is readily seen that $g(a) > 0$ and $g(b - na) < 0$. By the Intermediate Value Theorem, there exists a $c \in (a, b - na)$ such that $g(c) = 0$. This implies

$$
\begin{aligned}
f(c) + f(c + a) + \cdots + f(c + na) &= c + (c + a) + \cdots + (c + na) \\
&= (n + 1)c + \frac{n(n + 1)}{2}\alpha \\
&= (n + 1)\left(c + \frac{n}{2}\alpha\right),
\end{aligned}
$$

which is what we want.  ∎

**3.** The answer is yes. To see this, noting that $\det(A + B) = \det(A^t + B^t)$, we can write

$$
\begin{aligned}
\det(A)\det(A + B) &= \det(A)\det(A^t + B^t) = \det(I + AB^t) \\
&= \det(I + BA^t) = \det(B)\det(A^t + B^t) \\
&= \det(B)\det(A + B).
\end{aligned}
$$

Consequently, in view of $\det(A) + \det(B) = 0$, we have

$$
\det(A)\det(A + B) = \det(B)\det(A + B) = -\det(A)\det(A + B),
$$

yielding $2\det(A)\det(A + B) = 0$. But $\det(A) = \pm 1$ because $A$ is orthogonal. Thus, $\det(A + B) = 0$, as desired.  ∎

**4.** Let $x, y \in R$ be arbitrary. We can write

$$
x^{n+2}y^{n+2} = (xy)^{n+2} = (xy)^{n+1}xy = x^{n+1}y^{n+1}xy,
$$

yielding

$$
x^{n+1}(xy^{n+1} - y^{n+1}x)y = 0.
$$

As the above equality holds for all $x, y \in R$, changing $x$ to $x+1$ in the equality, we obtain

$$
(1 + x)^{n+1}(xy^{n+1} - y^{n+1}x)y = 0,
$$

or

$$
\left(\binom{n+1}{0} + \binom{n+1}{1}x + \cdots + \binom{n+1}{n+1}x^{n+1}\right)(xy^{n+1} - y^{n+1}x)y = 0.
$$

Multiplying both sides of the equality by $x^n$ and using the aforementioned equality, we get

$$
x^n(xy^{n+1} - y^{n+1}x)y = 0.
$$

Analogously, changing $x$ to $x + 1$ in this equality, we see that

$$
x^{n-1}(xy^{n+1} - y^{n+1}x)y = 0.
$$

Continuing in this way, we finally obtain

$$
(xy^{n+1} - y^{n+1}x)y = 0. \tag{$*$}
$$

On the other hand, by the hypothesis, we have

$$
x^{n+1}y^{n+1} = (xy)^{n+1} = (xy)^n xy = x^n y^n xy,
$$

yielding

$$
x^n(xy^n - y^n x)y = 0.
$$

Likewise, from this, we obtain

$$(xy^n - y^n x)y = 0.$$

Multiplying both sides of the above by $y$ from the left, we get

$$yxy^{n+1} - y^{n+1}xy = 0,$$

from which, in view of $(*)$, we see that

$$(yx - xy)y^{n+1} = 0.$$

Once again, preforming an argument involving changing $y$ to $y + 1$ and multiplying both sides of the obtained equality by an appropriate power of $y$, we eventually conclude that $yx - xy = 0$. That is, $xy = yx$ for all $x, y \in R$, which is what we want. ∎

**5.** Suppose that a person is chosen at random. Define the following events.

$A$ (resp. $B$) := the event that the person is guilty (resp. innocent).

$+$ (resp. $-$) := the event that the person responds positively (resp. negatively).

$E$ := the event that the person responds positively to four questions and negatively to one question.

It follows from the hypothesis that $P(A) = \frac{1}{3}$, $P(B) = \frac{2}{3}$, $P(+|A) = 0.8$, $P(+|B) = 0.4$, $P(-|A) = 0.2$, and $P(-|B) = 0.6$. Using Bayes' Theorem, we see that the desired probability is equal to

$$
\begin{aligned}
P(A|E) &= \frac{P(A \cap E)}{P(E)} = \frac{P(A)P(E|A)}{P(A)P(E|A) + P(B)P(E|B)} \\
&= \frac{\frac{1}{3}\binom{5}{4}(0.8)^4(0.2)}{\frac{1}{3}\binom{5}{4}(0.8)^4(0.2) + \frac{2}{3}\binom{5}{4}(0.4)^4(0.6)} = \frac{8}{16}.
\end{aligned}
$$

Thus, $P(A|E) = \frac{8}{16}$ is the desired probability. ∎

**6.** We claim that for each $j = 1, \ldots, k$, there are subsets $A_1, \ldots, A_j$ satisfying the following inequalities.

$$1 \le |A_1 \cap \cdots \cap A_j| \le r - (j - 1)(I(F) - 1) \qquad (*)$$

Obviously, the assertion follows from $(*)$ by letting $j = k$. We prove $(*)$ by induction on $j$. If $j = 1$, the assertion is trivial; in fact, any element $A_1$ of $F$ satisfies $(*)$ when $j = 1$. Suppose that $(*)$ holds for $j$, where $j \le k - 1$. We prove the assertion for $j + 1$. To this end, it follows from the hypothesis that $A_1 \cap \cdots \cap A_j$ has nonempty intersection with any element of $F$. This implies $|A_1 \cap \cdots \cap A_j| \ge I(F)$. Let $S$ be a subset of $A_1 \cap \cdots \cap A_j$ with $I(F) - 1$ elements. Thus, there exists an $A_{j+1} \in F$ such that $S \cap A_{j+1} = \emptyset$. So we can write

$$
\begin{aligned}
1 &\le |A_1 \cap \cdots \cap A_j \cap A_{j+1}| \le |(A_1 \cap \cdots \cap A_j) \cap S^c| \\
&= |(A_1 \cap \cdots \cap A_j) \setminus S| = |A_1 \cap \cdots \cap A_j| - (I(F) - 1) \\
&\le r - (j - 1)(I(F) - 1) - (I(F) - 1) = r - j(I(F) - 1),
\end{aligned}
$$

proving the induction assertion, finishing the proof. ∎

## 2.29. Twenty-Ninth Competition

**2.29.1. First Day. 1. First solution:** We can write

$$I := \int_0^a f(x)dx \int_0^a \frac{dx}{f(x)}$$

$$= \int_0^a \int_0^a f(x)\frac{1}{f(y)}dxdy = \int_0^a \int_0^a f(y)\frac{1}{f(x)}dxdy,$$

from which, we obtain

$$2I = I+I = \int_0^a \int_0^a \left(\frac{f(x)}{f(y)} + \frac{f(y)}{f(x)}\right)dxdy$$

$$= \int_0^a \int_0^a \frac{f(x)^2 + f(y)^2}{f(x)f(y)}dxdy.$$

On the other hand, obviously, we have $\frac{f(x)^2+f(y)^2}{f(x)f(y)} \geq 2$ for all $0 \leq x,y \leq a$. Thus,

$$2I \geq \int_0^a \int_0^a 2dxdy = 2a^2,$$

yielding

$$I \geq a^2,$$

as desired.

**Second solution:** By the Cauchy-Schwarz Inequality, we can write

$$a^2 = \left|\int_0^a \sqrt{f(x)} \times \frac{1}{\sqrt{f(x)}}dx\right|^2 \leq \left(\int_0^a f(x)dx\right)\left(\int_0^a \frac{dx}{f(x)}\right),$$

which is what we want. ∎

**2.** The "if part" is trivial. We prove the "only if part" of the assertion by contradiction. To this end, suppose $x = \frac{a}{b}$ for some $a,b \in \mathbb{N}$. It follows from the contradiction hypothesis that there exists a $k \in \mathbb{N}$ such that $n_{k+1} - 1 > b$. Set

$$x_k = \sum_{i=1}^k \frac{1}{n_1 \cdots n_i}.$$

We can write

$$\frac{a}{b} - x_k = x - x_k = \frac{1}{n_1 \cdots n_{k+1}} + \frac{1}{n_1 \cdots n_{k+1}n_{k+2}} + \cdots$$

$$< \frac{1}{n_1 \cdots n_k}\left(\frac{1}{n_{k+1}} + \frac{1}{n_{k+1}^2} + \frac{1}{n_{k+1}^3} + \cdots\right) = \frac{1}{n_1 \cdots n_k}\frac{1}{n_{k+1} - 1}.$$

Consequently,

$$0 < \frac{a}{b}n_1 \cdots n_k - x_k n_1 \cdots n_k < \frac{1}{n_{k+1} - 1},$$

whence

$$0 < an_1 \cdots n_k - bx_k n_1 \cdots n_k < \frac{b}{n_{k+1} - 1} < 1,$$

which is a contradiction because $an_1 \cdots n_k - bx_k n_1 \cdots n_k \in \mathbb{N}$. Thus, the assertion follows by contradiction. ∎

**3.** Note that $10^{\phi(m)} \equiv 1 \pmod{m}$ for all positive integers $m$ satisfying $\gcd(m, 10) = 1$, where $\phi$ denotes Euler's totient function. It follows that $m \mid 10^{k\phi(m)} - 1$ whenever $k, m \in \mathbb{N}$ with $\gcd(m, 10) = 1$. Moreover, if $\gcd(m, 3) = 1$, then

$$m \left| \frac{10^{k\phi(m)} - 1}{10 - 1} = 1 + 10 + \cdots + 10^{k\phi(m)-1} \right. .$$

Thus, if $\gcd(m, 30) = 1$, then $m\left| \overbrace{11 \ldots 1}^{k\phi(m) \text{ times}} \right.$ for all $k \in \mathbb{N}$. This proves the assertion. ∎

**4.** Proceed by way of contradiction. Suppose $I$ is a nonzero right ideal in $R$ whose elements all have square zero. It follows that $(x + y)^2 = 0$, and hence $xy = -yx$ for all $x, y \in R$. Thus, $(xI)(xI) = 0$, from which, we see that

$$(RxI)(RxI) \subseteq (Rx)I(xI) = R(xI)(xI) = 0,$$

for all $x \in R$. In other words, the two-sided ideal $RxI$ is nilpotent, which, in view of the hypothesis, yields $RxI = 0$. This shows that $xI$ is contained in the left annihilator of $R$ in $R$, denoted by l.ann$(R)$. But l.ann$(R)$ is nilpotent. Thus, $xI \subseteq$ l.ann$(R) = \{0\}$, yielding $xI = \{0\}$ for all $x \in R$. This easily implies $I^2 = \{0\}$, from which, we obtain $(RI)(RI) \subseteq RI^2 = \{0\}$. But $RI$ is a two-sided ideal of $R$. So it follows from the hypothesis that $RI = \{0\}$. This implies that $I$ is contained in the right annihilator of $R$ in $R$, denoted by r.ann$(R)$. But r.ann$(R)$ is obviously a nilpotent two-sided ideal. Thus, $I \subseteq$ l.ann$(R) = \{0\}$, yielding $I = \{0\}$, a contradiction. This proves the assertion. ∎

**5.** Define the function $f : X \longrightarrow Y$ by

$$f(A) = \{k \in \mathbb{Z} : 2k \in A\}.$$

It suffices to show that for any $B \in Y$, there exists a $A \in X$ such that $f(A) = B$. To this end, for given $B \in Y$, set

$$A := \{2k : k \in B\}.$$

It is obvious that $A \in X$ and $f(A) = B$, as desired. ∎

**6.** Note first that since $S \leq \mathbb{Z}_2^n$ is a vector space over $\mathbb{Z}_2$, $d$ is equal to the minimum of the number of ones occurring in the nonzero elements of $S$. It is plain that $S$ has $2^k$ elements. Use a matrix $A \in M_{(2^k-1)\times n}(\mathbb{Z}_2)$ to denote the nonzero elements of $S$, so that every nonzero element of $S$ corresponds to a row of $A$. We may assume that the first $k$ rows of $A$ form a basis for $S$. Denote this $k \times n$ submatrix of $A$ by $B$. It is plain that the number of one entries of the matrix $A$ is at least $d(2^k - 1)$. By showing that this number is at most $n2^{k-1}$, we conclude that $d(2^k - 1) \leq n2^{k-1}$, proving the assertion. We claim that the number of ones in any column of $A$ is either $0$ or $2^{k-1}$. Since every row of $A$ is a linear combination of the rows of $B$, the number of ones in any column of $A$ is computed as follows. First, if there exists no "one" in column $i$ ($1 \leq i \leq n$) of $B$, then neither does exist a "one" in the same column of $A$.

Next, let $t$ be the number of ones in column $i$ $(1 \leq i \leq n)$ of $B$. Since any linear combination of the rows of $B$ gives rise to a row of $A$, and since the coefficients come from $\mathbb{Z}_2$, it follows that the number of ones in column $i$ of $A$, denoted by $1_i$, is equal to the number of all subsets of $\{1, 2, \ldots, n\}$ with $k$ elements each of which contains an odd number of one entries from column $i$ of $B$. Thus, $1_i$ is equal to the number of all odd numbered subsets of a set with $t$ elements times the number of all subsets of a set with $k - t$ elements. In other words,

$$1_i = 2^{t-1} 2^{k-t} = 2^{k-1}.$$

Now, as $A$ has $k$ columns, the number of all ones is at most $n2^{k-1}$, finishing the proof. ∎

**2.29.2. Second Day. 1. First solution:** Let $f(z) = \sum_{n=0}^{+\infty} a_n z^n$. Set $g(z) = \overline{f(\bar{z})}$. It is plain that

$$g(z) = \sum_{n=0}^{+\infty} \bar{a}_n z^n,$$

and that $g$ is analytic on $D$. It follows that $f - g$ is analytic on $D$ and moreover $(f - g)(\frac{1}{n}) = f(\frac{1}{n}) - g(\frac{1}{n}) = 0$ for all $n \geq 2$. Hence, the set of zeros of the analytic function $f - g$ has a limit point in $D$. It thus follows from the Uniqueness Theorem for analytic functions that $f(z) = g(z)$ for all $z \in D$, whence $a_n = \bar{a}_n$ for all $n \in \mathbb{N}$. Therefore, $a_n \in \mathbb{R}$ for all $n \in \mathbb{N}$. This, in view of $a_n = \frac{f^{(n)}(0)}{n!}$, yields $f^{(n)}(0) \in \mathbb{R}$ for all $n \in \mathbb{N}$, as desired.

**Second solution:** Let $f(z) = \sum_{n=0}^{+\infty} a_n z^n$, where $a_n = \frac{f^{(n)}(0)}{n!}$ for all $n \in \mathbb{N} \cup \{0\}$. It suffices to show that $a_n \in \mathbb{R}$ for all $n \in \mathbb{Z}$ with $n \geq 0$. We prove this by induction on $n$. If $n = 0$, the assertion is trivial because $a_0 = f(0) = \lim_n f(\frac{1}{n})$. Assuming that $a_0, \ldots, a_{n-1} \in \mathbb{R}$, to prove $a_n \in \mathbb{R}$, set

$$g(z) = \sum_{i=0}^{+\infty} a_{i+n} z^i = \begin{cases} \frac{1}{z^n}\left(f(z) - \sum_{i=0}^{n-1} a_i z^i\right) & z \neq 0, \\ a_n & z = 0. \end{cases}$$

It is readily verified that $g$ is an analytic function satisfying $g(\frac{1}{n}) \in \mathbb{R}$ for all natural numbers $n \geq 2$. Thus, $a_n = g(0) = \lim_n g(\frac{1}{n}) \in \mathbb{R}$, which is what we want. ∎

**2.** (i) The assertion can be stated as follows. For a given $\varepsilon > 0$, every point of the space $X$ can be reached from any other point of the space within a finite number of steps each of which having a length less than $\varepsilon$. With this in mind, for given $\varepsilon > 0$, fix an $x \in X$ and set

$$C_\varepsilon(x) := \{y \in X : \exists x_1 = x, x_2, \ldots, x_n = y \in X \ni \forall i < n : d(x_i, x_{i+1}) < \varepsilon\}.$$

To prove the assertion, it suffices to show that $C_\varepsilon(x) = X$. This, in view of the hypothesis that $X$ is connected, follows as soon as we show that $C_\varepsilon(x)$ is a clopen subset, i.e., both a closed and open subset, of $X$. First, suppose $z \in \overline{C_\varepsilon(x)}$. It follows that there exists a sequence $(y_n)_{n=1}^{+\infty}$ in $C_\varepsilon(x)$ such that $\lim_n y_n = z$. Thus, there exists an $N \in \mathbb{N}$ such that $d(y_n, z) < \varepsilon$ whenever

$n \geq N$. On the other hand, since $y_n \in C_\varepsilon(x)$, we have $B_\varepsilon(y_n) := \{t \in X : d(t, y_n) < \varepsilon\} \subseteq C_\varepsilon(x)$. Consequently, $z \in C_\varepsilon(x)$ because $z \in B_\varepsilon(y_n)$. Thus, $C_\varepsilon(x)$ is a closed subset of $X$. Next, suppose $y \in C_\varepsilon(x)$ is arbitrary. It follows that $B_\varepsilon(y) \subseteq C_\varepsilon(x)$. This implies that $C_\varepsilon(x)$ is an open subset of $X$. Therefore, $C_\varepsilon(x) = X$ because $X$ is connected.

(ii) Plainly, $X = [0, 1) \cup (1, 2]$ is an example showing the converse of (i) does not necessarily hold in general.

(iii) To prove the assertion by contradiction, suppose $X$ is disconnected. It follows that there is a clopen subset $M$ of $X$. Since $X$ is compact and both $M$ and $M^c = X \setminus M$ are closed subsets of $X$, they both are compact subsets of $X$. But $M \cap M^c = \emptyset$. Hence, $d(M, M^c) := \inf\{d(x, y) : x \in M, y \in M^c\} > 0$. Pick $x \in M$ and $y \in M^c$. It is now obvious that $x$ cannot be reached from $y$ within a finite number of steps each of which having a length less than $\varepsilon = d(M, M^c)$. Because, otherwise $\varepsilon = d(M, M^c) \leq d(x_i, x_{i+1}) < \varepsilon$, where $x_i \in M$ and $x_{i+1} \in M^c$ are the consecutive foot steps in a walk from $x$ to $y$ with a finite number of steps each of which having a length less than $\varepsilon$. This is a contradiction. Thus, the assertion follows. ∎

**3.** (i) For given $g \in G$, define $\varphi_g : K \longrightarrow K$ by $\varphi_g(x) = g^{-1}xg$. Since $g \in N_G(K)$, the map $\varphi_g$ is well-defined because $g^{-1}xg \in K$ for all $x \in K$. This implies $\varphi_g \in \mathrm{Aut}(K)$. In other words, the map $\varphi : N_G(K) \longrightarrow \mathrm{Aut}(K)$ defined by $\varphi(g) = \varphi_g$ is well-defined. It is easily verified that $\varphi$ is a homomorphism of groups and that

$$
\begin{aligned}
\ker \varphi &= \{g \in N_G(K) |\ \varphi(g) = id\} \\
&= \{g \in N_G(K) |\ \forall x \in K : g^{-1}xg = x\} \\
&= \{g \in N_G(K) |\ \forall x \in K : gx = xg\} = C_G(K).
\end{aligned}
$$

Now, the assertion is a quick consequence of the First Isomorphism Theorem for groups.

(ii) By (i), in view of the hypothesis that $K \trianglelefteq G$, the group $\frac{G}{C_G(K)}$ is isomorphic to a subgroup of $\mathrm{Aut}(K)$. Since $K$ is cyclic, $\mathrm{Aut}(K)$ is abelian, and hence so is $\frac{G}{C_G(K)}$, whence $G' \leq C_G(K)$. It follows that $G \leq C_G(K) \leq G$ because $G' = G$. Thus, $G = C_G(K)$, implying $K \leq Z(G)$, which is what we want. ∎

**4.** To prove the "if part", it suffices to show that any two invariant subspaces of any nilpotent transformation $N$ on $F^n$ with $N^{n-1} \neq 0$ are comparable with respect to inclusion. Note first that $N^n = 0$ and that there exists a vector $\alpha \in F^n$ such that $N^{n-1}\alpha \neq 0$ because $N^{n-1} \neq 0$. It follows that the set $\{N^{n-1}\alpha, \ldots, N\alpha, \alpha\}$ is a basis for $F^n$. To see this, suppose $c_1 N^{n-1}\alpha + \cdots + c_{n-1}N\alpha + c_n\alpha = 0$ for some $c_i \in F$ $(1 \leq i \leq n)$. Taking $N^{n-1}$ of both sides of this equality yields $c_n N^{n-1}\alpha = 0$, implying $c_n = 0$. Thus, $c_1 N^{n-1}\alpha + \cdots + c_{n-1}N\alpha = 0$. Now, taking $N^{n-2}$ of both sides of the equality yields $c_{n-1} = 0$. Continuing in this manner, we obtain $c_i = 0$ for all $1 \leq i \leq n$. Thus, the set $\{N^{n-1}\alpha, \ldots, N\alpha, \alpha\}$ is a basis for $F^n$ because $\dim F^n = n$. Set

$$\mathcal{M}_0 = \langle 0 \rangle = \{0\}, \quad \mathcal{M}_i = \langle N^{n-1}\alpha, \ldots, N^{n-i}\alpha \rangle.$$

We have

$$\mathcal{M}_0 = \{0\} < \mathcal{M}_1 = \langle N^{n-1}\alpha \rangle < \cdots < \mathcal{M}_n = F^n.$$

It is plain that $\mathcal{M}_i$ is an invariant subspace of $N$ for all $1 \leq i \leq n$. This shows that any nilpotent transformation with $N^{n-1} \neq 0$ is triangularizable. Now, suppose that $\mathcal{M}$ is an invariant subspace for $N$. We prove the assertion by showing that $\mathcal{M} = \mathcal{M}_j$ for some $1 \leq j \leq n$. To this end, let $1 \leq j \leq n$ be the greatest natural number for which there exists a nonzero vector $x \in \mathcal{M}$ such that

$$x = c_1 N^{n-1}\alpha + \cdots + c_j N^{n-j}\alpha, \tag{$*$}$$

where $c_j \neq 0$. We show that $\mathcal{M} = \mathcal{M}_j$. Taking $N^{j-1}$ of both sides of $(*)$, we get $c_j N^{n-1}\alpha = N^{j-1}x \in \mathcal{M}$, whence $N^{n-1}\alpha \in \mathcal{M}$ because $c_j \neq 0$. Now, using $N^{n-1}\alpha \in \mathcal{M}$ and taking $N^{j-2}$ of both sides of $(*)$, we see that $N^{n-2}\alpha \in \mathcal{M}$. Continuing this way, we conclude that $N^{n-j}\alpha \in \mathcal{M}$. Thus, $\mathcal{M}_j = \langle N^{n-1}\alpha, \ldots, N^{n-j}\alpha \rangle \subseteq \mathcal{M}$. On the other hand, $\mathcal{M} \subseteq \mathcal{M}_j$ because $N^{n-j}\alpha \in \mathcal{M}$. Therefore, $\mathcal{M} = \mathcal{M}_j$, as desired.

To prove the "only if part", note first that the matrix $A$ can have at most one eigenvalue. Because, otherwise the corresponding eigenspaces of two distinct eigenvalues are not comparable with respect to inclusion, which is in contradiction with the hypothesis. So, with no loss of generality, we may assume that zero is the only eigenvalue of $A$, for $A - \lambda I$ and $A$ share the same lattice of invariant subspaces. In other words, we may assume that $A$ is nilpotent. We claim that there exists a nonzero vector $\alpha \in F^n$ such that the set $\{A^{n-1}\alpha, \ldots, A\alpha, \alpha\}$ is a basis for $F^n$. Suppose to the contrary that the set $\{A^{n-1}\alpha, \ldots, A\alpha, \alpha\}$ is not a basis for $F^n$. Let $\mathcal{M} := \langle A^{n-1}\alpha, \ldots, A\alpha, \alpha \rangle$. It follows that the subspace $\mathcal{M}$ is a nontrivial invariant subspace of $A$. Pick $\beta \in F^n \setminus \mathcal{M}$. It follows from the contradiction hypothesis that the subspace $\mathcal{M}' := \langle A^{n-1}\beta, \ldots, A\beta, \beta \rangle$ is also a nontrivial invariant subspace of $A$. Obviously, the subspaces $\mathcal{M}$ and $\mathcal{M}'$ are not comparable, a contradiction. Therefore, there exists a nonzero vector $\alpha \in F^n$ such that the set $\{A^{n-1}\alpha, \ldots, A\alpha, \alpha\}$ is a basis for $F^n$. Since $A$ is assumed to be nilpotent, we see that $A^n = 0$ but $A^{n-1} \neq 0$ because $A^{n-1}\alpha \neq 0$. This completes the proof. ∎

**5.** Use $A(m, n)$ to denote the average number of times that the players should play till one of them runs out of coins. If the game is played once, $A$ wins with the probability of $\frac{1}{2}$ changing the number of the coins of $A$ and $B$ to $m+1$ and $n - 1$, respectively; and, likewise, $B$ wins with the probability of $\frac{1}{2}$ changing the number of coins of $A$ and $B$ to $m - 1$ and $n + 1$, respectively. This yields the following recurrence relation on $m, n \in \mathbb{N}$.

$$A(m, n) = 1 + \frac{1}{2}A(m - 1, n + 1) + \frac{1}{2}A(m + 1, n - 1)$$

The boundary conditions are $A(0, n) = A(m, 0) = 0$. To solve this boundary recurrence equation, view $A$ as a function of one variable, say $n$. It follows from the recurrence equation that on the line $m + n = c$, where $c$ is a constant, we have

$$A(m - 1, n + 1) + A(m + 1, n - 1) - 2A(m, n) = -2. \qquad (*)$$

Thus, $A(m, n)$ must be a quadratic function. As $A(0, n) = A(m, 0) = 0$, we obtain $A(m, n) = cmn$. Substituting this into $(*)$, we obtain

$$-2c = c(m - 1)(n + 1) + c(m + 1)(n - 1) - 2cmn = -2,$$

implying $c = 1$. Consequently, $A(m, n) = mn$. Therefore, on average, the game is played as many times as the product of the coins owned by $A$ and $B$ altogether. ∎

**6.** It suffices to show that $[0, 1] \subseteq C - C$. To this end, let $y \in [0, 1]$ be arbitrary. If

$$\frac{1 - y}{2} = 0.y_1 y_2 \ldots$$

denotes the ternary expansion of $\frac{1-y}{2}$, where $y_i \in \{0, 1, 2\}$, then we can write

$$\frac{1 - y}{2} = 0.a_1 a_2 \ldots + 0.b_1 b_2 \ldots,$$

where $a_i, b_i \in \{0, 1\}$ and $y_i = a_i + b_i$ for all $i \in \mathbb{N}$. So we have

$$1 - y = 2(0.a_1 a_2 \ldots) + 2(0.b_1 b_2 \ldots),$$

implying

$$y = 0.a_1' a_2' \cdots + 0.b_1' b_2' \cdots := a + b,$$

where $a_i' = 2 - 2a_i$ and $b_i' = 2b_i$, and hence $a_i', b_i' \in \{0, 2\}$ for all $i \in \mathbb{N}$. Therefore, $a, b \in C$ because there is no "one" in their ternary expansions. This finishes the proof. ∎

## 2.30. Thirtieth Competition

**2.30.1. First Day. 1. First solution:** It suffices to evaluate the following limit

$$\lim_{t \to +\infty} \int_{1385}^{2006} f(tx)dx.$$

Setting $tx = u$, we have

$$\lim_{t \to +\infty} \int_{1385}^{2006} f(tx)dx = \lim_{t \to +\infty} \frac{1}{t} \int_{1385t}^{2006t} f(u)du.$$

As $\lim_{t \to +\infty} t = +\infty$, using L'Hopital's rule, we can write

$$\lim_{t \to +\infty} \int_{1385}^{2006} f(tx)dx = \lim_{t \to +\infty} \Big( 2006 f(2006t) - 1385 f(1385t) \Big) = 621.$$

Thus,

$$\lim_{n \to +\infty} \int_{1385}^{2006} f(nx)dx = 621,$$

as desired.

**Second solution:** We can write

$$\lim_{n \to +\infty} \int_{1385}^{2006} f(nx)\,dx = \lim_{n \to +\infty} \left( \int_{1385}^{2006} \big(f(nx) - 1\big)\,dx + 621 \right) = 621,$$

provided that $\lim_{n \to +\infty} \int_{1385}^{2006} \big(f(nx) - 1\big)\,dx = 0$. It follows from the hypothesis that for given $\varepsilon > 0$, there is an $N > 0$ such that $|f(x) - 1| < \frac{\varepsilon}{621}$ whenever $x > N$. Thus, for all $n > \frac{N}{1385}$ and $x \in [1385, 2006]$, we have

$$|f(nx) - 1| < \varepsilon.$$

This yields

$$\left| \int_{1385}^{2006} \big(f(nx) - 1\big)\,dx \right| \le \int_{1385}^{2006} |f(nx) - 1|\,dx \le \varepsilon,$$

for all $n > \frac{N}{1385}$. That is, $\lim_{n \to +\infty} \int_{1385}^{2006} \big(f(nx) - 1\big)\,dx = 0$, settling the assertion. ∎

**2.** We need the following lemma.

**Lemma.** *Let $\frac{\alpha}{\pi} \notin \mathbb{R} \setminus \mathbb{Q}$, where $\pi = 3.14159\ldots$. Then, the set $\{e^{in\alpha} : n \in \mathbb{N}\}$ is dense in the unit circle.*

**Proof.** Use $\mathbb{T}$ to denote the unit circle. Set $S = \{e^{in\alpha} : n \in \mathbb{N}\}$. Plainly, the set $S$ consists of distinct points, for $\frac{\alpha}{\pi} \notin \mathbb{R} \setminus \mathbb{Q}$, and it is a multiplicative semigroup. As $\mathbb{T}$ is compact, we see that $S$ has a limit point in $\mathbb{T}$. Thus, there is a strictly increasing sequence $(n_k)_{k=1}^{+\infty}$ of natural numbers such that $\lim_k e^{in_k \alpha}$ exists. This implies that $\lim_k e^{im_k \alpha} = 1$, where $m_k = n_{k+1} - n_k$. Now, let $\{e^{it} : t \in (a, b)\}$ be an arbitrary open arc of the unit circle, where $a, b \in \mathbb{R}$ and $a < b$. It follows that there is an $\ell \in \mathbb{N}$ such that $e^{im_k \alpha} \in \{e^{it} : t \in (-\delta, \delta)\}$ for all $k \ge \ell$, where $\delta = \frac{b-a}{2}$. Consequently, $\{e^{inm_\ell \alpha} : n \in \mathbb{N}\} \cap \{e^{it} : t \in (a, b)\} \ne \emptyset$. This shows that $S$ is dense in $\mathbb{T}$, which is what we want. ☐

**Remark.** A proof similar to that of the above lemma shows that *every additive subgroup of $\mathbb{R}$ is either dense or isolated, i.e., it has no limit points.* A quick consequence of this is the following. *Let $\alpha \in \mathbb{R} \setminus \mathbb{Q}$. Then, the set $\{m + n\alpha : m, n \in \mathbb{Z}\}$ is dense in $\mathbb{R}$.*

First, we claim that there is a sequence $(k_p)_{p=1}^{+\infty}$ of natural numbers such that $\lim_{p \to +\infty} a_j^{k_p} = 1$ for all $1 \le j \le m$. Proceed by induction on $m$. If $m = 1$, then $\lim_{n \to +\infty} a_1^n = c$. Firstly, $c \ne 0$ because $|c| = \lim_{n \to +\infty} |a_1|^n = 1$. Secondly, by showing that $a_1 = 1$, we prove the claim in this case. From $c = \lim_{n \to +\infty} a_1^n = \lim_{n \to +\infty} a_1^{n+1} = ca_1$, we obtain $c(1 - a_1) = 0$, which yields $a_1 = 1$ because $c \ne 0$. Assume that the assertion holds for $m - 1$. It follows from the induction hypothesis that there is a sequence $(k_\ell)_{\ell=1}^{+\infty}$ such that $\lim_{\ell \to +\infty} a_j^{k_\ell} = 1$ for all $1 \le j \le m - 1$. Without loss of generality, if necessary by passing to a subsequence of $(k_\ell)_{\ell=1}^{+\infty}$, we can assume that $\lim_{\ell \to +\infty} a_j^{k_\ell} = 1$ for all $1 \le j \le m - 1$ and $\lim_{\ell \to +\infty} a_m^{k_\ell} = b_m$, where $b_m \in \mathbb{C}$ and $|b_m| = 1$. Set

$$A = \big\{(a_1^k, \ldots, a_{m-1}^k, a_m^k) : k \in \mathbb{N}\big\}.$$

Let $A'$ be the set of limit points of $A$. It is obvious that $(1, \ldots, 1, b_m^k) \in A'$ for all $k \in \mathbb{N}$. As $|b_m| = 1$, it is easily seen from the lemma above that there exists a subsequence $(k_r)_{r=1}^{+\infty}$ such that $\lim_{r \to +\infty} b_m^{k_r} = 1$. Consequently, $(1, \ldots, 1, 1) \in (A')' \subseteq A'$. Thus, there is a subsequence $(k_p)_{p=1}^{+\infty}$ such that $\lim_{p \to +\infty}(a_1^{k_p}, \ldots, a_{m-1}^{k_p}, a_m^{k_p}) = (1, \ldots, 1, 1)$, and hence $\lim_{p \to +\infty} a_j^{k_p} = 1$ for all $1 \le j \le m$, proving the claim. It thus follows that

$$\lim_{p \to +\infty} \sum_{j=1}^{m} a_j^{k_p+1} = c = \lim_{p \to +\infty} \sum_{j=1}^{m} a_j^{k_p} = \sum_{j=1}^{m} 1 = m,$$

yielding $\sum_{j=1}^{m} a_j = m$ because $\lim_{p \to +\infty} a_j^{k_p} = 1$ for all $1 \le j \le m$. Now, we can write

$$\sum_{j=1}^{m} \mathrm{Re}(a_j) = \mathrm{Re}\left(\sum_{j=1}^{m} a_j\right) = m,$$

yielding $\sum_{j=1}^{m} \mathrm{Re}(a_j) = m$. This implies $\mathrm{Re}(a_j) = 1$, for $\mathrm{Re}(a_j) \le 1$ for all $1 \le j \le m$. As the $a_j$'s are on the unit circle, we see that $a_j = 1$ for all $1 \le j \le m$, which is what we want. ∎

**3.** It is worth noting that the hypothesis that $R$ is commutative is redundant. Consider the following five elements of $R/J$

$$J, \quad 1 + J, \quad a + J, \quad a^2 + J, \quad a^3 + J.$$

It follows from the hypothesis that at least two of the above elements are equal. Thus, one of the following $10 = \binom{5}{2}$ cases will occur

$$J = 1 + J, \quad J = a + J, \quad J = a^2 + J, \quad J = a^3 + J,$$
$$1 + J = a + J, \quad 1 + J = a^2 + J, \quad 1 + J = a^3 + J,$$
$$a + J = a^2 + J, \quad a + J = a^3 + J, \quad a^2 + J = a^3 + J.$$

Consequently, at least one of the following elements of $R$ belongs to $J$

$$1, \quad a, \quad a^2, \quad a^3, \quad a - 1, \quad a^2 - 1, \quad a^3 - 1 = a, \quad a^2 - a, \quad a^3 - a, \quad a^3 - a^2.$$

In view of the identity $a(a-1)(a+1) = 1$, it is readily verified that the above elements are invertible elements of $R$, and hence $J = R$, as desired. ∎

**4.** Since $p$ is an odd prime, by Fermat's Little Theorem, we have $2^{p-1} \equiv 1 \pmod{p}$. Let $s$ be the smallest natural number for which $2^s \equiv 1 \pmod{p}$. It follows that $s \mid p - 1 = 2q$, and hence $s \in \{1, 2, q, 2q\}$, for $q$ is prime. To prove the assertion by contradiction, assume that $s < 2q$. If $s = 1$, then $p \mid 1$, which is a contradiction. If $s = 2$, we obtain $p = 3$, yielding $q = 1$, which is a contradiction. Finally, if $s = q$, then the equation $x^2 \equiv 2^p \pmod{p}$ is solvable for $x$ in $\mathbb{Z}$. Thus $\left(\frac{2^q}{p}\right) = 1$, where $\left(\frac{a}{p}\right)$, with $a, p \in \mathbb{Z}$, denotes the Legendre symbol, which is defined to be $\pm 1$ depending on whether or not the equation $x^2 \equiv a \pmod{p}$ is solvable for $x$ in $\mathbb{Z}$. On the other hand,

$$\left(\frac{2^q}{p}\right) = \left(\frac{2}{p}\right)^q = (-1)^{\frac{q(p^2-1)}{8}} = (-1)^{\frac{q^2(q+1)}{2}} = -1,$$

which is a contradiction. Note that the last equality in the above is implied by $q \equiv 1 \pmod 4$. Therefore, we obtain a contradiction in any event. So the assertion follows. ∎

**5.** Let $X = X_1 + \cdots + X_{2m}$, where $X_i$'s $(1 \leq i \leq 2m)$ are independent random variables assuming the values zero and one with the probability of $\frac{1}{2}$. It is plain that

$$\mu = \mu_X = m, \quad \sigma = \sigma_X^2 = \frac{m}{2}.$$

Using the Chebyshev Inequality with $\lambda = \sqrt{2}$, we have

$$P(|X - m| \geq \sqrt{m}) \leq \frac{1}{2},$$

implying

$$P(|X - m| < \sqrt{m}) \geq \frac{1}{2}.$$

On the other hand,

$$\frac{1}{2} \leq P(|X - m| < \sqrt{m}) = \sum_{|k| < \sqrt{m}} \binom{2m}{m+k} (\frac{1}{2})^{2m},$$

yielding

$$\sum_{|k| < \sqrt{m}} \binom{2m}{m+k} \geq 2^{2m-1},$$

which is what we want.                                                        ∎

**6.** Use $p_1, \ldots, p_n$ to denote the members of the financial group. Suppose that the number of coins owned by person $p_i$ after doing business for $j$ times is $a_{ij}$, where $i = 1, \ldots, n$ and $j \in \mathbb{N}$. Without loss of generality, assume that the $(j+1)$st business is done among $p_1, p_2, p_3$, and $p_4$. It follows that

$$a_{3j} + a_{4j} - a_{1j} - a_{2j} + 2k > 0.$$

Multiplying both sides of the above inequality by $2k$, we easily conclude that

$$(a_{3j} + k)^2 + (a_{4j} + k)^2 + (a_{1j} - k)^2 + (a_{2j} - k)^2 > a_{1j}^2 + a_{2j}^2 + a_{3j}^2 + a_{4j}^2.$$

Consequently,

$$\sum_{i=1}^{n} a_{i(j+1)}^2 > \sum_{i=1}^{n} a_{ij}^2.$$

On the other hand, the sum of the number of coins owned by $p_1, \ldots, p_n$ is a fixed number. Therefore, a finite number of businesses can be done among these people, proving the assertion.                                      ∎

**2.30.2. Second Day. 1. First solution:** As $X$ is separable, $X$ has a countable base, say $\mathcal{B} = \{B_i\}_{i=1}^{+\infty}$, e.g., the family of all open balls with centers at the points of a countable dense subset of $X$ and with positive rational radii is easily seen to be a countable base for $X$. Let $D_1 = \{a \in X : \lim_{x \to a} f(x) < f(a)\}$ and $D_2 = \{a \in X : \lim_{x \to a} f(x) > f(a)\}$. It suffices to show that $D_1 \cup D_2$ is at most countable. To this end, for $a \in D_1$ (resp. $a \in D_2$) choose an $r_a \in \mathbb{Q}$ with $\lim_{x \to a} f(x) < r_a < f(a)$ (resp. $\lim_{x \to a} f(x) > r_a > f(a)$) and $B_{i_a} \in \mathcal{B}$ with $a \in B_{i_a}$ such that $f(x) < r_a$ (resp. $f(x) > r_a$) for all $x \in B_{i_a} \setminus \{a\}$. Define a function $\iota_1 : D_1 \longrightarrow \mathbb{Q} \times \mathcal{B}$ (resp. $\iota_2 : D_2 \longrightarrow \mathbb{Q} \times \mathcal{B}$) by $\iota_1(a) = (r_a, B_{i_a})$ (resp. $\iota_2(a) = (r_a, B_{i_a})$). If $(r_a, B_{i_a}) = (r_b, B_{i_b})$ for some $a, b \in X$, then $a = b$, for otherwise $f(x) < r_a = r_b$ (resp. $f(x) > r_a = r_b$) whenever $x \in B_{i_a} \setminus \{a\}$. But $B_{i_a} = B_{i_b}$. Thus, $f(b) < r_a = r_b < f(b)$ (resp. $f(b) > r_a = r_b > f(b)$), which is a contradiction. This proves that $\iota_1$ (resp. $\iota_2$) is one-to-one. Therefore, $D_1$ and $D_2$ are countable, and hence so is $D_1 \cup D_2$, which is what we want.

**Second solution:** For a given $f : X \longrightarrow \mathbb{R}$, use $D$ and $L$ to denote the the set of points at which $f$ is discontinuous and has a limit, respectively. By showing that $D \cap L$ is at most countable, we prove the assertion. Define $\omega : X \longrightarrow [0, +\infty)$ by

$$\omega(x) = \lim_{\delta \to 0^+} \left( \sup f\big(B_\delta(x)\big) - \inf f\big(B_\delta(x)\big) \right),$$

where $B_\delta(x) = \{y \in X : d(y, x) < \delta\}$ denotes the open ball with center at $x$ and radius $\delta > 0$. Note that $D = \bigcup_{n=1}^{+\infty} D_n$, where $D_n = \{x \in X : \omega(x) > \frac{1}{n}\}$. Consequently, $D \cap L = \bigcup_{n=1}^{+\infty} (D_n \cap L)$. It thus suffices to prove that $D_n \cap L$ is at most countable for all $n \in \mathbb{N}$. To this end, for a given $a \in D_n \cap L$, we have $\lim_{x \to a} f(x) = l_a$ for some $l_a \in \mathbb{R}$. It follows that there exists a $\delta_{n_a} > 0$ such that $|f(x) - l_a| < \frac{1}{2n}$ whenever $x \in B_{\delta_{n_a}}(a) \setminus \{a\}$. This implies $|f(x) - f(y)| < \frac{1}{n}$ for all $x, y \in B_{\delta_{n_a}}(a) \setminus \{a\}$. Hence, $\omega(x) \leq \frac{1}{n}$ for all $x \in B_{\delta_{n_a}}(a) \setminus \{a\}$. In other words, $x \notin D_n$ for all $x \in B_{\delta_{n_a}}(a) \setminus \{a\}$. This implies that $D_n \cap L$ is at most countable for all $n \in \mathbb{N}$, for otherwise $D_n \cap L$ would have (uncountably many) limit points because $X$ is separable. This obviously is in contradiction with $x \notin D_n$ for all $x \in B_{\delta_{n_a}}(a) \setminus \{a\}$ and $a \in D_n \cap L$, finishing the proof. ∎

**2. First solution:** If $M = +\infty$, the assertion is trivial. So, we may, without loss of generality, assume that $M < +\infty$. Let $\mathbb{D} = \{z \in \mathbb{C} : |z| < 1\}$ denote the unit disk around zero. Define $g : \mathbb{D} \longrightarrow \mathbb{C}$ by $g(z) = \frac{f(Rz) - a_0}{M + |a_0|}$. Plainly, $g$ is analytic on $\mathbb{D}$ and $g(0) = 0$. By the Maximum Modulus Principle, $|f(Rz)| \leq \sup_{|z|=R} |f(z)| = M$ for all $z \in \mathbb{D}$. It follows that

$$|g(z)| \leq \frac{|f(Rz)| + |a_0|}{M + |a_0|} \leq \frac{M + |a_0|}{M + |a_0|} = 1,$$

for all $z \in \mathbb{D}$. Thus, by the Schwarz Lemma, $|g(z)| \leq |z|$ for all $z \in \mathbb{D}$. Now, if $f(z_0) = 0$ for some $z_0 \in \mathbb{C}$, then $\frac{z_0}{R} \in \mathbb{D}$ because $|z_0| < R$. Consequently, $|g(\frac{z_0}{R})| \leq |\frac{z_0}{R}|$. But

$$\left| g\Big(\frac{z_0}{R}\Big) \right| = \left| \frac{f\big(R\frac{z_0}{R}\big) - a_0}{M + |a_0|} \right| = \frac{|a_0|}{M + |a_0|} \leq \frac{|z_0|}{R},$$

yielding $|z_0| \geq \frac{R|a_0|}{M + |a_0|}$, finishing the proof.

**Second solution:** Again, we may, without loss of generality, assume that $M < +\infty$. Let $0 < r < R$ be arbitrary. By the Cauchy Inequality, $|a_n| \leq \frac{M(r)}{r^n}$ for all $n \in \mathbb{N}$. In view of the Maximum Modulus Principle, we have $M(r) \leq M$ for all $r < R$, and hence we can write

$$|f(z) - a_0| \leq \sum_{n=1}^{+\infty} |a_n| |z|^n \leq M(r) \sum_{n=1}^{+\infty} \Big(\frac{|z|}{r}\Big)^n = \frac{M|z|}{r - |z|},$$

for all $z \in \mathbb{C}$ with $|z| < R$. On the other hand,

$$\frac{M|z|}{r - |z|} < |a_0| \iff |z| < \frac{r|a_0|}{M + |a_0|}.$$

Consequently, $f(z) \neq 0$ whenever $|z| < \frac{r|a_0|}{M+|a_0|}$. Since $r < R$ is arbitrary, we see that $f(z) \neq 0$ whenever $|z| < \frac{R|a_0|}{M+|a_0|}$. This proves the assertion. ∎

**3.** Let $H$ be the set of all elements of $G$ whose orders are finite. Obviously, $H$ is not empty because $1 \in H$. Let $x, y \in H$ be arbitrary. It follows that there is an $n \in \mathbb{N}$ such that $x^n = y^n = 1$. Consequently,

$$
\begin{aligned}
(xy^{-1})^n G' &= (xy^{-1}G')^n = \left((xG')(y^{-1}G')\right)^n \\
&= (xG')^n (y^{-1}G')^n = (x^n G')(y^{-n}G') = G',
\end{aligned}
$$

where the equality $\left((xG')(y^{-1}G')\right)^n = (xG')^n(y^{-1}G')^n$ follows from the hypothesis that $\frac{G}{G'}$ is abelian. Thus, $(xy^{-1})^n \in G'$, from which, we obtain an $m \in \mathbb{N}$ such that $(xy^{-1})^{nm} = 1$. That is, $xy^{-1} \in H$, and hence $H \leq G$, as desired. ∎

**4.** We prove the following, of which the assertion is a quick consequence. *Let $D$ be a division ring and $F$ a subfield of its center. If $A \in M_n(D)$ is algebraic over the subfield $F$ and $\mathrm{rank}(A) = \mathrm{rank}(A^2)$, then there is an idempotent $E$ in the $F$-algebra generated by $A$ which projects $D^n$ onto the range of $A$ along $\ker(A)$, i.e., $D^n = \mathrm{im}(A) \oplus \ker(A)$ and $E(x + y) = x$, where $x \in \mathrm{im}(A)$ and $y \in \ker(A)$.*

As is usual, the members of $M_n(D)$ can be viewed as linear transformations acting on the left of $D^n$ via the usual matrix multiplication. To show that $D^n = \mathrm{im}(A) \oplus \ker(A)$, note that by the Rank-Nullity Theorem, we have $\dim\left(\mathrm{im}(A)\right) + \dim\left(\ker(A)\right) = \dim(D^n) = n$. On the other hand,

$$\dim\left(\mathrm{im}(A) + \ker(A)\right) = \dim(\mathrm{im}(A)) + \dim(\ker(A)) - \dim\left(\mathrm{im}(A) \cap \ker(A)\right).$$

Thus, to prove $D^n = \mathrm{im}(A) \oplus \ker(A)$, it suffices to show that $\mathrm{im}(A) \cap \ker(A) = \{0\}$. To this end, let $B := A|_{\mathrm{im}(A)} : \mathrm{im}(A) \longrightarrow \mathrm{im}(A)$. We can write

$$
\begin{aligned}
\dim\left(\mathrm{im}(A^2)\right) &= \mathrm{rank}(A^2) = \mathrm{rank}(A) \\
&= \dim\left(\mathrm{im}(A)\right) = \dim\left(\ker(B)\right) + \dim\left(\mathrm{im}(B)\right) \\
&= \dim\left(\mathrm{im}(A) \cap \ker(A)\right) + \dim\left(\mathrm{im}(A^2)\right),
\end{aligned}
$$

yielding $\dim\left(\mathrm{im}(A) \cap \ker(A)\right) = 0$, whence $\mathrm{im}(A) \cap \ker(A) = \{0\}$.

For the rest, let $f(A) = A^m + a_{m-1}A^{m-1} + \cdots + a_1 A + a_0 I = 0$, where $m$ is minimal and $a_0, a_1, \ldots, a_{m-1} \in F$. If $a_0 \neq 0$, then we have nothing to prove. If $a_0 = 0$, then $(A^{m-1} + a_{m-1}A^{m-2} + \cdots + a_1 I)A = 0$. We claim that $a_1 \neq 0$. If not, then $(A^{m-2} + a_{m-1}A^{m-3} + \cdots + a_2 I)A^2 = 0$. Now, since $\mathrm{im}(A) = \mathrm{im}(A^2)$ because $\mathrm{rank}(A) = \mathrm{rank}(A^2)$, it follows that $(A^{m-2} + a_{m-1}A^{m-3} + \cdots + a_2 I)A = 0$, a contradiction. So $a_1 \neq 0$. Set $G = a_1^{-1}A^{m-1} + a_1^{-1}a_{m-1}A^{m-2} + \cdots + a_1^{-1}a_2 A + I$ and note that $D^n = \ker(A) \oplus \mathrm{im}(A)$. Then $E = I - G$ is the desired idempotent. Because for each $x = Ax' \in \text{im}(A)$ and $y \in \ker(A)$, we can write

$$
\begin{aligned}
E(x + y) &= Ex + Ey \\
&= \left(Ax' - a_1^{-1} f(A)x'\right) - \\
&\quad \left(a_1^{-1} A^{m-1} + a_1^{-1} a_{m-1} A^{m-2} + \cdots + a_1^{-1} a_2 A\right)y \\
&= (Ax' - 0) - 0 = Ax' = x.
\end{aligned}
$$

So the proof is complete. ∎

**5.** It is plain that the sets of odd and even numbers, denoted by $\mathbb{O}$ and $\mathbb{E}$, respectively, is a partition of the natural numbers such that neither $\mathbb{O} \oplus \mathbb{O}$ nor $\mathbb{E} \oplus \mathbb{E}$ contains any prime. To prove the uniqueness, suppose that $\mathbb{N} = \mathbb{A} \cup \mathbb{B}$ and that neither $\mathbb{A} \oplus \mathbb{A}$ nor $\mathbb{B} \oplus \mathbb{B}$ contains any prime. Without loss of generality, assume that $1 \in \mathbb{A}$. By showing that $\mathbb{A} = \mathbb{O}$ and $\mathbb{B} = \mathbb{E}$, we finish the proof. Proceed by induction. As $1 \in \mathbb{A}$ and $1 + 2 = 3$, $2 \in \mathbb{B}$. Assume that

$$\{1, 3, \ldots, 2k - 1\} \subseteq \mathbb{A},$$

and

$$\{2, 4, \ldots, 2k\} \subseteq \mathbb{B}.$$

It follows from Bertrand's conjecture that for $n = 2k + 1$, there is a prime $p$ such that

$$2k + 1 < p \le 4k + 2.$$

We have $2k + 1 < p \le 4k + 2$ because $k \ge 1$. But $p - (2k + 1)$ is an even number and $0 < p - (2k+1) \le 2k$. Hence, $p - (2k+1) \in \mathbb{B}$, whence $2k+1 \in \mathbb{A}$. Likewise, from $0 < p - (2k + 2) \le 2k + 1$, we see that $2k + 2 \in \mathbb{B}$. So the proof is complete by way of induction. ∎

**6.** Let $\overline{T}$ denote the closure of $T$. It is plain that the distance between any two points of $\overline{T}$ is less than or equal to $\alpha$. So, we may, with no loss of generality, assume that $T$ is closed. Let $x, y \in T$ be such that the distance between $x$ and $y$ is maximal. Use $\overrightarrow{xy}$ to denote the shortest arc joining $x$ and $y$ in the clockwise direction. Construct $T'$ from $T$ as follows. If $z \in T \cap \overrightarrow{xy}$, then $z \in T'$. If $z \in T$ and $z \notin \overrightarrow{xy}$, then $z' \in T'$, where $z'$ denotes the antipodal point of $z$. In this case, we have $z' \notin T$, for otherwise the distance between $z$ and $z'$ is equal to one half whereas $0 < \alpha < \frac{1}{2}$. Also, $z' \in \overrightarrow{xy}$, because otherwise the distance between $z$ and $x$ or $z$ and $y$ will be greater than the distance between $x$ and $y$, which is impossible. Thus, $T' \subseteq \overrightarrow{xy}$. On the other hand, the sum of the lengths of the arcs whose union is $T$ is equal to the sum of the lengths of the arcs whose union is $T'$. This obviously proves the assertion. ∎

## 2.31. Thirty-First Competition

**2.31.1. First Day. 1.** (a) and (b) The answer is no. Color the circle by two colors, say black (b) and white (w), as shown in the figure. That is, (b) and (w) are, respectively, assigned to the upper and lower half-open semicircles. It is obvious that any equilateral triangle inscribed in the circle has a white vertex

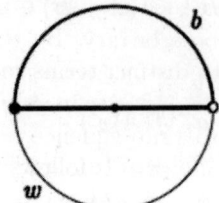

Figure 14

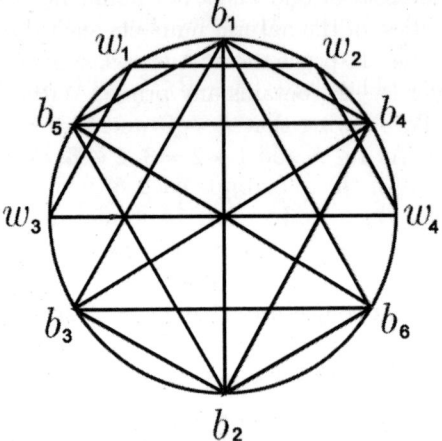

Figure 15

and two black vertices or vise versa. Also, the hypotenuse of any right triangle inscribed in the circle is a diameter of the circle whose vertices are of the two colors.

(c) The answer is yes. Proceed by contradiction. Assume that there is a coloring of the circle for which there is no monocolored isosceles triangle inscribed in the circle. First choose a monocolored chord, say with white ends, which is not a diameter. Use $w_1w_2$ to denote the chord. The points on the circle that together with this chord form an isosceles triangle lie on the perpendicular bisector of $w_1w_2$, which is a diameter of the circle. By the contradiction hypothesis, the two ends of this diameter are both black. Use $b_1b_2$ to denote this diameter. Likewise, the two ends of the diameter perpendicular to $b_1b_2$ must be both white. Use $w_3w_4$ to denote this diameter. Now consider the two chords $w_1w_3$ and $w_2w_4$. It follows that the ends of the diameters perpendicular to $w_1w_3$ and $w_2w_4$, respectively, must be all black. Use $b_3b_4$ and $b_5b_6$ to denote these diameters. It thus follows that the black-colored triangles $b_2b_4b_5$, $b_2b_3b_6$, $b_1b_4b_5$, and $b_1b_3b_6$ are all isosceles. This is a contradiction, and hence completes the proof.  ∎

2. Using the Sequence Lemma, it is readily seen that

$$A' + B \subseteq (A + B)', \quad A + B' \subseteq (A + B)',$$

implying that $(A' + B) \cup (A + B') \subseteq (A + B)'$. To see $(A + B)' \subseteq (A' + B) \cup (A + B')$, let $x \in (A + B)'$ be arbitrary. By the Sequence Lemma, there exists a sequence $(a_n + b_n)_{n=1}^{+\infty}$ with distinct terms such that $x = \lim_n (a_n + b_n)$, where the sequences $(a_n)_{n=1}^{+\infty}$ and $(b_n)_{n=1}^{+\infty}$ are in $A$ and $B$, respectively. As $A$ is bounded, if necessary, by passing to a subsequence, we may assume that there is an $a \in \overline{A} := A \cup A'$ such that $a = \lim_n a_n$. It follows that $\lim_n b_n = \lim_n ((a_n + b_n) - a_n) = x - a$, and hence $b := x - a \in B$ because $B$ is closed. There are two cases to consider. If $a \in A'$, then $x = a + b \in (A' + B) \subseteq (A' + B) \cup (A + B')$, which is what we want. If $a \in A \setminus A'$, then, if necessary by discarding a finite number of indices, we may assume that $a_n = a$ for all $n \in \mathbb{N}$. This implies that $b_n$'s $(n \in \mathbb{N})$ are all distinct. Moreover, $\lim_n b_n = x - a \in B'$. Thus, $x = a + (x - a) \in (A + B') \subseteq (A' + B) \cup (A + B')$, as desired. So, in any case, $x \in (A' + B) \cup (A + B')$. Therefore, $(A + B)' \subseteq (A' + B) \cup (A + B')$. This completes the proof. ∎

**3.** Let $H_1, \ldots, H_n$ be all of the subgroups each of which has index 2 in $G$. It thus follows that each $H_i$ $(i = 1, \ldots, n)$ is normal in $G$, and hence so is $H_1 \cap \cdots \cap H_n$ in $G$. Also, $H_1 \cap \cdots \cap H_n$ is of finite index in $G$ because so is every $H_i$ $(i = 1, \ldots, n)$ in $G$. Set $A := \frac{G}{H_1 \cap \cdots \cap H_n}$. We claim that $A$ is the desired group. Suppose that $B$ is a subgroup of $A$ of index 2. It follows that $B := \frac{K}{H_1 \cap \cdots \cap H_n}$, where $H_1 \cap \cdots \cap H_n \subseteq K \trianglelefteq G$. By the Third Isomorphism Theorem for groups , we have

$$\frac{A}{B} = \frac{\frac{G}{H_1 \cap \cdots \cap H_n}}{\frac{K}{H_1 \cap \cdots \cap H_n}} \cong \frac{G}{K}.$$

Thus, the index of $K$ in $G$ is 2. Therefore, $K = H_i$ for some $1 \leq i \leq n$. Consequently, $A$ has exactly $n$ subgroup of index 2 which are

$$\frac{H_1}{H_1 \cap \cdots \cap H_n}, \frac{H_2}{H_1 \cap \cdots \cap H_n}, \ldots, \frac{H_n}{H_1 \cap \cdots \cap H_n}.$$

(Note that as $H_i$'s are mutually disjoint, so are the aforementioned groups.) It remains to show that $A$ is abelian. For each $i = 1, \ldots, n$, the group $\frac{G}{H_i}$ is abelian because $H_i$ has index 2 in $G$. Thus, $G' \subseteq H_i$ for all $1 \leq i \leq n$, from which we obtain $G' \subseteq H_1 \cap \cdots \cap H_n$. Therefore, $A$ is abelian, as desired. ∎

**4.** No, one cannot find such two dice. To see this, proceed by contradiction. Consider two such dice $A$ and $B$. For each $i = 1, \ldots, 6$, let $A_i$ and $B_i$, respectively, denote the probability of getting $i$ for the two dice $A$ and $B$. For each $j = 2, \ldots, 12$, denote by $P_j$, the probability of getting a sum of $j$. Consequently, $P_j = \sum_{k=1}^{j} A_k B_{j-k}$. We have $P_2 = A_1 B_1$, $P_{12} = A_6 B_6$, and $P_7 = A_1 B_6 + A_2 B_5 + A_3 B_4 + A_4 B_3 + A_5 B_2 + A_6 B_1$. So, we can write

$$P_7 \geq A_1 B_6 + A_6 B_1 \geq 2\sqrt{A_1 B_6 A_6 B_1} = 2\sqrt{A_1 B_1 A_6 B_6}$$

$$= 2\sqrt{P_2 P_{12}} \geq 2\sqrt{\frac{2}{33} \cdot \frac{2}{33}} = \frac{4}{33},$$

which is impossible because $P_i \in \left(\frac{2}{33}, \frac{4}{33}\right)$ for all $2 \leq i \leq 12$. ∎

**5.** Note that the set of all balls whose centers are rational points of $\mathbb{R}^2$ and whose radii are positive rationals is countable. Use $\{B_1, B_2, B_3, \ldots\}$ to denote this set. Choose a set $\{x_i\}_{i=1}^{+\infty}$ such that $x_i \in B_i$ for all $i \in \mathbb{N}$. As the set $D := \{x_i\}_{i=1}^{+\infty}$ is evidently dense in the plane, it suffices to show that $D$ can be so chosen that no three points of $D$ are collinear. To this end, use induction on $n$ to choose $x_n$ in such a way that for all $i < j < n$, the points $x_i, x_j, x_n$ are not collinear. This obviously can be done because no ball is a union of a finite number of lines. ∎

**6.** Let $p(x) = x^n + a_1 x^{n-1} + \cdots + a_n$ be the characteristic polynomial of $A$ and let $\lambda_1, \ldots, \lambda_n$ be the eigenvalues of $A$, which are the roots of $p(x) = 0$. As, for each $i = 1, \ldots, n$, the polynomial $p(x)$ is divisible by $x - \lambda_i$, we can write

$$\frac{p(x)}{x - \lambda_i} = x^{n-1} + (\lambda_i + a_1)x^{n-2} + \cdots + (\lambda_i^{n-1} + a_1\lambda_i^{n-2} + \cdots + a_{n-2}\lambda_i + a_{n-1}).$$

We note that $p'(x) = \frac{p(x)}{x - \lambda_1} + \cdots + \frac{p(x)}{x - \lambda_n}$. It thus follows that

$$
\begin{aligned}
p'(x) = {}& nx^{n-1} + \left(\mathrm{tr}(A) + na_1\right)x^{n-2} + \cdots \\
& + \left(\mathrm{tr}(A^{n-1}) + a_1\mathrm{tr}(A^{n-2}) + \cdots + a_{n-2}\mathrm{tr}(A) + na_{n-1}\right).
\end{aligned}
$$

On the other hand, $p'(x) = nx^{n-1} + (n-1)a_1 x^{n-2} + \cdots + a_{n-1}$. Thus,

$$
\begin{aligned}
a_1 &= -\mathrm{tr}(A) \\
a_1\mathrm{tr}(A) + 2a_2 &= -\mathrm{tr}(A^2) \\
a_1\mathrm{tr}(A^2) + a_2\mathrm{tr}(A) + 3a_3 &= -\mathrm{tr}(A^3) \\
\vdots\ &=\ \vdots \\
a_1\mathrm{tr}(A^{n-2}) + \cdots + a_{n-2}\mathrm{tr}(A) + (n-1)a_{n-1} &= -\mathrm{tr}(A^{n-1}) \\
a_1\mathrm{tr}(A^{n-1}) + \cdots + a_{n-1}\mathrm{tr}(A) + na_n &= -\mathrm{tr}(A^n)
\end{aligned}
$$

Note that the last equality in the above is obtained from $p(\lambda_1) + \cdots + p(\lambda_n) = 0$. It now follows from Cramer's Rule that

$$
a_n = \frac{\det\begin{pmatrix}
1 & 0 & 0 & \cdots & \cdots & -\mathrm{tr}(A) \\
\mathrm{tr}(A) & 2 & 0 & \cdots & \cdots & -\mathrm{tr}(A^2) \\
\mathrm{tr}(A^2) & \mathrm{tr}(A) & 3 & 0 & \cdots & -\mathrm{tr}(A^3) \\
\vdots & \vdots & \ddots & \ddots & \vdots & \vdots \\
\mathrm{tr}(A^{n-2}) & \mathrm{tr}(A^{n-3}) & \cdots & \mathrm{tr}(A) & n-1 & -\mathrm{tr}(A^{n-1}) \\
\mathrm{tr}(A^{n-1}) & \mathrm{tr}(A^{n-2}) & \cdots & \cdots & \mathrm{tr}(A) & -\mathrm{tr}(A^n)
\end{pmatrix}}{\det\begin{pmatrix}
1 & 0 & 0 & \cdots & \cdots & 0 \\
\mathrm{tr}(A) & 2 & 0 & \cdots & \cdots & 0 \\
\mathrm{tr}(A^2) & \mathrm{tr}(A) & 3 & 0 & \cdots & 0 \\
\vdots & \vdots & \ddots & \ddots & \ddots & \vdots \\
\mathrm{tr}(A^{n-2}) & \mathrm{tr}(A^{n-3}) & \cdots & \mathrm{tr}(A) & n-1 & 0 \\
\mathrm{tr}(A^{n-1}) & \mathrm{tr}(A^{n-2}) & \mathrm{tr}(A^{n-3}) & \cdots & \mathrm{tr}(A) & n
\end{pmatrix}}.
$$

Note that the denominator of the fraction above is equal to $n!$. Thus, in view of $a_n = (-1)^n \det A$, the assertion follows by substitution.  ∎

**2.31.2. Second Day. 1.** For the first part, just note that

$$\left(\frac{n-1}{2}+1\right) + \left(\frac{n-1}{2}+2\right) + \cdots + \left(\frac{n-1}{2}+n\right) = n^2.$$

For the rest, suppose that $m, k$ are positive integers such that $(m+1)+(m+2)+\cdots+(m+12) = k^2$. It follows that $12m+78 = k^2$. Consequently, $n$ is even, and hence $k = 2t$ for some $t \in \mathbb{N}$. By substitution, we obtain $2t^2 - 6m = 39$, which is impossible because the left hand side is an even integer.  ∎

**2.** Performing the substitution $u = x^\alpha$ and $y = y^\beta$ and depending on the signs of $\alpha$ and $\beta$, we need to investigate the limit of

$$\frac{u^2 v^2}{u^3 + v^3}$$

as $u$ and $v$ tend to 0 or $+\infty$. There are four cases to consider.

(a) $\alpha > 0, \beta > 0$.

In this case, we need to investigate $\lim_{u,v\to 0+} \frac{u^2 v^2}{u^3+v^3}$. But by the Sandwich Theorem, this limit exists and equals zero because

$$0 \le \frac{u^2 v^2}{u^3 + v^3} = \left(\left(\frac{u}{v}\right)^{\frac{3}{2}} + \left(\frac{v}{u}\right)^{\frac{3}{2}}\right)^{-1} \sqrt{uv} \le \frac{\sqrt{uv}}{2},$$

for all $u, v > 0$.

(b) $\alpha > 0, \beta < 0$.

In this case, we need to investigate $\lim_{u\to 0+, v\to+\infty} \frac{u^2 v^2}{u^3+v^3}$. Again, by the Sandwich Theorem, the limit exists in this case and equals zero because

$$0 \le \frac{u^2 v^2}{u^3 + v^3} = \frac{u^2}{\frac{u^3}{v^2} + v} \le \frac{u^2}{v},$$

for all $u, v > 0$.

(c) $\alpha < 0, \beta > 0$.

Just as in (b), one can readily see that $\lim_{u\to+\infty, v\to 0+} \frac{u^2 v^2}{u^3+v^3} = 0$.

(d) $\alpha < 0, \beta < 0$.

By showing that there is a path on which the limit is infinity, we conclude that the limit does not exist in this case. To this end, assuming that $u = v$, we have

$$\lim_{u,v\to+\infty, u=v} \frac{u^2 v^2}{u^3 + v^3} = \lim_{u\to+\infty} \frac{u}{2} = +\infty,$$

as desired.

Alternatively, the cases (a), (b), and (c) above could be proven using the following inequalities

$$0 \le \frac{u^2 v^2}{u^3 + v^3} \le \min\left(\frac{u^2}{v}, \frac{v^2}{u}\right) \le \max(u, v),$$

for all $u, v > 0$.  ∎

**3.** Let $x = (x_1, \ldots, x_n)$ and $y = (y_1, \ldots, y_n)$. Define the function $f : A^n \to A^n$ by $f(z) = z'$, where $z = (z_1, \ldots, z_n)$, $z' = (z'_1, \ldots, z'_n)$, and

$$
z'_j = \begin{cases}
x_j & z_j = x_j \neq y_j, \\
y_j & z_j = y_j \neq x_j, \\
z_j & \text{otherwise,}
\end{cases}
$$

for all $1 \leq j \leq n$. It is readily seen that

$$
d(z, x) = d(f(z), y), \ d(z', y), \ d(z, y) = d(f(z), x) = d(z', x).
$$

Therefore, $f$ is a one-to-one correspondence between the sets $C$ and $D$, which is what we want.                                                                 ∎

**4.** First, we prove that "*if $F$ is a field, then the ring $F[x]$ has infinitely many maximal ideals.*" To see this, noting that the ideal generated by an irreducible monic polynomial is a maximal ideal and that such ideals are distinct whenever the generating polynomials are, it suffices to show that there are infinitely many irreducible monic polynomials in $F[x]$. To this end, motivated by Euclid's proof of the infinitude of primes, suppose by contradiction that for some $n \in \mathbb{N}$, there are only $n$ irreducible monic polynomials in $F[x]$, say $f_1, \ldots, f_n$. But the polynomial $f_1 \cdots f_n + 1$ must have an irreducible monic divisor. This implies that some $f_i$ ($1 \leq i \leq n$) divides $f_1 \cdots f_n + 1$. Consequently, $f_i$ must divide 1, which is a contradiction. Thus, there are infinitely many irreducible monic polynomials in $F[x]$.

Now, to prove the assertion, let $M$ be a maximal ideal of $R$. By the preceding paragraph, the ring $\frac{R}{M}[x]$ has infinitely many maximal ideals. Since the mapping $\phi : R[x] \to \frac{R}{M}[x]$, defined by $f(a_n x^n + \cdots + a_1 x + a_0) = (a_n + M)x^n + \cdots + (a_1 + M)x + (a_0 + M)$ is an epimorphism, i.e., a surjective homomorphism, of rings, the ring $R[x]$ has has infinitely many maximal ideals as well. This is because for every maximal ideal $I$ of $\frac{R}{M}[x]$, $\phi^{-1}(I)$ is also a maximal ideal of $R[x]$ and that $\phi^{-1}(I) \neq \phi^{-1}(J)$ whenever $I \neq J$. This completes the proof. ∎

**5.  First solution:** The hypothesis can be rephrased as follows. We have $AX = 0$, where

$$
A = \begin{pmatrix}
0 & \pm 1 & \cdots & \pm 1 \\
\pm 1 & \ddots & \ddots & \vdots \\
\vdots & \ddots & \ddots & \pm 1 \\
\pm 1 & \cdots & \pm 1 & 0
\end{pmatrix}_{2n \times 2n}, \quad
X = \begin{pmatrix} x_1 \\ \vdots \\ \vdots \\ x_{2n} \end{pmatrix}_{2n \times 1}, \quad
0 = \begin{pmatrix} 0 \\ \vdots \\ \vdots \\ 0 \end{pmatrix}_{2n \times 1}.
$$

That is, the diagonal entries of $A$ are all zero and the off diagonal entries of $A$ are all either 1 or $-1$. It suffices to prove that $A$ is invertible. Since the determinant of $A$ is an integer, we prove the assertion by showing that

$\det A \overset{2}{\equiv} 1$. We can write

$$\det A = \det \begin{pmatrix} 0 & \pm 1 & \cdots & \pm 1 \\ \pm 1 & \ddots & \ddots & \vdots \\ \vdots & \ddots & \ddots & \pm 1 \\ \pm 1 & \cdots & \pm 1 & 0 \end{pmatrix} \overset{2}{\equiv} \det \begin{pmatrix} 0 & 1 & \cdots & 1 \\ 1 & \ddots & \ddots & \vdots \\ \vdots & \ddots & \ddots & 1 \\ 1 & \cdots & 1 & 0 \end{pmatrix}$$

$$= \text{the number of derangements of } 2n \text{ elements}$$

$$= 2n! \left( \frac{1}{0!} - \frac{1}{1!} + \frac{1}{2!} - \cdots + (-1)^{2n} \frac{1}{(2n)!} \right)$$

$$\overset{2}{\equiv} 1.$$

Recall that a derangement on $n$ elements is a permutation on $n$ elements having no fixed points. Also, a standard argument employing the inclusion-exclusion principle shows that the number of derangements on $n$ elements is equal to

$$n! \sum_{k=0}^{n} \frac{(-1)^k}{k!}.$$

**Second solution:** We prove the following, of which the assertion is a quick consequence. *Let $G$ be an abelian torsion free group, i.e., the identity is the only element of the group having finite order. Let $n, k \in \mathbb{N}$ with $n > k \geq 2$ and $x_i$'s $(1 \leq i \leq n)$ be elements of $G$ such that removing any of them the remaining ones can be partitioned into $k$ subsets with equal sums. Prove that $x_i$'s are all equal. Therefore, either $x_i$'s are all zero or they are all equal to a nonzero element, in which case $n \overset{k}{\equiv} 1$.*

Since the subgroup generated by $x_i$'s $(1 \leq i \leq n)$ is finitely generated, in view of the Fundamental Theorem of Finitely Generated Abelian Groups, we may, without loss of generality, assume that $G = \mathbb{Z}$. Also, if necessary, by adding an appropriate positive integer to all $x_i$'s $(1 \leq i \leq n)$, we may assume that $x_i \in \mathbb{N}$ for each $i = 1, \ldots, n$. First from the hypothesis, we easily obtain $x_j \overset{k}{\equiv} \sum_{i=1}^{n} x_i$ for each $j = 1, \ldots, n$. Next, write $x_i$'s in the basis $k$. Plainly, there is an $N \in \mathbb{N}$ such that for all $1 \leq j \leq n$, we can write

$$x_j = \sum_{i=0}^{N} a_{ij} k^i,$$

where $a_{ij} \in \mathbb{N}$ with $0 \leq a_{ij} < k$ and $0 \leq i \leq N$. Note that $x_j \overset{k}{\equiv} \sum_{i=1}^{n} x_i$ for each $j = 1, \ldots, n$ obtains $a_{0j} = a_{01}$ for each $j = 1, \ldots, n$. We use induction on $N$ to prove that that $x_i$'s are all equal. If $N = 0$, the assertion is easy because in this case $x_j = a_{0j} = a_{01} = x_1$ for all $1 \leq j \leq n$, as desired. Assuming that the assertion holds for $N$, we prove it for $N + 1$. To this end, noting that $a_{0j} = a_{01}$ for each $j = 1, \ldots, n$, define

$$y_j := \frac{x_j - a_{01}}{k} = \sum_{i=1}^{N+1} a_{ij} k^{i-1},$$

where $1 \le j \le n$. It readily follows from the hypothesis that removing any of the $y_i$'s $(1 \le i \le n)$ the remaining ones can be partitioned into $k$ subsets with equal sums. Thus, the induction hypothesis applies to $y_i$'s $(1 \le i \le n)$, and hence $y_i$'s are all equal. Therefore, so are $x_i$'s $(1 \le i \le n)$, which is what we want. ∎

**6.** (a) Let $A = (t_1 q_1, 1 - t_1)$ and $B = (t_2 q_2, 1 - t_2)$, where $t_1, t_2 \in (0, 1]$ and $q_1, q_2 \in \mathbb{Q}$ are distinct. Let $r$ be an irrational number between $q_1, q_2 \in \mathbb{Q}$. It follows that the line $\ell$ joining $M$ and $(r, 0)$ partitions the plane into the line $\ell$ and two open half-planes, say $P_1$ and $P_2$. It is obvious that $T \setminus \{M\} \subset P_1 \cup P_2$ and that $\{A, B\} \not\subset P_1$ whereas $\{A, B\} \subset T \setminus \{M\} \subset P_1 \cup P_2$. It thus follows that any continuous path from $A$ to $B$ in the set $T$ must intersect the line $\ell$ at the point $M$ because every other point of the line $\ell$ does not belong to the set $T$. This proves the assertion.

(b) Let $f : T \to T$ be a continuous function and $f(M) = M' = (t_0 q_0, 1 - t_0)$ for some $t_0 \in [0, 1]$ and $q_0 \in \mathbb{Q}$. If $t_0 = 0$, there is nothing to prove, for the point $M$ would be a fixed point of $f$. If not, define the function $g : [0, 1] \to T$ by $g(t) = f(t t_0 q_0, 1 - t t_0)$. Plainly, $g$ is continuous. Define the set $D$ as follows

$$D = \Big\{ (t t_0 q_0, 1 - t t_0) \in T : t \in [0, 1] \Big\} = \Big\{ t(t_0 q_0, 1 - t_0) + (1 - t)(0, 1) : t \in [0, 1] \Big\}.$$

It is plain that $D$ is the line segment joining the points $M = (0, 1)$ and $M' = (t_0 q_0, 1 - t_0)$. Note that $D$ is homeomorphic to the compact interval $[0, 1]$. We prove the assertion by showing that $f$ has a fixed point in $D$. If $f$ takes no point of $D$ to the point $M$, the assertion is evident. That is because it would then follow from (a) that $f(D)$ is contained in $D$, and hence $f$ has a fixed point in $D$, for $D$ is homeomorphic to the compact interval $[0, 1]$. So assume that there is a $t \in [0, 1]$ such that $g(t) = M$. Define

$$t_m = \inf \Big\{ t \in [0, 1] : g(t) = M \Big\}.$$

Since $g$ is continuous and $g(0) = f(M) \ne M$, we see that $t_m > 0$ and moreover $g(t_m) = M$. Also, it follows from (a) that the image of $[0, t_m]$ under $g$ is contained in $D$. Use $\pi_2 : \mathbb{R}^2 \to \mathbb{R}$ to denote the projection on the $y$-axis. We can write

$$\pi_2 \Big( g(0) \Big) = 1 - t_0 < 1 - 0 t_0 = 1$$

and

$$\pi_2 \Big( g(t_m) \Big) = \pi_2(M) = 1 > 1 - t_m t_0.$$

Consequently, there is a $t_1 \in (0, t_m)$ such that $\pi_2 \Big( g(t_1) \Big) = 1 - t_1 t_0$. Since $g(t_1) \in D$, we obtain

$$f(t_1 t_0 q_0, 1 - t_1 t_0) = (t_1 t_0 q_0, 1 - t_1 t_0),$$

which means $f$ has a fixed point, as desired. ∎

# Part 3

# Problem Index

# First Competition

## Analysis.
*Problem 1.* (routine) Real analysis. Derivatives. Baire functions.
*Problem 2.* Matrix analysis.
*Problem 3.* Real analysis. Limits. Continuity.

## Algebra.
*Problem 1.* (routine) Ring theory. Commutative rings with identity. Prime ideals. Maximal ideals.
*Problem 2.* (routine) Group theory. Abelian groups.
*Problem 3.* Baby set theory. Isomorphic sets.

## General.
*Problem 1.* (routine) Triangles.
*Problem 2.* (routine) Decimal expansion.
*Problem 3.* (routine) Complex numbers.
*Problem 4.* (routine) Real numbers. Inequalities.

## Differential Equations.
*Problem 1.* (routine) Second order differential equations.

## Probability and Statistics.
*Problem 1.* (routine) Independent random variables.

## Topology.
*Problem 1.* (routine) Point-set topology of real numbers. (Ir)rational numbers.

# Second Competition

## Analysis.
*Problem 1.* Real analysis. Continuous functions. Limits of sequences.
*Problem 2.* (routine) Real analysis. Integral norm of continuous functions.
*Problem 3.* Point-set topology of real numbers. Cantor set.

## Algebra.
*Problem 1.* (routine) Finite fields.
*Problem 2.* (routine) Ring theory. (Right/Left) Quasi-regular elements.
*Problem 3.* (routine) Homomorphisms of rings. Polynomial rings in one variable.

## General.
*Problem 1.* (routine) Arithmetic. Congruences.
*Problem 2.* (routine) Real numbers. Inequalities.
*Problem 3.* Real analysis. Second order derivatives. Mechanics. Particle acceleration.
*Problem 4.* (routine) Counting. The product rule of combinatorics.

## Probability and Statistics.
*Problem 1.* (routine) Covariance of random variables.

## Topology.
*Problem 1.* (routine) Point-set topology. Connectedness.

## Differential Equations.
*Problem 1.* (routine) First order differential equations.

## Third Competition

### Analysis.
*Problem 1.* Real analysis. Derivatives.
*Problem 2.* (routine) Real analysis. Continuous functions shifting forward sequences of real numbers.
*Problem 3(a).* Roots of complex polynomials.
*Problem 3(b).* Continuity of the roots of complex polynomials.
*Problem 3(c).* Continuous algebraic functions. Holder functions.

### Algebra.
*Problem 1.* (routine) Modules. Dimension of vector spaces.
*Problem 2.* (routine) Group theory. Homomorphisms characterizing abelian groups.
*Problem 3.* (routine) Semigroups. Groups.

### General.
*Problem 1.* (routine) Real functions of two variables.
*Problem 2.* (routine) Binomial coefficients.
*Problem 3.* Plane geometry.

## Fourth Competition

### Analysis.
*Problem 1.* Real analysis. Continuity. Injectivity.
*Problem 2.* Sequences of nonnegative real numbers.

### Algebra.
*Problem 1.* (routine) Ring theory. Commutative rings with identity. Radicals of ideals. Prime ideals.
*Problem 2.* (routine) Linear algebra. Center of the algebra of all linear transformations.
*Problem 3.* Ring theory. Artinian integral domains are fields.

### General.
*Problem 1.* Integral calculus.
*Problem 2.* Conditional probability.

## Fifth Competition

### Analysis.
*Problem 1.* Real analysis. Composition of Riemann integrable functions.
*Problem 2.* Real analysis. Continuous functions not assuming any value more than twice.
*Problem 3.* Point-set topology of real numbers. Congestions points.

## Algebra.

*Problem 1.* (routine) Linear algebra. Eigenvalues of certain linear transformations.

*Problem 2.* Group theory. Groups having two elements satisfying certain relations.

*Problem 3.* Ring theory. Rings whose elements satisfy certain polynomial equations.

## General.

*Problem 1.* Number theory. Divisibility.

*Problem 2.* Roots of complex polynomials and those of their derivatives. Closed convex hulls.

*Problem 3.* Plane geometry. Conic sections. Optics.

## Sixth Competition

## Analysis.

*Problem 1.* (routine) Analytic geometry. Rational points.

*Problem 2.* (routine) Metric spaces. Uniform convergence of sequences of continuous functions.

*Problem 3.* Real analysis. Limits of improper integrals.

## Algebra.

*Problem 1.* Group theory. Finite groups. Subgroups.

*Problem 2.* Matrix theory.

*Problem 3.* Ring theory. Rings.

## General.

*Problem 1.* Integer numbers.

*Problem 2.* Integer matrices.

*Problem 3.* (routine) Plane geometry.

## Seventh Competition

## Analysis.

*Problem 1.* Point set topology of real numbers.

*Problem 2.* Real analysis. Continuity.

*Problem 3.* Limits of sequences of of real numbers.

## Algebra.

*Problem 1.* Ring theory. Nilpotent elements. Idempotent elements.

*Problem 2.* Group theory. Normal subgroups.

*Problem 3.* Linear algebra. Linear transformations.

## Eighth Competition

## Analysis.

*Problem 1.* (routine) Real analysis. Continuous extension.

*Problem 2.* Convergence of series of real numbers.

*Problem 3.* (routine) Riemann-Stieltjes integral. Convergence of sequences of Riemann-Stieltjes integrals.

## Algebra.
*Problem 1.* (routine) Real matrices. The general linear group of order 2.
*Problem 2.* Ring Theory. Infinite integral domains.
*Problem 3.* (routine) Real matrices.

## General.
*Problem 1.* (routine) Minimum without using derivative.
*Problem 2.* (routine) Arithmetic. Divisibility.
*Problem 3.* Probability. Average function value.

# Ninth Competition

## Analysis.
*Problem 1.* (routine) Point-set topology of real numbers. Equivalence relation.
*Problem 2.* Real analysis. Continuity of inverse functions.
*Problem 3.* (routine) Real analysis. Increasing continuous function.
*Problem 4.* Real analysis. Riemann integrable function.
*Problem 5.* Real analysis. Continuity. Injectivity. Surjectivity. Square filling curves.

## Algebra.
*Problem 1.* Group theory. Finite groups. Subgroups of prime indexes. Normal subgroups.
*Problem 2.* Field of integers mod 3, $\mathbb{Z}_3$. Polynomial ring with coefficients in $\mathbb{Z}_3$. Finite fields.
*Problem 3.* (routine) Polynomial ring with coefficients in $\mathbb{Z}_7$. Greatest common divisor.
*Problem 4.* Field theory. Quadratic extensions.
*Problem 5.* Linear algebra. Linear transformations.

## General.
*Problem 1.* Determinants.
*Problem 2.* (routine) Algebraic equations.
*Problem 3.* Real analysis. Continuous functions. Integrals.
*Problem 4.* Diophantine equations. Number of solutions.
*Problem 5.* Puzzles.
*Problem 6.* Probability.
*Problem 7.* Determinants.

# Tenth Competition

## Analysis.
*Problem 1.* Real analysis. Derivatives. Limits.
*Problem 2.* Real analysis. Higher derivatives. Definite integrals.
*Problem 3.* (routine) Real analysis. Limits of definite integrals. Convergence of function sequences.

## Algebra.
*Problem 1.* Group theory. Additive group of real numbers. Maximal subgroups.
*Problem 2.* (routine) Ring theory. Ideals.

*Problem 3.* Field theory. Field extensions.
*Problem 4.* (routine) Matrix theory over general fields. Nilpotent matrices.

## General.
*Problem 1.* Probability. Random variables.
*Problem 2.* Point-set topology of the plane.
*Problem 3.* (routine) Differential and integral calculus. Differential equations. Integral equations.
*Problem 4.* (routine) Arithmetic. Divisibility.

## Eleventh Competition

## Analysis.
*Problem 1.* Real analysis. Derivatives.
*Problem 2.* Real analysis. Convergence of function sequences. Functional equations. Uniform convergence.
*Problem 3.* Real analysis. Higher order derivatives. Convergence of series.

## Algebra.
*Problem 1.* Ring theory. Ideals. (Right) identity elements.
*Problem 2.* Field theory. Algebraic numbers. Finite field extensions.
*Problem 3.* Group theory. Finite groups.

## General.
*Problem 1.* Probability. Determinants.
*Problem 2.* Real numbers. Algebraic equations.
*Problem 3.* Baby set theory. Group theory.

## Twelfth Competition

## Analysis.
*Problem 1.* Real analysis. Continuity. Injectivity. Surjectivity.
*Problem 2.* Real analysis. Uniform convergence of series.
*Problem 3.* (routine) Integral calculus. Definite integrals.

## Algebra.
*Problem 1.* Group theory. Finite groups. Sylow theory.
*Problem 2.* Group theory. Finite abelian groups. Homomorphisms.
*Problem 3.* Ring theory.
*Problem 4.* Ring theory. Principal ideals.
*Problem 5.* Matrix theory over general fields. Determinant. Trace.
*Problem 6.* Algebraic numbers.

## General.
*Problem 1.* Plane Geometry.
*Problem 2.* Combinatorics. Pigeonhole principal.
*Problem 3.* Binomial coefficients.

## Thirteenth Competition

## Analysis.
*Problem 1.* Real analysis. Riemann integrable functions. Additive functions.

*Problem 2.* Convergence of series of real numbers.

*Problem 3.* Real analysis. Convergence of function sequences. Definite integrals.

## Algebra.
*Problem 1.* Ring theory. Integral domains.
*Problem 2.* Ring theory. Right (Left) ideals.
*Problem 3.* Group theory.
*Problem 4.* Group theory. General linear groups. Permutation groups.
*Problem 5.* (routine) Matrix theory. Eigenvectors. Eigenvalues.

## General.
*Problem 1.* (routine) Plane geometry.
*Problem 2.* Rational polynomials.
*Problem 3.* Space geometry.

## Fourteenth Competition

## Analysis.
*Problem 1.* Real analysis. Differentiability.
*Problem 2.* Real analysis. Uniform continuity. Improper integrals. Limits.
*Problem 3.* Real analysis. Continuous functions.

## Algebra.
*Problem 1.* Group theory. Finite $p$-groups. Derived subgroups.
*Problem 2.* Group theory. Finite groups. Normal subgroups. Center of groups.
*Problem 3.* Ring theory. Greatest commons divisors. Least common multiples.
*Problem 4.* (routine) Ring theory. Polynomial rings of two variables. Irreducibility.
*Problem 5.* Finite-dimensional vector spaces. Direct sums.

## General.
*Problem 1.* Combinatorics.
*Problem 2.* Combinatorics. Counting.
*Problem 3.* Real numbers.

## Fifteenth Competition

## Analysis.
*Problem 1.* Real analysis. Convexity.
*Problem 2.* (routine) Real analysis. Continuous functions. Uniform convergence of function sequences.
*Problem 3.* Real analysis. Real functions.

## Algebra.
*Problem 1.* Group theory. Finite groups. Sylow $p$-subgroups.
*Problem 2.* Ring theory. (Right/Left) ideals. Prime ideals. Nilpotent elements.

*Problem 3.* (routine, as stated, but its nontrivial counterpart is nonroutine!) Matrix theory.

**General.**
*Problem 1.* Elementary number theory.
*Problem 2.* Integral calculus.
*Problem 3.* Combinatorics. Probability. Binomial coefficients.

## Sixteenth Competition

**Analysis.**
*Problem 1.* (routine) Real analysis. Continuous functions. Uniform convergence.
*Problem 2.* Real analysis. Periodic functions. Riemann-Stieltjes integrals.
*Problem 3.* Real analysis. Uniform continuity.

**Algebra.**
*Problem 1.* Group theory. Finite nonabelian groups. Inner automorphisms.
*Problem 2.* Ring theory. Nilpotent elements. Idempotent elements.
*Problem 3.* Matrix theory. Elementary number theory.

## Seventeenth Competition

**Analysis.**
*Problem 1.* Real analysis. Derivatives.
*Problem 2.* (routine) Differential and integral calculus. Real polynomials. Divisibility.
*Problem 3.* Real continuous functions on metric spaces.
*Problem 4.* Real analysis. Continuous functions.

**Algebra.**
*Problem 1.* Group theory. Finite groups.
*Problem 2.* Ring theory. Polynomial rings in two variables with coefficients from matrix rings over fields.
*Problem 3.* Linear algebra. Linear transformations. Rank. Nullity.

## Eighteenth Competition

**Analysis.**
*Problem 1.* Real analysis. Power series.
*Problem 2.* Real analysis. Periodic functions. Limits of definite integrals.
*Problem 3.* Real analysis. Open maps.

**Algebra.**
*Problem 1.* Group theory. Finite groups. Normal subgroups.
*Problem 2.* Ring theory. Ascending chains of right ideals.
*Problem 3.* Vector spaces. Direct sums.

## Nineteenth Competition

**Analysis.**
*Problem 1.* (routine) Real analysis. Differential and integral calculus. Zeros of continuous functions.
*Problem 2.* Real analysis. Limits.
*Problem 3.* Real analysis. Continuous nowhere differentiable functions.

**Algebra.**
*Problem 1.* Group theory. Normal subgroups. Posets.
*Problem 2.* Ring theory. Minimal left ideals. Ideals.
*Problem 3.* (routine) Matrix theory. Rank.
*Problem 4.* Matrix theory. Rank. Determinant. Trace.

## Twentieth Competition

*Problem 1.* Vector spaces. Complex numbers. Linear independence.
*Problem 2.* Real analysis. Sequences of real numbers. Limits.
*Problem 3.* Group theory. Automorphism groups. Solvability.
*Problem 4.* Real analysis. Functional equations. Uniform continuity.
*Problem 5.* Ring theory. Commutative rings with identity. Zero divisors.
*Problem 6.* Real analysis. Periodic functions. Convergence of improper integrals.

## Twenty-First Competition

**Analysis.**
*Problem 1.* Real analysis. Point-set topology. Limits.
*Problem 2.* Real analysis. Continuous functions. Limits. Integrals.
*Problem 3.* Real analysis. One-sided limits. Riemann integrable functions.

**Algebra.**
*Problem 1.* Ring theory. Small ideals.
*Problem 2.* Group theory. Center of groups. Derived subgroups.
*Problem 3.* Matrix theory. Trace. Symmetric matrices. Characteristic polynomials.

## Twenty-Second Competition

**Analysis.**
*Problem 1.* Real analysis. Derivatives.
*Problem 2.* Real continuous functions on metric spaces. Series. Uniform convergence.
*Problem 3.* Complex analysis. Sequences of analytic functions. Integrals.

**Algebra.**
*Problem 1.* Group theory. Group index.
*Problem 2.* Ring theory. Commutative rings. Maximal ideals.
*Problem 3.* Linear algebra. Characteristic polynomials. Rank.

**General.**
*This section has forty routine multiple choice problems.*

## Twenty-Third Competition

**Analysis.**

*Problem 1.* Real analysis. Continuous bounded functions. Uniform norm. Compactness. Uniform continuity.

*Problem 2.* Real analysis. Trigonometric series.

*Problem 3.* Real analysis. Convergence of function sequences. Differentiability.

**Algebra.**

*Problem 1.* Group theory. Inner automorphism groups. Nonabelian groups.

*Problem 2.* Ring theory. (Right) Ideals. Division rings.

*Problem 3.* (routine) Matrix theory. Zero trace.

## Twenty-Fourth Competition

**First Day.**

*Problem 1.* Real analysis. Derivatives.

*Problem 2.* Real analysis. Real valued harmonic functions. Complex analysis.

*Problem 3.* Number theory. Primes. Divisibility. Congruences.

*Problem 4.* (routine) Group theory. Infinite abelian groups.

*Problem 5.* Combinatorics. Counting.

*Problem 6.* Probability. Conditional probability.

**Second Day.**

*Problem 1.* Euclidean spaces. Metric spaces. Open balls.

*Problem 2.* (routine) Real analysis. Differential and integral calculus. Continuous functions.

*Problem 3.* Matrix theory. Rank inequalities.

*Problem 4.* (routine) Ring theory. (Left) ideals.

*Problem 5.* Combinatorics.

*Problem 6.* Combinatorics. Game theory.

## Twenty-Fifth Competition

**First Day.**

*Problem 1.* Group theory. Finite groups. Group centers.

*Problem 2.* Linear algebra. Idempotents. Image. Kernel.

*Problem 3.* Real analysis. Derivatives. Higher order derivatives.

*Problem 4.* Real analysis. Derivatives.

*Problem 5.* Probability.

*Problem 6.* Combinatorics. Triangles with integer sides. Partitions of numbers.

**Second Day.**

*Problem 1.* Ring theory. Integral domains. Multiplicative group of units.

*Problem 2.* Number theory. Prime numbers. Divisibility. Congruences.

*Problem 3.* Continuous functions on metric spaces.

*Problem 4.* Functions of one complex variable.

*Problem 5.* Combinatorics. Matrices with $\{-1, 1\}$ entries.

*Problem 6.* Finite sets. Relations.

## Twenty-Sixth Competition

**First Day.**
*Problem 1.* Real analysis. Continuous functions. Derivatives.
*Problem 2.* Complex analysis. Analytic functions.
*Problem 3.* (routine) Number theory. Primes. Congruences.
*Problem 4.* Ring theory. Commutative rings with identity. Maximal ideals.
*Problem 5.* Combinatorics.
*Problem 6.* Combinatorics. Finite sets.

**Second Day.**
*Problem 1.* Real analysis. Continuous functions.
*Problem 2.* Real analysis. Continuous functions.
*Problem 3.* Group theory. Maximal subgroups.
*Problem 4.* Matrix theory. Eigenvalues. Eigenvectors.
*Problem 5.* Combinatorics. Probability. Counting.
*Problem 6.* Discrete mathematics.

## Twenty-Seventh Competition

**First Day.**
*Problem 1.* Matrix analysis.
*Problem 2.* Complex analysis. Analytic functions.
*Problem 3.* Sequences of real numbers. Recurrence relations.
*Problem 4.* Set theory.
*Problem 5.* Matrices. Product of matrices.
*Problem 6.* Group theory. Normal subgroups.

**Second Day.**
*Problem 1.* Real analysis. Derivatives.
*Problem 2.* Point set topology of real numbers.
*Problem 3.* Analytic geometry.
*Problem 4.* Combinatorics. Finite sets.
*Problem 5.* Number theory. Congruences. Divisibility.
*Problem 6.* Ring theory. Boolean rings. Isomorphisms.

## Twenty-Eighth Competition

**First Day.**
*Problem 1.* Complex analysis.
*Problem 2.* Euclidean spaces. Continuous functions.
*Problem 3.* Number theory. Arithmetic functions. Euler's totient function.
*Problem 4.* Group theory. Symmetric groups. Group center.
*Problem 5.* Vector spaces over $\mathbb{Z}_2$.
*Problem 6.* Posets.

**Second Day.**
*Problem 1.* (routine) Integral calculus.
*Problem 2.* Real analysis. Continuous functions.

*Problem 3.* Complex matrices. Orthogonal matrices. Determinants.
*Problem 4.* Ring theory. Unital rings whose elements satisfy certain relations.
*Problem 5.* Probability. Conditional probability.
*Problem 6.* Combinatorics. Set theory.

## Twenty-Ninth Competition

**First Day.**
*Problem 1.* (routine) Integral calculus.
*Problem 2.* Rational numbers. Series of rational numbers.
*Problem 3.* (routine) Number theory. Divisibility.
*Problem 4.* Ring theory. Nilpotent ideals. Right ideals.
*Problem 5.* (routine) Baby set theory.
*Problem 6.* Combinatorics. Finite-dimensional vector spaces.

**Second Day.**
*Problem 1.* Complex analysis. Analytic functions.
*Problem 2.* Connected metric spaces. Compact metric spaces.
*Problem 3.* Group theory. Normalizer subgroup. Centralizer subgroup. Automorphism group. Cyclic groups. Derived subgroups. Group center.
*Problem 4.* Matrix theory. Triangular matrices. Invariant subspaces. Nilpotent matrices.
*Problem 5.* Probability. Expectation values.
*Problem 6.* Point-set topology of real numbers. Cantor set.

## Thirtieth Competition

**First Day.**
*Problem 1.* (routine) Integral calculus.
*Problem 2.* Limits of sequences of complex numbers.
*Problem 3.* Ring theory.
*Problem 4.* Number theory. Primes. Congruences. Divisibility.
*Problem 5.* Binomial coefficients. Inequalities.
*Problem 6.* Natural numbers.

**Second Day.**
*Problem 1.* Real functions on separable metric spaces. Points of discontinuity.
*Problem 2.* Complex analysis. Analytic functions. Power series.
*Problem 3.* Group theory. Derived subgroups. Elements of finite order.
*Problem 4.* Matrix theory. Rank. Image. Kernel. Direct sum. Idempotents.
*Problem 5.* Number theory. Primes.
*Problem 6.* Point-set topology. Circles. Arc lengths.

## Thirty-First Competition

**First Day.**
*Problem 1.* Plane geometry. Triangles. Coloring circles by two colors.
*Problem 2.* Point-set topology of $\mathbb{R}^n$. Minkowski sums. Limit points.
*Problem 3.* Group theory. Subgroups of index two. Finite abelian groups.
*Problem 4.* Probability. Dice throwing.

*Problem 5.* Point-set topology of $\mathbb{R}^2$. Collinearity.

*Problem 6.* Matrix theory. Determinant. Trace.

## Second Day.

*Problem 1.* Number theory. Complete squares.

*Problem 2.* Calculus of two real variables. Limits.

*Problem 3.* Set theory. One-to-one correspondence.

*Problem 4.* Ring theory. Polynomial rings over commutative rings with identity. Maximal ideals.

*Problem 5.* Real numbers. Partitions of sets with equal sums.

*Problem 6.* Metric spaces. Fixed point theory.

APPENDIX A

# Historical Introduction And The Winners List

## A.1. Iranian University Students Mathematics Competitions – a historical introduction

The history of the Iranian Mathematics Competitions for university students[1] is an integral part of the contemporary history of mathematics in Iran. On March 31st, 1972, the first general assembly of the Iranian Mathematical Society (IMS) was held at the former National University (now Shahid Beheshti University). Upon a motion proposed by Mehdi Behzad, then president of the society –which at that time was called the Society's secretary– the need for a mathematics competition aimed at university students was approved. Herewith a committee, consisting of Houshang Attarchi, Mehdi Behzad, Fereidoun Ghahramani, Mohammad-Ali Gheyni, and Bahman Vahidi, was first appointed to prepare a guideline for the competitions and to make preparations for holding these competitions, which were intended for university students. This committee developed a guideline and required that the competitions must be held annually and simultaneously with the Annual Iranian Mathematics Conference (AIMC). The primary goals of the competitions were to discover, encourage, and nurture mathematical talents throughout the country and to create a friendly and scientific rivalry among students and the Iranian universities. In those years, the mathematics courses in most of the universities across the country were heavily influenced by the traditional and old curriculum as opposed to the modern one. Only a handful of universities, such as the University of Tehran, Sharif (then Aryamehr) University of Technology, (former Pahlavi) University of Shiraz, and Mossaheb Institute of Mathematics offered courses such as Set Theory, Modern Algebra, Modern Analysis, and Combinatorics. The secondary goal of the competitions was to motivate the students to view mathematics in new ways to maintain and sustain a more diverse perspective of mathematics, thereby putting pressure on universities to change their educational systems.

The following year, on March 30th, 1973, simultaneous with the Fourth AIMC held at the University of Tehran, the First National Mathematics Competition was held with the participation of 22 students from five universities across the country. Half of the questions were in the areas of Mathematical Analysis and Algebra and the rest were in other areas categorized as General,

---

[1]This note, together with a list of the winners which follows this note, is largely a translation of a historical introduction written by Rashid Zaare-Nahandi for the Persian version of this book.

i.e., innovative questions, Probability and Statistics, Differential Equations, and Topology. The second competition took place in March 1974 at the (former Pahlavi) University of Shiraz with the participation of 51 students from eleven universities. The students' participation grew steadily in the following years so that in the thirtieth competition more than 180 students from 40 universities took part.

Ever since 1973, the competitions were held annually except for 1979 with closure of universities because of the revolution and 1981-1983 when universities were closed due to the cultural revolution. Students' and universities' reception of the competitions caused a great increase in the number of participating students. Yet, when the competitions were held on the same date and location as the AIMC's, the organizational matters of the event, in particular grading the student's papers, had become so overwhelming that it would take a number of months to announce the results. This would sometimes lead to discontent and objections, which occasionally resulted in changes to the final scores. However, it was only due to the tireless and diligent efforts of the questions committees in those years that the competitions survived and were kept alive despite all the difficulties and shortcomings. In view of all this, the IMS was persuaded to hold the competitions on a date and in a location different from those of the AIMC's. Consequently, from 1996-2003 the competitions took place over a three-day period. Since 2004 they have been taking place over a four-day period at one of the Iranian universities during the Spring of every year.

In June 1999, the IMS executive committee and board of trustees gathered at Tafresh University to reconsider the bylaws and regulations of the IMS. At that time, a committee was established to substantially change the regulations concerning the National Mathematics Competitions. From 2000, the competitions were held according to the new regulations. It must, however, be said that minor changes and improvements were made at some stages due to the experiences obtained from previous competitions.

According to the current guidelines of the competitions, each university or higher educational institution can send only one team consisting of at most five students and a leader, possibly along with a deputy leader, to the competitions. Each university is responsible for selecting its own team, presumably through holding preliminary tests among its students. The contests take place in two sessions, with each session lasting three and a half hours. The students answer twelve questions of which four are in the area of analysis, four are in the area of algebra, and another four are in other areas such as discrete mathematics, probability, number theory, etc, categorized as innovative questions. Attempts are made to propose problems whose solutions require not only mathematical knowledge but innovation.

The questions of the competitions are now designed and selected by a questions committee consisting of a chairperson, a supervisor, and three other people as the heads of the following sections: analysis, algebra, and innovative questions. The heads of the three sections can each choose a vice head to assist them. The questions committee is appointed by the Executive Committee of the IMS for a period of two to three years. The committee is responsible for designing questions and holding the competitions. It is also responsible for appointing a grading team, whose members are mostly chosen from the winners of the previous competitions. In the course of a number of months, the committee selects 24 questions which are divided into two sets of 12. The contests are held in two sessions. For each sitting, 6 questions are selected out of the 12 questions by a jury consisting of the leaders of the participating teams. The participants are ranked based on their individual scores. The top 5 individual scorers win gold medals, the next 10 top contestants are awarded silver medals, and up to the next top 15 receive bronze medals. It is worth mentioning that since April 2004, i.e., from the 28th competition on, the problems together with their proposed solutions have been posted online, on the IMS website currently only in Persian. This is done for two reasons. First, the universities can simultaneously hold the competitions for their interested students. Second, this way the problems and their proposed solutions are preserved on the Internet permanently.

Now after thirty-one competitions, it seems that most of the goals of the competitions have been achieved. A glance at the list of the winners of competitions reveals that many of them have gone on to become distinguished mathematicians and are now working at prominent universities and institutions throughout the world. Also, a glance at the questions of these thirty-one competitions shows that the competitions and the Iranian mathematical community, on the whole, have evolved to a greater level of sophistication in terms of content, quantity and quality.

In 1995, the executive committee of the IMS approved the preparation of a book containing the problems and solutions to the problems of the first twenty competitions. It took the IMS several months to collect the problems. About one third of the proposed solutions of all problems were available in the IMS files. Rashid Zaare-Nahandi informed Bamdad Yahaghi about the decision made by the IMS and asked whether Yahaghi was interested in preparing such book. Yahaghi showed interest and readiness and it took him a year of hard work to prepare a preliminary copy of the manuscript. Under the recommendation of the IMS, the analysis and the algebra part of the manuscript was then refereed by Karim Seddighi and Mohammad Reza Darafsheh, respectively. This was simultaneous with Yahaghi's departure for Canada to continue his PhD studies at Dalhousie University. Unfortunately, the original plan by the IMS to publish this book within a year could not be realized.

In 2005, shortly after the author's return to Iran, his interest in the project on which he had spent such considerable time and energy was rekindled. He began to revise, update, and rewrite the book in both Persian and English. It must be said that the IMS did not have the proposed solutions to the problems of Competitions 21, 23, 24, and 27. It took Yahaghi almost two years to prepare both the English and Persian versions of the book. We must point out that this book could not have come into existence without the efforts of the colleagues who have contributed to these competitions. This book perhaps sheds light on the time and effort that have gone into organizing these competitions. It is hoped that this book succeeds in introducing these competitions to an international audience.

Following this introduction, the names of the members of the questions committees[2] and the date, location, and the winners of the competitions, together with their educational and employment affiliations, are quoted.

---

[2] Up until 1998, the executive committee of the IMS used to appoint only one person as the chairperson of the competition. This person was charged with the task of inviting other colleagues to submit questions and to mark the students' papers. As outlined before, as of 1999, the questions committee is appointed by the Executive Committee of the IMS. And it is this committee that is responsible for designing questions and holding the competitions.

## A.2. Winners of the Competitions

### The First Competition, University of Tehran, March 1973
*Questions Committee: Mehdi Behzad.*

1. *Elizabeth Ebrahimzadeh*, Sharif (former Aryamehr) University of Technology.
   PhD (University of California, Berkeley, 1984),
   Professor (California State University, Sacramento, USA).
2. *Mohammad Reza Darafsheh*, University of Tehran.
   PhD (University of Birmingham, UK, 1978),
   Professor (University of Tehran, Iran).
3. *Hashem Madadi Almousavi*, Ferdowsi University of Mashhad.
   Deceased.
4. *Mirebrahim Hashemi Aghdam*, Sharif (former Aryamehr) University of Technology.
   –
5. *Yousef Bahrampour*, (former Pahlavi) University of Shiraz.
   PhD (University of Oregon, USA, 1983),
   Professor (University of Kerman, Iran).

### The Second Competition, Shiraz (former Pahlavi) University, March 1974
*Questions Committee: Mehdi Behzad.*

1. *Mohammad Ali Najafi*, Sharif (former Aryamehr) University of Technology.
   Dropped out PhD candidate (MIT, USA, 1978),
   Instructor (Sharif University of Technology, Iran).
2. *Firooz Khosraviyani*, University of Tehran.
   PhD (University of Wales, Aberystwyth, UK, 1981),
   Associate Professor (Texas A&M International University, USA).
3. *Mahdi Zekavat*, (former Pahlavi) University of Shiraz.
   PhD (University of Saskatchewan, Canada, 2000),
   Assistant Professor (Shiraz University, Iran).
4. *Hamid Hamed Akbari Tousi*, University of Tehran.
   –
5. *Shahram Arshad Riyazi*, former National University.
   –

### The Third Competition, (former Jondi Shapour) University of Ahwaz, March 1975
*Questions Committee: Mohammad Ali Gheyni.*

1. *Pirooz Vakili*, Sharif (former Aryamehr) University of Technology.
   PhD (Harvard University, USA, 1988),
   Associate Professor (Boston University, USA).
2. *Moslem Nikfar*, University of Tehran.
   PhD (Shiraz University, Iran, 1995),
   Assistant Professor (Isfahan University of Technology, Iran).

3. *Ali Asghar Babadi Margha*, University of Tehran.
   —

4. *Saeed Ghahramani*, Sharif (former Aryamehr) University of Technology.
   PhD (University of California, Berkeley, USA, 1983),
   Professor (Western New England College, USA).
5. *Homayoun Moin*, Sharif (former Aryamehr) University of Technology.
   MSc (University of Wisconsin, Madison, 1979),
   Editor (Iran University Press Center, Tehran, Iran).

## The Fourth Competition, University of Tabriz (former Azarabadegan), March 1976
*Questions Committee: Vahab Davarpanah.*

1. *Pirooz Vakili*, Sharif (former Aryamehr) University of Technology.
   See the Third Competition results.
2. *Homayoun Moin*, Sharif (former Aryamehr) University of Technology.
   See the Third Competition results.
3. *Naser Hosseini*, (former Pahlavi) University of Shiraz.
   PhD (University of Miami, USA, 1986),
   Associate Professor (University of Kerman, Iran).
4. *Shahla Marvizi Ahdout*, Sharif (former Aryamehr) University of Technology.
   PhD (Massachusetts Institute of Technology, USA, 1981),
   Associate Professor (Long Island University, USA).
5. *Ali Karimi*, Tarbiat Moallem (Teacher Training) University.
   —

## The Fifth Competition, Sharif (former Aryamehr) University of Technology, March 1977
*Questions Committee: Mohammad Reza Nouri-Moghadam.*

1. *Hamid Kazemi*, Sharif (former Aryamehr) University of Technology.
   Deceased.
2. *Mohsen Maesumi Fakhar*, Sharif (former Aryamehr) University of Technology.
   PhD (New York University, USA, 1990),
   Associate Professor (Lamar University, USA).
3. *Masoud Khalkhali*, Sharif (former Aryamehr) University of Technology.
   PhD (Dalhousie University, Canada, 1991),
   Professor (University of Western Ontario, Canada).
4. *Ebrahim Saátchi*, (former Azarabadegan) University of Tabriz.
   —

5. *Safa Nourbakhash*, former National University.
   —

## The Sixth Competition, The University of Isfahan, March 1978
*Questions Committee: Magerdich Toumanian.*

1. *Nasser Boroojerdian*, University of Tehran.

PhD (University of Tehran, 1994),
Assistant Professor (Amirkabir University of Technology, Iran).
2. *Hamid Kazemi*, Sharif (former Aryamehr) University of Technology.
   Deceased.
3. *Aliasghar Alikhani-Koopaei*, University of Isfahan.
   PhD (University of California, Santa Barbara, USA, 1985),
   Associate Professor (Penn State Berks, USA).
4. *Mehdi Salehi Nejad*, Ferdowsi University of Mashhad.
   –

5. *Ali Rejali*, University of Isfahan.
   PhD (University of Sheffield, UK, 1989),
   Professor (University of Isfahan, Iran).

## The Seventh Competition, Ferdowsi University of Mashhad, March 1980
*Questions Committee: Akbar Hassani.*

1. *Fraydoun Rezakhanlou*, University of Tehran.
   PhD (New York University, USA, 1989),
   Professor (University of California, Berkeley, USA).
2. *Mehdi Alavi Shoushtari*, University of Ahwaz.
   Deceased.
3. *Seyed Esmail Seyedabadi*, Sharif University of Technology.
   –

4. *Rajabali Kamyabi-Gol*, Ferdowsi University of Mashahd.
   PhD (University of Alberta, Canada, 1997),
   Associate Professor (Ferdowsi University of Mashhad, Iran).
5. *Ali Asghar Jodayree Akbarfam* , University of Tabriz.
   PhD (University of Ottawa, Canada, 1989),
   Associate Professor (University of Tabriz, Iran).

## The Eighth Competition, Shiraz University, March 1984
*Questions Committee: Asadollah Niknam.*

1. *Mohammad Hassan Jahanbakht*, University of Isfahan.
   MSc (Sharif University of Technology, Iran).
   –

2. *Mojtaba Moniri*, University of Tehran.
   PhD (University of Minnesota-Minneapolis, USA, 1994),
   Associate Professor (Tarbiat Modares University, Iran).
3. *Mohammad Taghi Jahandideh*, Shiraz University.
   PhD (Dalhousie University, Canada, 1997),
   Assistant Professor (Isfahan University of Technology, Iran).
4. *Samad Ahmadi*, Tarbiat Moallem (Teacher Training) University.
   PhD (University of Kent at Canterbury, UK, 1998),
   Senior Lecturer (De Montfort University, UK).
5. *Kamal Aghigh*, University of Tabriz.

PhD (Panjab University, Chandigarh, India, 2002),
Assistant Professor (Khajeh Nasir Toosi University of Technology, Iran).

## The Ninth Competition, Tehran Teacher Training (Tarbiat Moallem) University, September 1985

*Questions Committee: Rahim Zaare-Nahandi.*

1. *Nasser Boroojerdian,* University of Tehran.
   See the Sixth Competition results.
2. *Mojtaba Moniri,* University of Tehran.
   See the Eighth Competition results.
3. *Majid Ashrafi,* Shiraz University.
   −

4. *Jamal Rooin,* Tarbiat Moallem (Teacher Training) University.
   PhD (Tarbiat Moallem (Teacher Training) University, Iran, 2001),
   Assistant Professor (IASBS of Zanjan, Iran).
5. *Ali Parsian,* University of Tehran.
   PhD (University of Tehran, Iran, 1997),
   Assistant Professor (Tafresh University, Iran).

## The Tenth Competition, University of Sistan and Baluchestan, March 1986

*Questions Committee: Karim Seddighi.*

1. *Reza Jahani-Nezhad,* University of Kashan.
   PhD (Sharif University of Technology, Iran, 1996),
   Assistant Professor (University of Kashan, Iran).
2. *Amir Akbary Majdabadno,* University of Tehran.
   PhD (University of Toronto, Canada, 1997),
   Associate Professor (University of Lethbridge, Canada).
3. *Shaahin Ajoodani Namini,* University of Tehran.
   PhD (California Institute of Technology, USA, 1998).
   −

4. *Hamid Reza Farhadi,* Tarbiat Moallem (Teacher Training) University.
   PhD (University of Manitoba, Canada, 1996),
   Assistant Professor (Sharif University of Technology, Iran).
5. *Masoud Amini,* Ferdowsi University of Mashhad.
   PhD (University of Illinois at Urbana-Champaign, USA, 1998),
   Associate Professor (Tarbiat Modares University, Iran).

## The Eleventh Competition, The University of Birjand, March 1987

*Questions Committee: Karim Seddighi.*

1. *Vahid Tarokh,* Sharif University of Technology.
   PhD (University of Waterloo, Canada, 1995),
   Professor (Harvard University, USA).
2. *Masoud Amini,* Ferdowsi University of Mashhad.
   See the Tenth Competition results.
3. *Shaahin Ajoodani Namini,* University of Tehran.

See the Tenth Competition results.
4. *Reza Karami*, Isfahan University of Technology.

–

## The Twelfth Competition, Guilan University, March 1988
*Questions Committee: Mohammad Ali Shahabi.*

1. *Shaahin Ajoodani Namini*, University of Tehran.
   See the Tenth Competition results.
2. *Shaahin Amiri Sharifi*, Sharif University of Technology.

–

3. *Gholam Hossein Esslamzadeh*, Shiraz University.
   PhD (University of Alberta, Canada, 1997),
   Associate Professor (Shiraz University, Iran).
4. *Bamdad R. Yahaghi*, Sharif University of Technology.
   PhD (Dalhousie University, 2002),
   Assistant Professor (Golestan University, Iran).
5. *Ali Iranmanesh*, Shiraz University.
   PhD (Tarbiat Modares University, Iran, 1995),
   Professor (Tarbiat Modares University, Iran).

## The Thirteenth Competition, University of Tehran, March 1989
*Questions Committee: Mohammad Ali Shahabi.*

1. *Kambiz Mahmoodian*, University of Tehran.
   PhD (University of Toronto, 1999),
   Assistant Professor (Sharif University of Technology).
2. *Saeed Zakeri*, University of Tehran.
   PhD (State University of New York at Stony Brook, USA, 1999),
   Assistant Professor (Queens College and Graduate Center of CUNY, USA).
3. *Mohammad Sal Moslehian*, Ferdowsi University of Mashhad.
   PhD (Ferdowsi University of Mashhad, Iran, 1999),
   Professor (Ferdowsi University of Mashhad, Iran).
4. *Bamdad R. Yahaghi*, Sharif University of Technology.
   See the Twelfth Competition results.
5. *Behrooz Mashayekhifard*, Ferdowsi University of Mashhad.
   PhD (Ferdowsi University of Mashhad, Iran, 1996),
   Associate Professor (Ferdowsi University of Mashhad, Iran).

## The Fourteenth Competition, The University of Isfahan, March 1990
*Questions Committee: Heydar Zahed Zahedani.*

1. *Shaahin Amiri Sharifi*, Sharif University of Technology.

–

2. *Hessam Hamidi Tehrani*, Sharif University of Technology.
   PhD (Columbia University, USA, 1997).

–

3. *Saeed Zakeri*, University of Tehran.
   See the Thirteenth Competition results.

4. *Hamid Mousavi*, Tarbiat Moallem (Teacher Training) University.
   PhD (Tarbiat Moallem (Teacher Training) University, Iran, 1999),
   Assistant Professor (University of Tabriz, Iran).
5. *Shahab Shahabi*, University of Tehran.
   PhD candidate (McGill University, Canada).

## The Fifteenth Competition, Ferdowsi University of Mashhad, March 1991
*Questions Committee: Mohammad Reza Darafsheh.*

1. *Hessam Hamidi Tehrani*, Sharif University of Technology.
   See the Fourteenth Competition results.
2. *Ali Rajai*, Sharif University of Technology.
   PhD (Princeton University, USA, 1998),
   –

3. *Shahriar Mokhtari Sharghi*, Sharif University of Technology.
   PhD (Columbia University, USA, 1998).
   –

4. *Pedram Safari*, Sharif University of Technology.
   PhD (Columbia University, USA, 2000),
   Lecturer (Harvard University, USA).
5. *Ataollah Togha*, University of Kerman.
   PhD (The George Washington University, USA, 2004),
   Assistant Professor (Bronx Community College of CUNY, USA).

## The Sixteenth Competition, Razi (Rhazes or Rasis) University of Kermanshah, March 1992
*Questions Committee: Mohammad Reza Darafsheh.*

1. *Shahriar Mokhtari Sharghi*, Sharif University of Technology.
   See the Fifteenth Competition results.
1. *Ataollah Togha*, University of Kerman.
   See the Fifteenth Competition results.
3. *Ali Sabetian*, Shiraz University.
   –

3. *Pedram Safari*, Sharif University of Technology.
   See the Fifteenth Competition results.
5. *Mehdi Nadjafikhah*, Iran University of Science and Technology.
   PhD (Iran University of Science and Technology, Iran, 1998),
   Assistant Professor (Iran University of Science and Technology, Iran).
5. *Ali Rajai*, Sharif University of Technology.
   See the Fifteenth Competition results.

## The Seventeenth Competition, Shahid Beheshti (former National) University, March 1993
*Questions Committee: Mohammad Reza Darafsheh.*

1. *Hossein Hajiabolhassan*, Sharif University of Technology.

PhD (Sharif University of Technology, Iran, 1999),
Associate Professor (Shahid Beheshti (former National) University, Iran).
2. *Payman L. Kassaei*, Sharif University of Technology.
PhD (MIT, USA, 1999),
Lecturer (King's College London, UK).
3. *Behrang Noohi*, Sharif University of Technology.
PhD (MIT, USA, 2000).

–

4. *Arash Rastegar*, Sharif University of Technology.
PhD (Princeton University, USA, 1998),
Assistant Professor (Sharif University of Technology, Iran).
5. *Ali Dadban*, University of Tehran.

–

## The Eighteenth Competition, Sharif University of Technology, March 1994
*Questions Committee: Jafar Zafarani.*

1. *Ramin Takloo-Bighash*, Sharif University of Technology.
PhD (The Johns Hopkins University, USA, 2001),
Assistant Professor (University of Illinois at Chicago, USA).
2. *Kasra Rafi*, Sharif University of Technology.
PhD (State University of New York at Stony Brook, USA, 2001),
L.E. Dickson Instructor (University of Chicago, USA).
3. *Behrang Noohi*, Sharif University of Technology.
See the Seventeenth Competition results.
4. *Payman L. Kassaei*, Sharif University of Technology.
See the Seventeenth Competition results.
5. *Aminollah Zargarian*, University of Tehran.

–

## The Nineteenth Competition, University of Kerman, March 1995
*Questions Committee: Ahmad Haghani.*

1. *Amir Jafari*, Sharif University of Technology.
PhD (Brown University, USA, 2003).

–

1. *Ali Lashgari Faghani*, Isfahan University of Technology.
PhD (Universität Zürich, Switzerland, 2001).

–

3. *Fatemeh Ayatollahzadeh Shirazi*, University of Tehran.
PhD (University of Tehran, Iran, 1999),
Assistant Professor (University of Tehran, Iran).
3. *Mohammad Reza Raoofi*, Isfahan University of Technology.
PhD (University of Indiana Bloomington, USA, 2004).

–

5. *Reza Naserasr*, Sharif University of Technology.

PhD (Simon Fraser University, USA, 2004),
PDF (Carleton University, Canada).
6. *Hossein Movasati*, Sharif University of Technology.
PhD (IMPA – Instituto de Matematica Pura e Aplicada, Brazil, 2002),
Adjoint Professor (IMPA, Brazil).
7. *Ahmad Mojiri*, Isfahan University of Technology.
PhD (University of Ottawa, Canada, 2003),
Assistant Professor (Texas A&M University-Texarkana, USA).

## The Twentieth Competition, Sharif University of Technology, February 1996
*Questions Committee: Omid Ali Karamzadeh and Yahya Tabesh.*

1. *Keivan Mallahi-Karai*, Sharif University of Technology.
PhD (Yale University, USA, 2006),
Visiting Lecturer (Jacobs University Bremen, Germany).
2. *Hossein Movasati*, Sharif University of Technology.
See the Nineteenth Competition results.
3. *Omid Naghshineh Arjmand*, Sharif University of Technology.
PhD Candidate (Sharif University of Technology, Iran).
4. *Ali Reza Amini Harandi*, Isfahan University of Technology.
PhD (University of Isfahan, Iran, 2005),
Assistant Professor (Shahrekord University, Iran).
5. *Ebrahim Samei*, Shahid Beheshti (former National) University.
PhD (University of Manitoba, Canada, 2005),
PDF (University of Waterloo, Canada).

## The Twenty-First Competition, University of Tehran, March 1997
*Questions Committee: Omid Ali Karamzadeh.*

1. *Kia Dalili*, Sharif University of Technology.
PhD (Rutgers University, New Brunswick, USA, 2005),
PDF (University of Missouri, USA).
2. *Maryam Mirzakhani*, Sharif University of Technology.
PhD (Harvard University, USA, 2004),
Professor (Princeton University, USA).
3. *Ebrahim Samei*, Shahid Beheshti (former National) University.
See the Twentieth Competition results.
4. *Hadi Jorati*, Sharif University of Technology.
PhD (Princeton University, USA, 2006),
Visiting Researcher (University of Helsinki, Finland).
5. *Hossein Abedi Andani*, Isfahan University of Technology.
PhD candidate (Tarbiat Modares University, Iran).
5. *Daryoush Kiani*, University of Tabriz.
PhD (Sharif University of Technology, Iran, 2004),
Assistant Professor (Amirkabir University of Technology, Iran).
5. *Payam Nesser Tayoub*, University of Tehran.
—

## The Twenty-Second Competition, University of Ahwaz, March 1998
*Questions Committee: Omid Ali Karamzadeh.*

1. *Maryam Mirzakhhani*, Sharif University of Technology.
   See the Twenty-First Competition results.
2. *Eaman Eftekhari*, Sharif University of Technology.
   PhD (Princeton University, USA, 2005),
   PDF (IPM, Iran).
3. *Payam Nasser Tayoub*, University of Tehran.
   –
4. *Mohammad Ahmadvand*, Bu-Ali Sina (Avecina) University of Hamedan.
   –
5. *Abolghassem Karimi*, Shahid Beheshti (former National) University.
   PhD (Shahid Beheshti (former National) University, Iran, 2006),
   Assistant Professor (Islamic Azad University of Gorgan, Iran).

## The Twenty-Third Competition, Sharif University of Technology, March 1999
*Questions Committee: Omid Ali Karamzadeh and Yahya Tabesh.*

1. *Hadi Salmasian*, Sharif University of Technology.
   PhD (Yale University, USA, 2006),
   PDF (Queen's University, Canada).
2. *Mohsen Bahramgiri*, Sharif University of Technology.
   PhD (MIT, USA, 2006).
   Assistant Professor (Sharif University of Technology, Iran).
3. *Mohammad Javaheri*, Sharif University of Technology.
   PhD (State University of New York at Stony Brook, USA, 2005),
   Visiting Assistant Professor (University of Oregon, USA).
4. *Bijan Ahmadi*, Shahid Beheshti (former National) University.
   PhD candidate (Sharif University of Technology, Iran).
5. *Kamal Azizi*, University of Tabriz.
   PhD (University of Tabriz, Iran, 2006),
   Assistant Professor (University of Tabriz, Iran).

## The Twenty-Fourth Competition, Khajeh Nasir Toosi University of Technology, May 2000
*Questions Committee: Saeed Azam, Rouzbeh Tusserkani, Ali Reza Jamali (Chair), and Hossein Mohebi.*

1. *Omid Amini*, Sharif University of Technology.
   PhD (École Polytechnique, France, 2007).
   –
2. *Kasra Alishahi*, Sharif University of Technology.
   PhD candidate (Sharif University of Technology, Iran).
3. *Mazyar Mirrahimi*, Sharif University of Technology.
   PhD (Ecole des Mines de Paris, France, 2006).
   –
4. *Seyed Reza Moghaddasi*, Sharif University of Technology.

PhD (Sharif University of Technology, Iran, 2005),
Assistant Professor (Sharif University of Technology, Iran).
5. *Masood Aryapoor*, Sharif University of Technology.
   PhD candidate (Yale University, USA).
6. *Fereydoun Derakhshani*, University of Tehran.
   –

7. *Bijan Ahmadi*, Shahid Beheshti (former National) University.
   See the Twenty-Third Competition results.
8. *Afshin Amini*, Shiraz University
   PhD candidate (Shiraz University, Iran).

## The Twenty-Fifth Competition, IKIU of Qazvin, May 2001
*Questions Committee: Saeed Azam, Rouzbeh Tusserkani, Ali Reza Jamali (Chair), and Hossein Mohebi.*

1. *Amir Mohammadi*, Sharif University of Technology.
   PhD candidate (Yale University, USA).
2. *Salman Abolfathe Beigi*, Sharif University of Technology.
   PhD candidate (MIT, USA).
3. *Hamid Reza Dorbidi*, Sharif University of Technology.
   PhD candidate (Sharif University of Technology, Iran).
4. *Babak Amini*, Shiraz University.
   PhD candidate (Shiraz University, Iran).
5. *Afshin Amini*, Shiraz University.
   See the the the Twenty-Fourth Competition results.
6. *Rasoul Azizi*, Isfahan University of Technology.
   PhD Candidate (The University of Texas at El Paso, USA).
6. *Javad Ebrahimi Boroujeni*, Sharif University of Technology.
   PhD candidate (Simon Fraser University, Canada).

## The Twenty-Sixth Competition, IASBS of Zanjan, May 2002
*Questions Committee: Saeed Azam, Rouzbeh Tusserkani, Ali Reza Jamali (Chair), and Hossein Mohebi.*

1. *Salman Abolfathe Beigi*, Sharif University of Technology.
   See the Twenty-Fifth Competition results.
2. *Ali Shourideh*, Sharif University of Technology.
   PhD candidate (University of Minnesota, USA).
3. *Amin Aminzadeh Gohari*, Sharif University of Technology.
   PhD candidate (University of California, Berkeley, USA).
3. *Javad Ebrahimi Boroojeni*, Sharif University of Technology.
   See the Twenty-Fifth Competition results.
5. *Majid Hadian*, Sharif University of Technology.
   PhD candidate (Universität Bonn, Germany).
6. *Rahbar Rasouli*, University of Tehran.
   MSc (University of Tehran, Iran, 2006),
   –

7. *Amir Moradifam*, Iran University of Science and Technology.

PhD candidate (UBC, Canada).

## The Twenty-Seventh Competition, Bu-Ali Sina (Avecina) University of Hamedan, May 2003

*Questions Committee: Gholam Hossein Eslamzadeh, Hossein Hajiabolhassan, Mohammad Reza Pournaki, and Mohammad Taghi Dibaei (Chair).*

1. *Mohsen Sharifi Tabar*, Sharif University of Technology.
   PhD candidate (Sharif University of Technology, Iran).
2. *Ali Shourideh*, Sharif University of Technology.
   See the Twenty-Sixth Competition results.
3. *Mohammad Farajzadeh Tehrani*, Sharif University of Technology.
   PhD candidate (Princeton University, USA).
4. *Payam Valadkhan*, Sharif University of Technology.
   Student (Sharif University of Technology, Iran).
5. *Hamid Hassanzadeh*, Tarbiat Moallem (Teacher Training) University.
   PhD candidate (Tarbiat Moallem (Teacher Training) University, Iran).
6. *Rahbar Rasooli*, University of Tehran.
   See the Twenty-Sixth Competition results.
7. *Maryam Khosravi*, Tarbiat Moallem (Teacher Training) University.
   PhD candidate (Tarbiat Moallem (Teacher Training) University, Iran).
8. *Amir Moradifam*, Iran University of Science and Technology.
   See the Twenty-Sixth Competition result.

## The Twenty-Eighth Competition, Sharif University of Technology, May 2004

*Questions Committee: Gholam Hossein Eslamzadeh, Hossein Hajiabolhassan, Majid Mirza-Vaziri, Abdolrasool Pourabbas, Mohammad Reza Pournaki, and Rashid Zaare-Nahandi (Chair).*

1. *Iman Setayesh*, Sharif University of Technology.
   PhD candidate (Princeton University, USA).
2. *Omid Haji Mirsadeghi*, Sharif University of Technology.
   Student (Sharif University of Technology, Iran).
3. *Armin Morabbi*, Sharif University of Technology.
   Student (Sharif University of Technology, Iran).
4. *Sajad Lakzian*, Amirkabir University of Technology.
   PhD candidate (Amirkabir University of Technology, Iran).
5. *Mohammad Kazem Anvari*, Ferdowsi University of Mashhad.
   PhD candidate (Ferdowsi University of Mashhad, Iran).

## The Twenty-Ninth Competition, University of Mazandaran in Babolsar, May 2005

*Questions Committee: Hossein Hajiabolhassan, Majid Mirza-Vaziri, Mojtaba Moniri, Mohammad Reza Pournaki, Mehdi Radjabalipour (Chair), Bamdad R. Yahaghi, Rashid Zaare-Nahandi (Supervisor), and Manouchehr Zaker.*

1. *Iman Setayesh*, Sharif University of Technology.
   See the Twenty-Eighth Competition results.

2. *Mohammad Farajzadeh Tehrani*, Sharif University of Technology.
   Student (Sharif University of Technology, Iran).
3. *Mohammad Abbas Rezaei*, Sharif University of Technology.
   PhD candidate (University of Chicago, USA).
4. *Mohammad Hossein Mousavi*, Sharif University of Technology.
   Student (Sharif University of Technology, Iran).
5. *Fatemeh Doroodian*, Amirkabir University of Technology.
   PhD candidate (Amirkabir University of Technology, Iran).
6. *Mahmoud Hassanzadeh*, University of Tehran.
   Student (University of Tehran, Iran).

## The Thirtieth Competition, Tafresh University, May 2006

*Questions Committee: Hossein Hajiabolhassan, Majid Mirza-Vaziri, Mojtaba Moniri, Mohammad Reza Pournaki, Mehdi Radjabalipour (Chair), Bamdad R. Yahaghi, Rashid Zaare-Nahandi (Supervisor), and Manouchehr Zaker.*

1. *Ali Akbar Daemi*, Sharif University of Technology.
   Student (Sharif University of Technology, Iran).
2. *Mohammad Gharakhani*, Sharif University of Technology.
   Student (Sharif University of Technology, Iran).
3. *Omid Haji Mirsadeghi*, Sharif University of Technology.
   Student (Sharif University of Technology, Iran).
4. *Mostafa Einollahzadeh Samadi*, Sharif University of Technology.
   Student (Sharif University of Technology, Iran).
5. *Behzad Mehrdad*, Sharif University of Technology.
   Student (Sharif University of Technology, Iran).

## The Thirty-First Competition, Ferdowsi University of Mashhad, May 2007

*Questions Committee: Mojtaba Gheerati, Hassan Shirdareh Haghighi, Omid Naghshineh Arjmand, and Fariborz Azarpanah (Chair).*

1. *Ali Akbar Daemi*, Sharif University of Technology.
   Student (Sharif University of Technology, Iran).
2. *Nasser Talebizadeh*, Sharif University of Technology.
   Student (Sharif University of Technology, Iran).
3. *Erfan Salavati*, Sharif University of Technology.
   Student (Sharif University of Technology, Iran).
4. *Nima Ahmadipour Anari*, Sharif University of Technology.
   Student (Sharif University of Technology, Iran).
5. *Mohammad Gharakhani*, Sharif University of Technology.
   See the Thirtieth Competition results.

# Index

Abel's Continuity Theorem[3], 193, 195

Abelian groups, C1 Alg. 2(r), C3 Alg. 2(r), C12 Alg. 2, C24 $D_1$ 4(r), C31 $D_1$ 3

Acceleration, 6

Algebraic, 18, 51

Algebraic equations, C9 Gen. 2(r), C11 Gen. 2

Algebraic functions, C3 Ana. 3

Algebraic numbers, 19, C11 Alg. 2, C12 Alg. 5

Algebras of all linear transformations, C4 Alg. 2(r)

Algorithm, 38

AM-GM Inequality, 60

Analytic functions, C26 $D_1$ 2, C27 $D_1$ 2, C29 $D_2$ 1, C30 $D_2$ 2

Analytic geometry, C6 Ana. 1(r), C27 $D_2$ 3

Archimedean property of real numbers, 57

Area, 21, 35, 47

Arithmetic, C2 Gen. 1(r), C8 Gen. 2(r), C10 Gen. 4(r)

Arzela's Theorem, 138

Automorphism, 19

Baire Category Theorem, 174, 222

Baire function, 2, 56

Banach space, 173, 204

Bayes' Theorem, 79, 201-203, 229

Bertrand's conjecture, 51, 241

Bertrand's principle, 51

Binomial coefficients, C3 Gen. 2(r), C12 Gen. 3, C15 Gen. 3, C30 $D_1$ 5

Binomial Theorem, 75, 77, 86, 93, 102

Bisection method, 37

Boolean ring, 85, 131, 223

Cantor set, 4, 49

Cantor set, C2 Ana. 3, C29 $D_2$ 6

Cauchy Inequality, 239

Cauchy product of two polynomials, 75

Cauchy-Riemann Equations, 198

Cauchy-Schwarz Inequality, 230

Cauchy's criterion for the convergence of improper integrals, 180

Cauchy's Integral Formula, 190, 197

Cauchy's Theorem, 126

Cayley-Hamilton Theorem, 132

Centralizer, 126, 161, 209, 217

Chain
ascending, 28, 98, 131, 170
descending, 10, 79

Characteristic, 16, 20

Chebyshev Inequality, 50, 238

Circle, 12, 12, 21, 23, 51

Class equation, 126

Clopen subset, 232

Closed convex hull, 87

Combinatorics, C12 Gen. 2, C14 Gen. 1, C14 Gen. 2, C15 Gen. 3, C24 $D_1$ 5, C24 $D_2$ 6, C25 $D_1$ 6, C25 $D_2$ 5, C25 $D_2$ 6, C26 $D_1$ 5, C26 $D_1$ 6, C26 $D_2$ 5, C27 $D_2$ 4, C28 $D_2$ 6, C29 $D_1$ 6

Commutator subgroup, 23, 178

Commutative rings with identity, C1 Alg. 1(r), C4 Alg. 1(r), C20 5, C26 $D_1$ 4, C31 $D_2$ 4

Compact, 7, 30, 39

Compactness, C23 Ana. 1

Comparison Test, 136, 166

Complete metric space, 41

Complete square, 52

Complex numbers, C1 Gen. 3(r)

Complex Analysis, C22 Ana. 3, C24 $D_1$ 2, C25 $D_2$ 4, C26 $D_1$ 2, C27 $D_1$ 2, C28 $D_1$ 1, C29 $D_2$ 1, C30 $D_2$ 2

---

[3]The letters C, $D_i$ ($i = 1, 2$), and r stand for 'Competition', 'Day', and routine, respectively, e.g., C26 $D_1$ 3(r) means the third question of Day 1 of the twenty-sixth competition which is categorized as routine. The abbreviations Ana., Alg., and Gen. respectively, stand for Analysis, Algebra, and General, e.g., C5 Ana. 2 means the second analysis question of the fifth competition.

Conditional probability, C4 Gen. 2

Congestion point, 10, 84

Congruences, C2 Gen. 1(r), C24 $D_1$ 3, C25 $D_2$ 2, C26 $D_1$ 3(r), C27 $D_2$ 5, C30 $D_1$ 4

Conic sections, C5 Gen. 3

Connected, 6, 36, 40, 49

Connected component, 7

Continuity, C1 Ana. 3, C4 Ana. 1, C7 Ana. 2, C9 Ana. 2, C9 Ana. 5, C12 Ana. 1, C26 $D_2$ 1

Continuous functions, C2 Ana. 1, C3 Ana. 2(r), C5 Ana. 2, C9 Ana. 3(r), C9 Gen. 3, C14 Ana. 3, C15 Ana. 2(r), C16 Ana. 1(r), C17 Ana. 3, C17 Ana. 4, C19 Ana. 3, C21 Ana. 2, C22 Ana. 2, C24 $D_2$ 2(r), C25 $D_2$ 3, C26 $D_1$ 1, C26 $D_2$ 2, C28 $D_1$ 2, C28 $D_2$ 2

Convergence
of sequences, C8 Ana. 3(r)
of series, C8 Ana. 2, C11 Ana. 3, C13 Ana. 2, C22 Ana. 2, C29 $D_1$ 2

Convergent, 14, 19, 21, 30,37
pointwise, 19
uniformly, 12, 24, 31

Convexity, C15 Ana. 1

Counting, C2 Gen. 4(r), C9 Gen. 4, C12 Gen. 2, C14 Gen. 2, C24 $D_1$ 5, C25 $D_1$ 6, C26 $D_2$ 5

Darboux's property, 56

Darboux's Theorem, 56, 56, 181, 189, 197, 221

Decimal expansion, C1 Gen. 2(r)

Dedekind's Extension of Abel's Theorem, 135, 136

Definite integrals, C9 Gen. 3, C10 Ana. 2, C10 Ana. 3(r), C12 Ana. 3(r), C13 Ana. 3, C18 Ana. 2

Dense, 15, 51, 52

Derangement, 247

Derivative, 2, 11, 15, 17, 19, 34, 39, 49

Derivatives, C1 Ana. 1(r), C3 Ana. 1, C10 Ana. 1, C11 Ana. 1, C15 Ana. 3, C17 Ana. 1, C22 Ana. 1, C24 $D_1$ 1, C25 $D_1$ 3, C25 $D_1$ 4, C26 $D_1$ 1, C27 $D_2$ 1

Derived subgroup, 23, 31, 51, 98, 178

Derived subgroups, C14 Alg. 1, C21 Alg. 2, C29 $D_2$ 3, C30 $D_2$ 3

Determinant, 16, 19, 113, 116, 127, 246

Determinants, C9 Gen. 1, C9 Gen. 7, C11 Gen. 1, C19 Alg. 4, C28 $D_2$ 3, C31 $D_1$ 6

Differential calculus, C17 Ana. 2(r), C19 Ana. 1(r), C24 $D_2$ 2(r)

Differential equations, C1 Differential Equations(r), C2 Differential Equations(r), C10 Gen. 3(r), C15 Ana. 3

Differentiability, C14 Ana. 1, C19 Ana. 3, C23 Ana. 3

Dihedral group, 196

Dini's Theorem, 125

Diophantine equations, C9 Gen. 4

Dirichlet's Theorem, 194

Disconnected, 2, 44

Divisibility, C5 Gen. 1, C8 Gen. 2(r), C10 Gen. 4(r), C17 Ana. 2(r), C24 $D_1$ 3, C25 $D_2$ 2, C27 $D_2$ 5, C29 $D_1$ 3(r), C30 $D_1$ 4

Division algorithm, 120

Division ring, 39, 65, 92, 120, 156, 165, 206, 240

Domain, 40

Eigenvalue, 11, 22, 44

Eigenvector, 22, 44

Eisenstein's Criterion, 126, 132

Element
idempotent, 13, 26, 46
identity, 5, 13
invertible, 5
left identity, 13
left quasi-regular, 5
maximal, 29
nilpotent, 13
right quasi-regular, 5
right identity, 8
quasi-regular, 5

Elementary number theory, C15 Gen. 1, C16 Alg. 3, C24 $D_1$ 3, C25 $D_2$ 2, C26 $D_1$ 3(r), C27 $D_2$ 5, C28 $D_1$ 3, C29 $D_1$ 3(r), C30 $D_1$ 4, C30 $D_2$ 5, C31 $D_2$ 1

Ellipse, 11, 88

Endomorphism, 5, 9

Equation
differential, 3, 6, 18, 32, 36
functional, 19
integral, 18

Euclidean
algorithm, 112
metric, 46, 205
norm, 2
plane, 9, 45, 140
ring, 66, 112

Euclid's proof of the infinitude of primes, 246

Euclidean spaces, C24 $D_2$ 1

Euler's totient function, 46, 64, 162, 231

Expansion,
binary, 28, 38

decimal, 26, 38
hexadecimal, 38
octal, 38
ternary, 4, 49
Extension, 14, 16, 18, 19
Galois, 111
normal, 111
separable, 111

Fermat's Little Theorem, 147, 198, 216, 237
Field, 5, 8, 10, 14, 16, 18, 19, 20, 23, 25, 27, 28, 30, 32, 39, 44, 48, 49, 51
Field extensions, C9 Alg. 4, C10 Alg. 3, C11 Alg. 2
Finite-dimensional vector spaces, C14 Alg. 5, C18 Alg. 3, C28 $D_1$ 5, C28 $D_2$ 6
Finite fields, C2 Alg. 1(r), C4 Alg. 3, C9 Alg. 2
Finite groups, C6 Alg. 1, C9 Alg. 1, C11 Alg. 3, C11 Gen. 3, C12 Alg. 1, C12 Alg. 2, C14 Alg. 2, C15 Alg. 1, C16 Alg. 1, C17 Alg. 1, C18 Alg. 1, C25 $D_1$ 1
Finite $p$-groups, C14 Alg. 1
Finitely generated (right/left) ideals, C13 Alg. 2
First Isomorphism Theorem
for groups, 111, 141, 155, 161, 178, 190, 233
for modules, 74, 179
for rings, 191
Fixed point theory, C31 $D_2$ 6
Flux, 36
Frobenius Inequality, 206
Function
analytic, 32, 43, 45, 49, 51
arithmetic, 46
auxiliary, 17
bounded, 23, 31, 34
choice, 38
complex valued, 7
continuous, 2, 4, 7, 9, 10, 12, 13, 14, 15, 16, 17, 19, 20, 21, 23, 26, 27, 28, 29, 30, 31, 34, 39, 42, 43, 44, 46, 48, 50,53
continuous bounded, 39
continuously differentiable, 14, 29, 42
convex, 24
decreasing, 30, 34
differentiable, 19, 21, 23, 27, 31, 34, 42, 43, 45
entire, 46
harmonic, 40
increasing, 15, 24, 34
integral part, 26

nonnegative, 4, 25
nowhere differentiable, 29
one-to-one, 9, 15, 19, 20, 46, 46
periodic, 26, 28, 30
real, 2, 3, 7, 8, 17, 24
surjective, 20, 42, 46, 48
uniformly continuous, 23, 26, 30, 34, 39
Functional equations, C20 4
Fundamental Theorem of Algebra, 224
Fundamental Theorems of Calculus,
The First, 100, 122, 171, 197
The Second, 100, 118
Fundamental Theorem of finite(ly generated) abelian groups, 63, 127, 161, 187, 247

Game theory, C24 $D_2$ 6
Gamma function, 36
Gauss-Lucas Theorem, 86
General linear groups, C8 Alg. 1(r), C13 Alg. 4
Greatest common divisor, 16, 23
Group
abelian, 2, 20, 40, 45, 52
commutative, 2, 20, 40
cyclic, 5, 20, 63, 65, 120, 130
finite, 16, 19, 23, 28, 41
finite abelian, 52
finite nonabelian, 26
finite $p$-, 23
infinite, 40
nonabelian, 26, 39
of inner automorphisms, 26, 39, 161, 178
of permutations, 22, 111, 190, 225
of units, 211
$p$, 23, 126, 147, 161
symmetric, 57, 111, 141, 155, 190
quaternionic, 210
simple, 28
simple abelian, 120
solvable, 30, 147, 178
Group homomorphisms, C12 Alg. 3

Hamel basis, 78
Higher order derivatives, C11 Ana. 3, C25 $D_1$ 3
Holder functions, C3 Ana. 3
Homomorphisms
of groups, C3 Alg. 2(r)
of rings, C2 Alg. 3(r)
Hyperbola, 88

Ideal
left, 22, 25, 41
maximal, 2, 14, 32, 43, 53

minimal left, 29
nilpotent, 48
prime, 2, 9, 25
principal, 20, 43
right, 22, 25, 28, 39, 48
small, 31
two-sided, 19, 29, 39, 41, 48
Improper integrals, C6 Ana. 3, C14 Ana.
    2, C20 6
Inclusion, 29, 49
Inclusion-exclusion principle, 158, 247
Inequalities, C1 Gen. 4(r), C2 Gen. 2(r),
    C28 $D_1$ 6, C30 $D_1$ 5
Injectivity, C4 Ana. 1, C9 Ana. 5
Inner automorphisms, C16 Alg. 1, C23
    Alg. 1
Instantaneous
    speed, 67
    velocity (vector), 67
Integrable functions, C5 Ana. 1, C9
    Ana. 4, C13 Ana. 1, C21 Ana. 3
Integral, 10, 26, 35, 38
Integral calculus, C4 Gen. 1, C12 Ana.
    3(r), C15 Gen. 2, C17 Ana. 2(r),
    C19 Ana. 1(r), C21 Ana. 2, C24 $D_2$
    2(r), C28 $D_2$ 1(r), C29 $D_1$ 1(r), C30
    $D_1$ 1(r)
Integral domain, 10, 14, 22, 42
Integral domains, C4 Alg. 3, C8 Alg. 2,
    C13 Alg. 1, C25 $D_2$ 1
Integration by parts, 69, 101, 125, 129,
    160
Intermediate value property, 56, 221
Intermediate Value Theorem, 56, 67, 70,
    83, 105, 128, 137, 142, 147, 164, 172,
    182, 206, 215, 219, 221, 228
Invariant subspace(s), 49
Inverse Probability Theorem, 79
Isomorphic sets, 2, C1 Alg. 3, C31 $D_2$ 3
Iteration method, 37

Jacobian, 36

Labeled graph, 149, 200
Lagrange's Theorem, 92, 120, 127
Largest prime power dividing a factorial,
    123
Least common multiple, 23
Lebesgue number, 106
Lebesgue's Integrability Criterion for
    Riemann integrals, 80, 106, 186
Lebesgue's Number Lemma, 106
Left annihilator, 231
Legendre symbol, 237
Leibniz's Theorem, 194
L'Hopital's rule, 172, 235
Limit, 2, 31, 50, 52

Limit Comparison Test, 126, 136
Limit point, 30, 52
Limits, C1 Ana. 3, C10 Ana. 1, C19
    Ana. 2, C21 Ana. 1, C21 Ana. 2,
    C31 $D_2$ 2
Limits of sequences, C2 Ana. 1, C7 Ana.
    3, C20 2, C30 $D_1$ 2
Line, 6, 12, 23, 37, 45, 45
Linearly dependent, 9, 30
Linear transformation, 5, 11, 13, 16, 27,
    32, 42
Linear transformations, C5 Alg. 1(r), C7
    Alg. 3, C9 Alg. 5, C17 Alg. 3, C22
    Alg. 3, C25 $D_1$ 2

Matrix, 2, 12, 14, 16, 19, 20, 22, 25, 29,
    31, 39, 41, 43, 44, 45, 47, 49, 51, 52
Matrix analysis, C1 Ana. 2, C27 $D_1$ 1
Matrix theory, C6 Ana. 2, C8 Alg. 3(r),
    C10 Alg. 4(r), C12 Alg. 5, C13 Alg.
    5(r), C15 Alg. 3, C16 Alg. 3, C19
    Alg. 3(r), C19 Alg. 4, C21 Alg. 3,
    C23 Alg. 3(r), C24 $D_2$ 3, C26 $D_2$ 5,
    C27 $D_1$ 5, C28 $D_2$ 3, C29 $D_2$ 4, C30
    $D_2$ 4, C31 $D_1$ 6
Maximal ideals, C1 Alg. 1(r), C22 Alg.
    2, C26 $D_1$ 4
Maximal subgroup, 18
Maximal subgroups, C10 Alg. 1, C26 $D_2$
    3
Maximum Modulus Principle, 216, 239
Maximum Modulus Theorem for
    harmonic functions, 197, 198
Mean Value Property, 197
Mean Value Theorem, 56, 70, 125, 137,
    188, 195, 210, 215
Mean Value Theorem for integrals, 185
Mean Value Theorem for second order
    derivatives, 67
Mechanics, C2 Gen. 3
Metric space
    compact, 42, 49
    complete, 41
    connected, 49
    separable, 51

Minkowski sum, 52
Module, 8
Morera's Theorem, 190
Möbius
    function, 162
    Inversion Formula, 224

Natural numbers, C30 $D_1$ 6
Nilpotent matrix, 18, 49
Noetherian ring, 165
Normalizer, 155

Normal subgroup, 13, 16, 23, 28, 45
Normal subgroups, C7 Alg. 2, C9 Alg. 1,
    C14 Alg. 2, C18 Alg. 1, C19 Alg. 1,
    C27 $D_1$ 6
Normed linear space, 160
Normed spaces of continuous functions,
    C2 Ana. 2(r)
Nowhere dense subset, 174, 221

Odd prime, 20, 42, 43
One-to-one correspondence, 2, 24, 48, 53
One-sided limits, C21 Ana. 3
Open map, 28
Operator norm, 218
Optical property of
    ellipses, 88
    hyperbolas, 88
Optics, C5 Gen. 3
Orthogonal matrix, 47

Partition, 42, 45, 51
Perfect subset, 221
Periodic functions, C18 Ana. 2, C20 6
Permutation, 12, 22
Permutation groups, C13 Alg. 4, C28 $D_1$
    4
Persian alphabet, 6
Persian calendar, 5
Picard's Great Theorem, 224
Plane, 8, 23, 37
Plane geometry, C3 Gen. 3, C5 Gen. 3,
    C6 Gen. 3(r), C12 Gen. 1, C13
    Gen. 1(r), C31 $D_1$ 1
Point-set topology, C1 Topology(r), C2
    Ana. 3, C2 Topology(r), C5 Ana. 3,
    C7 Ana. 1, C9 Ana. 1(r), C10 Gen.
    2, C21 Ana. 1, C24 $D_2$ 1, C27 $D_2$ 2,
    C29 $D_2$ 2, C29 $D_2$ 6, C30 $D_2$ 6, C31
    $D_1$ 2, C31 $D_1$ 5
Pointwise convergence, C11 Ana. 2, C23
    Alg. 3
Poisson random variable, 60, 61
Polynomial, 11, 16, 23, 27, 27, 51
    characteristic, 31, 32, 117, 132, 191,
    220, 244
    minimal, 132, 191, 221
    elementary symmetric, 152
Polynomial rings, C2 Alg. 3(r), C9 Alg.
    2, C9 Alg. 3(r), C14 Alg. 4(r), C17
    Ana. 2(r), C17 Alg. 1, C31 $D_2$ 4
Poset, 46
Power series, C18 Ana. 1, C30 $D_2$ 2
Prime ideals, C1 Alg. 1(r), C4 Alg. 1(r),
    C15 Alg. 2
Prime numbers, C25 $D_2$ 2, C26 $D_1$ 3(r),
    C30 $D_1$ 4, C30 $D_2$ 5
Prime ring, 211

Primitive Element Theorem, 111
Primitive root, 50
Principal ideals, C12 Alg. 4
Probability, 4, 10, 15, 17, 18, 19, 25, 33,
    40, 42, 44, 47, 52
Probability and Statistics, C1
    Probability and Statistics(r), C2
    Probability and Statistics(r), C4
    Gen. 2, C8 Gen. 3, C9 Gen. 6, C11
    Gen. 1, C15 Gen. 3, C24 $D_1$ 6, C25
    $D_1$ 5, C26 $D_2$ 5, C28 $D_2$ 5, C29 $D_2$
    5, C31 $D_1$ 4

Quasi-regular elements, C2 Alg. 2(r)
Quotient rings, C9 Alg. 2

Radical of an ideal, 9
Radicals of ideals, C4 Alg. 1(r)
Radius of convergence, 28, 51
Random variables, 4, 6, 18, 50
Random variables, C1 Probability and
    Statistics(r), C2 Probability and
    Statistics(r), C10 Gen. 1
Rank, 25, 27, 29, 32, 41, 51
Rank-Nullity Theorem, 92, 98, 113, 117,
    165, 176, 240
Rational Root Theorem, 143
Rational numbers, C29 $D_1$ 2
Real numbers, C14 Gen. 3, C31 $D_2$ 5
Real (valued) functions, C18 Ana. 3,
    C30 $D_2$ 1
Recurrence relations, C27 $D_1$ 3
Regular $n$-gon, 23
Riemann integrable, 10, 20, 21, 23, 31
Riemann's criterion for integrability, 81,
    106, 166
Riemann-Stieltjes integrals, C8 Ana.
    3(r), C16 Ana. 2
Right annihilator, 231
Ring
    commutative, 2, 9, 11, 20, 22, 30, 32,
    43, 47, 50, 53
    division, 39, 65, 92, 120, 156, 165, 206,
    240
    noetherian, 165
    uncountable,
    unital, 18, 20, 22, 25, 43, 47
    with identity, 2, 10, 11, 12, 13, 19, 29,
    30, 50, 53
Ring homomorphisms, C12 Alg. 4
Rolle's Theorem, 118, 221
Root, 7, 11, 19, 20, 22, 29, 37, 37, 45
Roots of complex polynomials, C3 Ana.
    3, C5 Gen. 2
Rouché's Theorem, 73

Sandwich Theorem, 245

Schroder-Bernstein Theorem, 58, 109

Schwarz Lemma, 239

Second Isomorphism Theorem for
groups, 97, 161

Second Mean Value Theorem for
integrals, 180

Semigroup, 8

Semigroups, C3 Alg. 3(r)

Sequence, 7, 9, 12, 13, 19, 21, 24, 26, 28,
30, 32, 34, 39, 45, 48
  decreasing, 7, 14, 20
  increasing, 7, 19, 48

Sequences of complex numbers, C30 $D_1$ 2

Sequences of real numbers, C4 Ana. 2

Series, 20, 21, 34, 37, 39, 48
  power, 28, 36, 51

Set
  of measure zero, 80
  partially ordered, a.k.a. poset, 46
  power, 2
  totally ordered, 82

Simple left module, 179

Space geometry, C13 Gen. 3

Speed, 6

Splitting field, 111

Square, 21

Squeeze Lemma, 119

Stolz's Theorems on limits
  the first theorem, 95
  the second theorem, 95, 185

Surjectivity, C9 Ana. 5, C12 Ana. 1

Sylow $p$-subgroup(s), 25, 130, 155

Sylow's Theorems
  the first theorem, 147
  the second theorem, 155
  the third theorem, 130

Sylow theory, C12 Alg. 1, C15 Alg. 1

Sylvester Inequality, 206

Sylvester's problem, 143

Symmetric difference of sets, 127


Tangential acceleration, 67

Taylor's Formula, 86

Third Isomorphism Theorem for groups,
175, 243

Tree, 149

Triangle, 3, 42, 51

Triangles, C1 Gen. 1(r)

Trigonometric series, C23 Ana. 2

Topological space, 6

Torsion free group, 247

Turán's Theorem, 200


Uniform Continuity, C14 Ana. 2, C16
Ana. 3, C20 4, C22 Ana. 2, C23
Ana. 1

Uniform convergence, C6 Ana. 2(r), C11
Ana. 2, C12 Ana. 2, C13 Ana. 3,
C15 Ana. 2(r), C16 Ana. 1(r), C22
Ana. 2, C22 Ana. 3

Uniqueness Theorem for analytic
functions, 219, 232


Vandermonde's determinant formula, 152

Variation of parameters, 60

Vector field, 36

Vector space, 4, 5, 9, 11, 13, 16, 18, 23,
27, 32, 42, 48


Well-ordering principle of natural
numbers, 57, 59, 169


Zero divisor(s), 30

Zorn's Lemma, 38, 78, 102, 156

# Texts and Readings in Mathematics

1. R. B. Bapat: Linear Algebra and Linear Models (Second Edition)
2. Rajendra Bhatia: Fourier Series (Second Edition)
3. C. Musili: Representations of Finite Groups
4. H. Helson: Linear Algebra (Second Edition)
5. D. Sarason: Complex Function Theory (Second Edition)
6. M. G. Nadkarni: Basic Ergodic Theory (Second Edition)
7. H. Helson: Harmonic Analysis (Second Edition)
8. K. Chandrasekharan: A Course on Integration Theory
9. K. Chandrasekharan: A Course on Topological Groups
10. R. Bhatia (ed.): Analysis, Geometry and Probability
11. K. R. Davidson: C* – Algebras by Example
12. M. Bhattacharjee et al.: Notes on Infinite Permutation Groups
13. V. S. Sunder: Functional Analysis — Spectral Theory
14. V. S. Varadarajan: Algebra in Ancient and Modern Times
15. M. G. Nadkarni: Spectral Theory of Dynamical Systems
16. A. Borel: Semisimple Groups and Riemannian Symmetric Spaces
17. M. Marcolli: Seiberg – Witten Gauge Theory
18. A. Bottcher and S. M. Grudsky: Toeplitz Matrices, Asymptotic
    Linear Algebra and Functional Analysis
19. A. R. Rao and P. Bhimasankaram: Linear Algebra (Second Edition)
20. C. Musili: Algebraic Geometry for Beginners
21. A. R. Rajwade: Convex Polyhedra with Regularity Conditions
    and Hilbert's Third Problem
22. S. Kumaresan: A Course in Differential Geometry and Lie Groups
23. Stef Tijs: Introduction to Game Theory
24. B. Sury: The Congruence Subgroup Problem
25. R. Bhatia (ed.): Connected at Infinity
26. K. Mukherjea: Differential Calculus in Normed Linear Spaces
    (Second Edition)
27. Satya Deo: Algebraic Topology: A Primer (Corrected Reprint)
28. S. Kesavan: Nonlinear Functional Analysis: A First Course
29. S. Szabó: Topics in Factorization of Abelian Groups
30. S. Kumaresan and G. Santhanam: An Expedition to Geometry
31. D. Mumford: Lectures on Curves on an Algebraic Surface (Reprint)
32. J. W. Milnor and J. D. Stasheff: Characteristic Classes (Reprint)
33. K. R. Parthasarathy: Introduction to Probability and Measure
    (Corrected Reprint)
34. A. Mukherjee: Topics in Differential Topology
35. K. R. Parthasarathy: Mathematical Foundations of Quantum
    Mechanics
36. K. B. Athreya and S. N. Lahiri: Measure Theory
37. Terence Tao: Analysis I (Second Edition)
38. Terence Tao: Analysis II (Second Edition)

39. W. Decker and C. Lossen: Computing in Algebraic Geometry
40. A. Goswami and B. V. Rao: A Course in Applied Stochastic
        Processes
41. K. B. Athreya and S. N. Lahiri: Probability Theory
42. A. R. Rajwade and A. K. Bhandari: Surprises and Counterexamples
in Real Function Theory
43. G. H. Golub and C. F. Van Loan: Matrix Computations (Reprint of the
Third Edition)
44. Rajendra Bhatia: Positive Definite Matrices
45. K. R. Parthasarathy: Coding Theorems of Classical and Quantum
Information Theory
46. C. S. Seshadri: Introduction to the Theory of Standard Monomials
47. Alain Connes and Matilde Marcolli: Noncommutative Geometry,
        Quantum Fields and Motives
48. Vivek S. Borkar: Stochastic Approximation: A Dynamical Systems
        Viewpoint
49. B. J. Venkatachala: Inequalities: An Approach Through Problems
50. Rajendra Bhatia: Notes on Functional Analysis
51. A. Clebsch (ed.): Jacobi's Lectures on Dynamics
        (Second Revised Edition)
52. S. Kesavan: Functional Analysis
53. V. Lakshmibai and Justin Brown: Flag Varieties: An Interplay of
        Geometry, Combinatorics, and Representation Theory
54. S. Ramasubramanian: Lectures on Insurance Models
55. Sebastian M. Cioaba and M. Ram Murty: A First Course in Graph Theory and
        Combinatorics